METAL IONS IN
BIOLOGICAL SYSTEMS

VOLUME 15

Zinc and Its Role in Biology and Nutrition

AN IMPORTANT MESSAGE TO READERS...

A Marcel Dekker, Inc. Facsimile Edition contains the exact contents of an original hard cover MDI published work but in a new soft sturdy cover.

Reprinting scholarly works in an economical format assures readers that important information they need remains accessible. Facsimile Editions provide a viable alternative to books that could go "out of print." Utilizing a contemporary printing process for Facsimile Editions, scientific and technical books are reproduced in limited quantities to meet demand.

Marcel Dekker, Inc. is pleased to offer this specialized service to its readers in the academic and scientific communities.

METAL IONS IN BIOLOGICAL SYSTEMS

Edited by

Helmut Sigel

Institute of Inorganic Chemistry
University of Basel
Basel, Switzerland

with the assistance of Astrid Sigel

VOLUME 15
Zinc and Its Role in Biology and Nutrition

CRC Press
Taylor & Francis Group
Boca Raton London New York

CRC Press is an imprint of the
Taylor & Francis Group, an **informa** business

First published 1983 by Marcel Dekker, Inc.

Published 2019 by CRC Press
Taylor & Francis Group
6000 Broken Sound Parkway NW, Suite 300
Boca Raton, FL 33487-2742

© 1983 by Taylor & Francis Group, LLC
CRC Press is an imprint of Taylor & Francis Group, an Informa business

First issued in paperback 2019

No claim to original U.S. Government works

ISBN 13: 978-0-367-45193-6 (pbk)
ISBN 13: 978-0-8247-7462-2 (hbk)
ISBN 13: 978-1-003-41809-2 (ebk)

DOI: 10.1201/9781003418092

**Visit the Taylor & Francis Web site at
http://www.taylorandfrancis.com**

**and the CRC Press Web site at
http://www.crcpress.com**

Library of Congress Cataloging in Publication Data
Main entry under title:

Zinc and its role in biology and nutrition. '

    (Metal ions in biological systems ; v. 15)
    Includes index.
    1. Zinc--Metalloenzymes. 2. Zinc in the body.
I. Sigel, Helmut. II. Sigel Astrid. III. Series.
QP532.M47 vol. 15   574.19214s   82-23632
[QP535.Z6]   [574.19'214]
ISBN 0-8247-1750-3

ISSN: 0161-5149

PREFACE TO THE SERIES

Recently, the importance of metal ions to the vital functions of living organisms, hence their health and well-being, has become increasingly apparent. As a result, the long-neglected field of "bioinorganic chemistry" is now developing at a rapid pace. The research centers on the synthesis, stability, formation, structure, and reactivity of biological metal ion-containing compounds of low and high molecular weight. The metabolism and transport of metal ions and their complexes is being studied, and new models for complicated natural structures and processes are being devised and tested. The focal point of our attention is the connection between the chemistry of metal ions and their role for life.

No doubt, we are only at the brink of this process. Thus, it is with the intention of linking coordination chemistry and biochemistry in their widest sense that the series METAL IONS IN BIOLOGICAL SYSTEMS reflects the growing field of "bioinorganic chemistry." We hope, also, that this series will help to break down the barriers between the historically separate spheres of chemistry, biochemistry, biology, medicine, and physics, with the expectation that a good deal of the future outstanding discoveries will be made in the interdisciplinary areas of science.

Should this series prove a stimulus for new activities in this fascinating "field" it would well serve its purpose and would be a satisfactory result for the efforts spent by the authors.

Fall 1973   Helmut Sigel

PREFACE TO VOLUME 15

Zinc is required for the growth and development of all species, including humans. It is now recognized that this metal ion is an integral component of many proteins and indispensable to their catalytic function and stability. Up to now about 160 zinc metalloenzymes have been identified (*Chem. Eng. News*, May 18, 1981, p. 46), and it is realized that they are involved in every aspect of metabolism, including the replication and translation of genetic material.

Hence it is not surprising that zinc shows up in nearly every volume of this series; for example, its role in enzymes was discussed in Volume 6 and the therapy of zinc deficiency and the pharmacological use of zinc were covered in Volume 14. In the present volume, which is completely devoted to this important metal ion, first the categories of zinc metalloenzymes are considered, together with models of the enzymic metal-ion binding sites. The substitution of zinc by other metal ions is described and the roles of zinc in DNA and RNA polymerases and in snake toxins are discussed. Other chapters deal with the spectroscopic properties of metallothionein and the interaction of zinc with erythrocytes. The volume closes with three chapters covering the nutritional aspects of zinc: its absorption and excretion, its influence on the activity of enzymes and hormones, and the zinc deficiency syndrome during parenteral nutrition in humans.

Helmut Sigel

CONTENTS

Chapter 6

CONTRIBUTORS

Numbers in parentheses indicate the pages on which the authors' contributions begin.

Ivano Bertini[*] Institute of General and Inorganic Chemistry, Faculty of Pharmacy, University of Florence, Florence, Italy (101)

Robert S. Brown Department of Chemistry, University of Alberta, Edmonton, Alberta, Canada (55)

Neville J. Curtis[†] Department of Chemistry, University of Alberta, Edmonton, Alberta, Canada (55)

Alphonse Galdes Center for Biochemical and Biophysical Sciences and Medicine, Harvard Medical School, Boston, Massachusetts (1)

Joan Huguet[‡] Department of Chemistry, University of Alberta, Edmonton, Alberta, Canada (55)

Stig Jarnum Medical Department P, Division of Gastroenterology, Rigshospitalet, Copenhagen, Denmark (415)

Jeremias H. R. Kägi Institute of Biochemistry, University of Zürich, Zürich, Switzerland (213)

Manfred Kirchgessner Institut für Ernährungsphysiologie der Technischen Universität München, Freising-Weihenstephan, Federal Republic of Germany (319/363)

Karin Ladefoged Medical Department P, Division of Gastroenterology, Rigshospitalet, Copenhagen, Denmark (415)

Present Affiliations:

[*]Institute of General and Inorganic Chemistry, Faculty of Mathematical, Physical and Natural Sciences, University of Florence, Florence, Italy

[†]Research School of Chemistry, Australian National University, Canberra, Australia

[‡]Institute of Organic Chemistry, University of Zürich, Zürich, Switzerland

Claudio Luchinat Institute of General and Inorganic Chemistry,
    Faculty of Pharmacy, University of Florence, Florence, Italy
    (101)

Joseph M. Rifkind Laboratory of Cellular and Molecular Biology,
    National Institute on Aging, Gerontology Research Center,
    Baltimore City Hospitals, Baltimore, Maryland (275)

Hans-Peter Roth Institut für Ernährungsphysiologie der Technischen
    Universität München, Freising-Weihenstephan, Federal Republic
    of Germany (363)

Anthony T. Tu Department of Biochemistry, Colorado State University,
    Fort Collins, Colorado (193)

Bert L. Vallee Center for Biochemical and Biophysical Sciences and
    Medicine, Harvard Medical School, Boston, Massachusetts (1)

Milan Vašák Institute of Biochemistry, University of Zürich, Zürich,
    Switzerland (213)

Edgar Weigand Institut für Ernährungsphysiologie der Technischen
    Universität München, Freising-Weihenstephan, Federal Republic
    of Germany (319)

Cheng-Wen Wu Department of Pharmacological Sciences, State University
    of New York at Stony Brook, Stony Brook, New York (157)

Felicia Ying-Hsiueh Wu Department of Pharmacological Sciences, State
    University of New York at Stony Brook, Stony Brook, New York
    (157)

CONTENTS OF OTHER VOLUMES

*Out of print

Other Volumes are in preparation.

Comments and suggestions with regard to contents, topics, and the
like for future volumes of the series would be greatly welcome.

METAL IONS IN BIOLOGICAL SYSTEMS

VOLUME 15

Zinc and Its Role in Biology and Nutrition

Chapter 1

CATEGORIES OF ZINC METALLOENZYMES

Alphonse Galdes and Bert L. Vallee
Center for Biochemical and Biophysical Sciences and Medicine
Harvard Medical School
Boston, Massachusetts

1. INTRODUCTION

Historically, the role of zinc in biological systems has been investigated in three phases. The first was nutritional and indicated
that zinc is essential for the development and growth of all species.
These studies were initiated by Raulin [1], who showed in 1869 that
zinc is required for the growth of *Aspergillus niger*, and culminated
100 years later in the demonstration that this metal is indispensable
for human growth and development [2].

1

The second phase was biochemical and led to the realization
that zinc is an integral component of many proteins and indispensable
to their catalytic function and structural stability. This phase
owes its origins to Keilin and Mann's [3] observation that carbonic
anhydrase, an enzyme essential for respiration in mammals, contains
zinc, which is required for its action. This was the first demonstration of a specific biochemical role for zinc. Fifteen years
later, a second enzyme, carboxypeptidase A, was similarly found to
contain an essential zinc atom [4]. Thereafter, zinc has been found
in numerous other enzymes, so that now over 20 functionally distinct
zinc-containing proteins have been identified, representing a total
of about 160 enzymes from different species. These comprise enzymes
in each of the six classes designated by the International Union on
Biochemistry (Table 1). In addition, a number of zinc-containing
proteins that do not have known enzymatic properties have also been
discovered. It is now realized that zinc-containing proteins are
involved in every aspect of metabolism. In particular, the replication and translation of the genetic material of all species is
affected by zinc enzymes, accounting for the absolute requirement
of this element by all forms of life.

The third phase in the evolution of zinc biology was the
recognition of clinical manifestations associated with abnormal zinc
metabolism in humans and other vertebrates. In spite of the ubiquitous occurrence of zinc, its deficiency or imbalance in humans is not
uncommon. Postalcoholic cirrhosis of the liver was the first human
pathological condition recognized to be related to zinc metabolism
[5], and the most recent was acrodermatitis enteropathica [6].

This chapter, which focuses on the second of these three
phases--the occurrence and function of zinc in enzymes--begins with
a brief survey of the role of zinc in metalloenzymes.

TABLE 1

Zinc Metalloenzymes, 1981

Name	Number	Source	Role[a]
Class I: oxidoreductases			
Alcohol dehydrogenase	9	Vertebrates, plants,	A, D
Alcohol dehydrogenase	1	Yeast	A
D-Lactate dehydrogenase	1	Barnacle	?
D-Lactate cytochrome reductase	1	Yeast	?
Superoxide dismutase	12	Vertebrates, plants, fungi, bacteria	D
Class II: transferases			
Aspartate transcarbamylase	1	*E. coli*	B
Transcarboxylase	1	*Propionibacterium*	?
Phosphoglucomutase	1	*shermanii*	?
RNA polymerase	10	Yeast	A
DNA polymerase	3	Wheat germ, bacteria, viruses	A
Reverse transcriptase	3	Sea urchin, *E. coli*, T$_4$ phage	A
Terminal dNT transferase	1	Oncogenic viruses	A
Nuclear poly(A) polymerase	2	Calf thymus	A
Mercaptopyruvate sulfur transferase	1	Rat liver, virus *E. coli*	?
Class III: hydrolases			
Alkaline phosphatase	8	Mammals, bacteria	A, D
Fructose-1,6-biphosphatase	2	Mammals	C
Phosphodiesterase (exonuclease)	1	Snake venom	A
Phospholipase C	1	*Bacillus cereus*	A
Nuclease P$_1$	1	*Penicillium cirtrinum*	?
α-Amylase	1	*B. subtilis*	B
α-D-Mannosidase	1	Jack bean	?
Aminopeptidase	10	Mammals, fungi, bacteria	A, C
Aminotripeptidase	1	Rabbit intestine	A
DD-Carboxypeptidase	1	*Streptomyces albus*	A
Procarboxypeptidase A	2	Pancreas	A
Procarboxypeptidase B	1	Pancreas	A
Carboxypeptidase A	4	Vertebrates, crustaceans	A
Carboxypeptidase B	4	Mammals, crustaceans	A

TABLE 1 (Continued)

Name	Number	Source	Role[a]
Carboxypeptidase (other)	5	Mammals, crustaceans, bacteria	A
Dipeptidase	3	Mammals, bacteria	A
Angiotensin-converting enzyme	3	Mammals	A
Neutral protease	16	Vertebrates, fungi, bacteria	A
Collagenase	4	Mammals, bacteria	A
Elastase	1	*Pseudomonas aeruginosa*	?
Aminocylase	1	Pig kidney	?
β-Lactamase II	1	*B. cereus*	A
Creatinase	1	*Pseudomonas putida*	?
Dihydropyrimidine aminohydrolase	1	Bovine liver	?
AMP deaminase	1	Rabbit muscle	?
Nucleotide pyrophosphatase	1	Yeast	A
Class IV: lyases			
Fructose-1,6-bisphosphate aldolase	4	Yeast, bacteria	A
L-Rhammnulose-1-phosphate aldolase	1	*E. coli*	A
Carbonic anhydrase	22	Animals, plants	A
δ-Aminolevulinic acid dehydratase	2	Mammalian liver, erythrocytes	A
Glyoxalase I	4	Mammals, yeast	A
Class V: isomerases			
Phosphomannose isomerase	1	Yeast	?
Class VI: ligases			
tRNA synthetase	3	*E. coli, Bacillus stearothermophilus*	A
Pyruvate carboxylase	2	Yeast, bacteria	?
Total	163		

[a]A denotes a catalytic role, B a structural, C a regulatory, and D a noncatalytic role. A question mark indicates that available information is insufficient to make an assignment.

1.1. Function of Zinc in Metalloenzymes

The function of zinc in metalloenzymes can be divided into four categories: catalytic, structural, regulatory (or modulatory), and noncatalytic.

Zinc is said to have a *catalytic* role when it is essential for and *directly* involved in catalysis by the enzyme; carbonic anhydrase, carboxypeptidase, thermolysin, and aldolases are examples of this type. The removal of catalytic zinc results in an inactive apoenzyme which, however, often retains the native tertiary structure. The manner in which zinc participates in catalysis is detailed further below.

Zinc plays a *structural* role when it is required *solely* for the structural stability of the protein, being necessary for activity only to the extent that the overall conformation of the enzyme affects its action. Structural zinc often (but not exclusively) stabilizes the quaternary structure of oligomeric holoenzymes. Thus a zinc atom serves to dimerize *Bacillus subtilis* α-amylase without affecting its enzymatic activity [7]. Similarly, zinc stabilizes the pentameric quaternary structure of aspartate transcarbamylase (see Sec. 3). The removal of zinc from this type of metalloenzyme (or from the dissociated subunits) prevents association.

A *regulatory* (or *modulatory*) role is indicated when the zinc regulates, but is not essential for, enzymatic activity (which is present in the absence of metal) or for the stability of the protein. Regulatory zinc may act either as an activator (e.g., bovine lens leucine aminopeptidase) or as an inhibitor (e.g., porcine kidney leucine aminopeptidase, fructose-1,6-biphosphatase).

A fraction of the zinc in certain metalloenzymes (e.g., equine and human alcohol dehydrogenases and *Escherichia coli* alkaline phosphatase) is neither involved directly in catalysis nor essential for the maintenance of the tertiary structure of the enzyme (although it

may stabilize it), so that its function is as yet obscure. In the
absence of specific knowledge on how the metal acts, it is noncom-
mittally referred to as *noncatalytic*.

A given metalloenzyme may contain multiple numbers and types
of zinc atoms. Thus equine alcohol dehydrogenase contains both a
catalytic and a noncatalytic zinc atom per subunit, and leucine
aminopeptidase contains a catalytic and a regulatory zinc atom per
subunit. In Table 1 we have categorized the zinc metalloenzymes
known up to the present according to these roles of the metal, A
denoting a catalytic, B a structural, C a regulatory, and D a noncatalytic role. The appearance of multiple letters identifies different zinc atoms playing different roles in the same enzyme. In some
instances the enzyme from different sources differ in this regard and
are therefore listed separately on this basis (e.g., alcohol dehydrogenase from yeast and vertebrates, respectively).

Metal exchange studies have shown that the replacement of
catalytic and regulatory zinc with other metal ions can profoundly
affect activity, whereas that of structural and noncatalytic zinc
atoms has only minor consequences. Moreover, the coordination geometry of catalytic zinc appears to differ significantly from that of
structural and noncatalytic zinc atoms (the coordination properties
of regulatory zinc atoms are as yet unknown). Thus x-ray analysis
shows catalytic zinc to be bound by three protein ligands and a water
molecule, whereas structural and noncatalytic zinc are fully coordinated by four protein ligands. The presence of a water molecule
bound to catalytic zinc, signifying an open coordination site, is
considered essential for the function of zinc in catalysis. Furthermore, the coordination geometry of catalytic zinc is highly distorted
and fluctuates between tetracoordinate and pentacoordinate (e.g.,
carbonic anhydrase; see below) properties, which reflect it to be
entatic [8], whereas the coordination geometry of noncatalytic zinc
is more regular and hence not entatic. Entasis may be considered the
keystone to catalysis by zinc metalloenzymes because it lowers the
energy barrier for the transition state and hence accelerates the
conversion of substrate to products.

In spite of intensive research, the precise mode of action of zinc is still not fully understood in any metalloenzyme. From the earliest days of the study of zinc enzymes, two types of mechanisms have been thought to account for the manner in which the metal affects catalysis. One, termed the *zinc-carbonyl* mechanism, proposes that the substrate binds *directly* to zinc and displaces the metal-bound H_2O molecule in the process. The zinc is then envisaged to act as a Lewis acid and to polarize the bound substrate, thereby facilitating nucleophilic attack (e.g., aldolase, peptidases). Thus, according to this hypothesis, the function of the zinc is to activate the *electrophile*. The second, known as the *zinc-hydroxide mechanism*, proposes that the substrate does *not* bind directly to the metal and that zinc mediates its function through the metal-bound water molecule, which is *not* displaced on binding substrate. The zinc is thought to lower the pK_a of the bound water molecule from ~14 to ~7. The resultant metal-bound hydroxide ion can then attack the substrate (e.g., carbonic anhydrase). Hence, in this hypothesis, the function of the zinc is to activate the *nucleophile*.

These two hypotheses need not be mutually exclusive, and a mechanism that integrates the zinc-carbonyl and zinc-hydroxide mechanisms can be envisaged. According to this integrated hypothesis, the substrate binds *directly* to the metal but does *not displace* the metal-bound water, resulting in a *pentacoordinate* intermediate. The function of the zinc then would be *both* to polarize the substrate *and* to activate the water molecule, which acts as a nucleophile. In addition, the metal, through its flexible coordination geometry, would act as a template to bring together the substrate and nucleophile; this suggestion is consistent with the entatic nature of the metal. This third mechanism, originally proposed for carbonic anhydrase [9], is thought to find support in recent crystallographic studies showing that CO_2 and H_2O are bound simultaneously to the catalytic zinc atom in this enzyme [10]. However, at present the evidence for this mechanism is not decisive.

In the balance of this chapter we consider representative examples of each of the roles of zinc in metalloenzymes by using

carbonic anhydrase and alcohol dehydrogenase to illustrate a catalytic role, aspartate transcarbamylase for a structural role, leucine
aminopeptidase for a regulatory role, and alcohol dehydrogenase for
a noncatalytic role.

2. CATALYTIC ROLE OF ZINC: CARBONIC ANHYDRASE

Carbonic anhydrase (EC 4.2.1.1, carbonate hydrolyase) catalyzes the
reversible hydration of CO_2. It was first purified from bovine red
blood cells by Keilin and Mann [3], who found that it contains 0.33%
zinc.

The erythrocytes of mammals invariably contain a high-activity
isoenzyme of carbonic anhydrase, generally designated and abbreviated
CAC or CA II, and in addition often contain another genetically distinct isoenzyme of low activity referred to as CAB or CA I. Carbonic
anhydrase activity is also found in a large number of other mammalian
tissues [11].

HCAB and HCAC, the low- and high-activity enzymes of human
erythrocytes, as well as BCAB, the major enzyme from bovine erythrocytes, are the carbonic anhydrases best characterized. The properties
reported in this section pertain to all three of these enzymes.

The molecular weight of these erythrocyte carbonic anhydrases
is about 30,000, and they contain one atom of zinc per molecule. The
complete amino acid sequences of HCAB [12], HCAC [13], and BCAB [14]
have been reported (Table 2). HCAB consists of 260 and HCAC of 259
amino acid residues with a sequence homology of about 60%. The homology of BCAB with HCAC is higher than that with HCAB. In fact, the 19
residues known to be within a radius of 8 Å about the active-site zinc
atom in HCAC are invariant in BCAB, whereas in HCAB four of these
residues differ (Table 2; see also Ref. 15). In all the carbonic
anhydrases that have been sequenced thus far, the zinc-coordinating
groups are invariably 3-histidines [15].

The crystal structures of HCAB [16] and HCAC [17,18] determined
at pH 8.5 are known at 2-Å resolution (Fig. 1). The overall three-

TABLE 2

Amino Acid Sequences for HCAB and HCAC[a]

	1			5					10
HCAB:	Ac-Ala-Ser-Pro-Asp-Trp-Gly-Tyr-Asp-Asp-Lys-
HCAC:	Ac --- -His-His- -Gly-Lys-His-

		15					20		
Asn-Gly-Gln-Pro-Glu-Trp-Ser-Lys-Leu-Tyr-
-Pro-Glu-His-   -His-   -Asp-Phe-

		25					30		
Pro-Ile-Ala-Asn-Gly-Asn-Asn-Gln-Ser-Pro-
-Lys-   -Glu-Arg-

		35					40		
Val-Asp-Ile-Lys-Thr-Ser-Glu-Thr-Lys-His-
-Asp-   -His-Thr-Ala-   -Tyr-

		45					50		
Asp-Thr-Ser-Leu-Lys-Pro-Ile-Ser-Val-Ser-
-Pro-   -Leu-

		55					60		
Tyr-Asn-Pro-Ala-Thr-Ala-Lys-Glu-Ile-Ile-
-Asp-Gln-   -Ser-Leu-Arg-   -Leu-

		65					70		
Asn-Val-Gly-His-Ser-Phe-His-Val-Asn-Phe-
-Asn-   -Ala-   -Asn-   -Glu-

		75					80		
Glu-Asp-Asn-Asn-Asp-Arg-Ser-Val-Leu-Lys-
Asp-   -Ser-Gln-   -Lys-Ala-

		85					90		
Gly-Gly-Pro-Phe-Ser-Asp-Ser-Tyr-Arg-Leu-
-Leu-Asp-Gly-Thr-

		95					100		
Phe-Gln-Phe-His-Phe-His-Trp-Gly-Ser-Thr-
Ile-   -Leu-

		105					110		
Asn-Glu-His-Gly-Ser-Glu-His-Thr-Val-Asp-
Asp-Gly-Gln-

		115					120		
Gly-Val-Lys-Tyr-Ser-Ala-Glu-Leu-His-Val-
Lys-Lys-   -Ala-   -Leu-

		125					130		
Ala-His-Trp-Asn-Ser-Ala-Lys-Tyr-Ser-Ser-
Val-   Thr-----   -Gly-Asp-

TABLE 2 (Continued)

```
                    135                140
        Leu-Ala-Glu-Ala-Ala-Ser-Lys-Ala-Asp-Gly-
        Phe-Gly-Lys-    -Val-Gln-Gln-Pro-
                    145                150
        Leu-Ala-Val-Ile-Gly-Val-Leu-Met-Lys-Val-
                    -Leu-    -Ile-Phe-Leu-
                    155                160
        Gly-Glu-Ala-Asn-Pro-Lys-Leu-Gln-Lys-Val-
            -Ser-   -Lys-   -Gly-
                    165                170
        Leu-Asp-Ala-Leu-Gln-Ala-Ile-Lys-Thr-Lys-
        Val-    -Val-    -Asp-Ser-
                    175                180
        Gly-Lys-Arg-Ala-Pro-Phe-Thr-Asn-Phe-Asp-
                -Ser-    -Asp-
                    185                190
        Pro-Ser-Thr-Leu-Leu-Pro-Ser-Ser-Leu-Asp-
            -Arg-Gly-              -Glu-
                    195                200
        Phe-Trp-Thr-Tyr-Pro-Gly-Ser-Leu-Thr-His-
        Tyr-                              -Thr-
                    205                210
        Pro-Pro-Leu-Tyr-Glu-Ser-Val-Thr-Trp-Ile-
                    -Leu-    -Cys-
                    215                220
        Ile-Cys-Lys-Glu-Ser-Ile-Ser-Val-Ser-Ser-
        Val-Leu-        -Pro-
                    225                230
        Glu-Gln-Leu-Ala-Gln-Phe-Arg-Ser-Leu-Leu-
                -Val-Leu-Lys-        -Lys-    -Asn-
                    235                240
        Ser-Asn-Val-Glu-Gly-Asp-Asn-Ala-Val-Pro-
        Phe-    -Gly-            -Glu-Pro-Glu-Glu-Leu-
                    245                250
        Met-Gln-His-Asn-Asn-Arg-Pro-Thr-Gln-Pro-
            -Val-Asp-    -Trp-        -Ala-
                    255                260
        Leu-Lys-Gly-Arg-Thr-Val-Arg-Ala-Ser-Phe-
            -Asn-    -Gln-Ile-Lys-                    -Lys
```

[a]Only the residues that differ between the two sequences are shown for HCAC.

Source: Refs. 12 and 15.

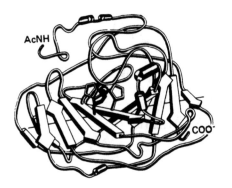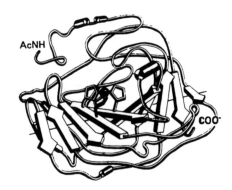

FIG. 1.   Schematic representation of the main chain folding in (A)
HCAC and (B) HCAB.   The arrows depict the β structure and the cylinders depict the helical segments in the peptide backbone.   The dark
ball in the middle of the molecule defines the position of the essential zinc atom ligated by three histidyl residues.   (From Refs. 16
and 18, with permission.)

dimensional structures of both enzymes are ellipsoidal and very
similar (Fig. 1), with an extensive β-pleated sheet transversing the
molecule.   The zinc atom is bound to this sheet near the center of
the molecule at the bottom of a cavity (Fig. 1).   The zinc coordination sphere is a distorted tetrahedron and consists of the N_τ atoms
of His 94(93) and His 96(95) and the N_π atom of His 119(118) in HCAB
(and HCAC, respectively) (Fig. 2).   A water molecule or a hydroxide
ion occupies the fourth coordination position (Fig. 2).   The metal-
bound solvent molecule is hydrogen bonded to Thr 199(197) (Fig. 2),
which in turn is hydrogen bonded to the buried Glu 106(105).   This
hydrogen-bonding system is a particularly prominent feature of the
active site of carbonic anhydrase.   The amino acid arrangement in the
active center is such that the cavity can be divided into essentially
hydrophobic and hydrophilic halves [19].

     In the far ultraviolet (UV), the circular dichroism (CD) and
optical rotatory dispersion (ORD) of HCAB and HCAC in solution are
invariant in the pH interval 4.0-10.5, indicating that the native
structure is maintained over this pH range [20,21]; outside this pH
range there is irreversible denaturation.   The zinc atom clearly
participates in maintaining the wide range of the enzyme's pH

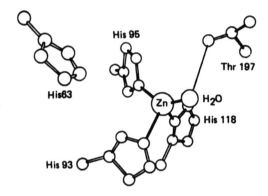

FIG. 2. Schematic representation of the active site of HCAC, showing
the metal-coordination geometry. (From Ref. 61, with permission.)

stability since apocarbonic anhydrase denatures below pH 5.5 [20].
Nevertheless, x-ray analysis of apo and zinc carbonic anhydrases
shows that their tertiary structures are very similar, indicating
that the metal atom is *not* essential in maintaining the overall
tertiary structure [22], much as its removal seems to destabilize
it and allow denaturation. Thus the density of the peak corresponding to the zinc atom is the only major feature in the electron-density
difference maps between apo and zinc HCAC crystals [22], and the
UV-ORD spectrum of the apoenzyme is very similar to that of native
HCAB [20]. Both CD studies of the relative susceptibilities of holo-
and apo-BCAB toward denaturation by guanidine hydrochloride [23] and
laser Raman scattering studies of HCAB [24] also demonstrate that
the tertiary structure of the apo- and holoenzymes are essentially
the same.

Chemical modification of carbonic anhydrase has implicated the
involvement of only a few amino acid side chains in catalysis. Amida-
tion of lysyl residues or nitration of tyrosyl residues does not
diminish activity [25,26], and although iodination of tyrosines in
HCAB does abolish activity, this seems to be due to a major conformational change [27]. Histidyl residues are the only ones that can be
modified specifically with concomitant loss of activity [28-32].

Chelating agents remove zinc from the enzyme with concomitant loss of activity, yielding a stable apoenzyme which can be reconstituted with Zn(II) and other divalent metal ions.

Substitution of Co(II) for zinc generates catalytic activities and pH-rate profiles similar to those of the native enzyme [33]. Substitution of Cd(II) for the zinc of HCAB [33,34] maximally results in 30% of the esterase activity of the native enzyme, but hydration of CO_2 is not restored. The apparent pK_a for esterase activity, 7.0 in the native enzyme, is shifted to 9.1 for the Cd(II)-substituted enzyme. Mn(II)-, Ni(II)-, Cu(II)-, and Hg(II)-substituted HCAB have less than 15% of the esterase activity of the native enzyme and no CO_2 hydration activity [33].

Optical spectroscopic [35-37] and x-ray crystallographic studies of HCAC [22] show that Zn(II), Co(II), Cu(II), and Mn(II) occupy the same binding site, whereas Hg(II) is displaced slightly in both HCAB and HCAC [22]. The apparent stability constants for these metal-enzyme complexes increase with pH and vary considerably for different metal ions (Table 3). These high-stability constants are due largely to the extremely slow rate of dissociation of the metal from the metalloprotein [38]. The half-life for the dissociation of zinc from native HCAB at neutral pH has been estimated to be 3 years!

TABLE 3

Stability Constants for Metallocarbonic Anhydrases

Metal	Log K, pH 5.5	pH-independent stability constant log K
Mn(II)	3.8	--
Co(II)	7.2	--
Ni(II)	9.5	--
Cu(II)	11.6	--
Zn(II)	10.5	15.0
Cd(II)	9.2	--
Hg(II)	21.5	--

Source: Ref. 36.

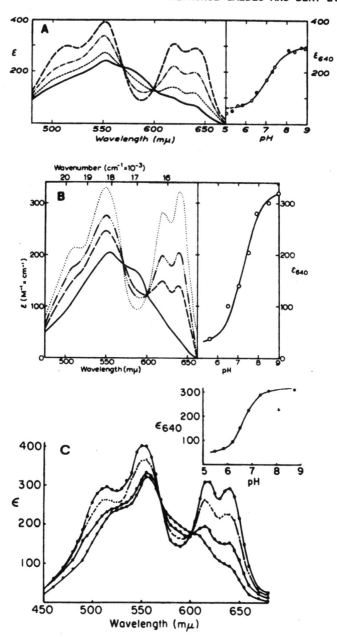

FIG. 3. Visible absorption spectra of cobalt carbonic anhydrase as a function of increasing pH, in order of increasing maximal absorption: (A) BCAB at pH 5.8, 6.5, 7.3 and 8.7, respectively; (B) HCAB at pH 5.6, 7.0, 7.5 and 9.0, respectively; (C) HCAC at pH 6.19, 6.50, 6.86 and 7.35, respectively. (From Refs. 35, 36, and 62, with permission.)

Detailed structural and functional information about carbonic anhydrase has been obtained by means of a variety of spectroscopic techniques, including optical absorption, natural and magnetic circular dichroism (CD and MCD), electron paramagnetic resonance (EPR), nuclear magnetic resonance (NMR), and perturbed angular correlation of γ rays (PAC). Most of these studies have been performed on the Co(II)- or Cd(II)-substituted carbonic anhydrases.

The absorption, CD, and MCD spectra of cobalt carbonic anhydrase are sensitive to interactions of ligands with the active center. At acidic pH the absorption spectrum of the cobalt enzyme has an intense absorption band ($\varepsilon \sim 200$) centered at about 550 nm (Fig. 3). Under these conditions the MCD spectrum exhibits a negative band centered at 600 nm and a small positive band at 520 nm [39]. As the pH is raised from 6 to 9 the absorption and MCD spectra split into four bands, centered at 520, 555, 615, and 645 nm (Fig. 3) [35,36,39]. The pK_a describing this spectral change is between pH 7 and 8 (Fig. 3), depending on the conditions [20,35-37], and is the same as the pK_a for the pH-rate profile of enzymatic activity (Fig. 4) [37,40-42]. The absorption and MCD spectra at acidic pH resemble those of tetrahedral Co(II) complex ions. Hence it has been postulated that the metal in this form of the enzyme exists in a distorted tetrahedral coordination geometry [39,43,44]. The spectra at alkaline pH do not closely resemble those of known Co(II) complex ions, but the widely split bands are very suggestive of pentacoordinate Co(II) ions and consistent with a highly distorted coordination geometry. Hence the metal in the alkaline form of the enzyme has been postulated to be in a pentacoordinate, or tetragonally distorted, tetrahedral geometry [39,43-45]. The group with pK_a 7-8 affecting both the spectral properties and the activity of the enzyme (Figs. 3 and 4) is generally assumed to be the metal-bound water molecule, which at alkaline pH is thought to dissociate into a metal-bound hydroxide ion [37,46,47] (see also below).

The binding of sulfonamide and anionic inhibitors also perturbs the absorption and MCD spectra of the cobalt enzyme and generates three distinct groups of absorption spectra, depending on the nature

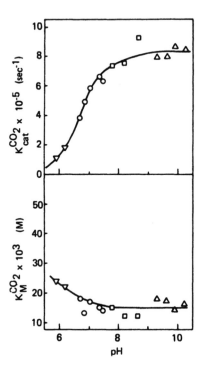

FIG. 4.   pH dependence of k_{cat} (upper graph) and K_m (lower graph) for the hydration of CO_2 catalyzed by BCAB.   (From Ref. 32, with permission.)

of the inhibitor (Fig. 5) [37,46,47].   These may be summarized as follows:

1.   Intense spectra, with $\varepsilon > 300$ in the visible region ($\lambda_{max} \sim 575$ nm), with $\varepsilon > 50$ in the infrared ($\lambda_{max} \sim 1000$ nm), and no absorption in the near infrared (770-700 nm) (sulfonamide, aniline, cyanide)

2.   Spectra with $\varepsilon \sim 200\text{-}300$ in the visible region and $\varepsilon \sim 10$ in the infrared, some of which absorb in the near infrared (chloride, bromide, benzoate)

3.   Low-intensity spectra with $\varepsilon \simeq 100$ in the visible and 10 in the infrared, and with characteristic absorption between 710 and 840 nm (acetate, nitrate, iodide)

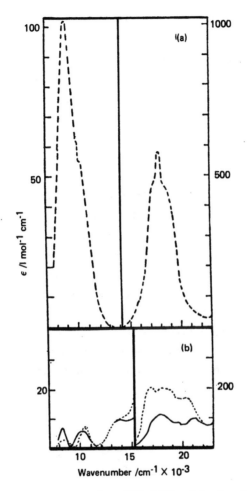

FIG. 5. Electronic spectra of Co(II)-BCAB with the following
inhibitors: (a) aniline at pH 6.0; (b) bromide (---) and acetate
(——) at pH 5.9. (From Ref. 45, with permission.)

These spectra have been interpreted to be indicative of tetracoordinate metal geometry (1), equilibrium mixtures between pentacoordinate and tetracoordinate species (2), and of pentacoordinate
metal geometry (3), respectively [45].

EPR [48] and PAC spectra [49] of Co(II)- and Cd(II)-substituted
HCAB, respectively, and x-ray crystallographic studies of zinc HCAB

[10] further establish that the metal in carbonic anhydrase has the
capacity to be five-coordinate.

Much evidence, including crystal structures and spectral data,
shows that monoanionic inhibitors react with carbonic anhydrase by
coordinating directly to the metal (Fig. 6) [22,50-54]. Sulfonamides
are a class of specific monoanionic inhibitors of carbonic anhydrase
[50,55] among which the aromatic sulfonamides are the most effective,
strongly suggesting that hydrophobic interactions with the protein
are important for their binding.

However, the metal atom is also essential for the binding of
sulfonamides, as these bind to the apoenzyme with an affinity 1000-
fold lower than that for the native enzyme [33]. Indeed, in common
with other inhibitors, the sulfonamides are thought to bind directly
to the metal atom (Fig. 6). Furthermore, only zinc and cobalt, the
two metals that result in carbonic anhydrase activity, induce sulfonamide binding [33]; evidently, some of the features of the active
site that potentiate activity also facilitate sulfonamide binding.

High-resolution ^{1}H NMR studies of HCAB and HCAC have led to the
assignment of most of the histidyl C_2 protons in these proteins [56,
57]. These studies are not thought to support the hypothesis [58]

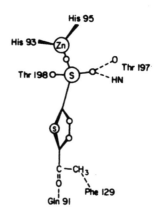

FIG. 6.   Schematic representation of the binding of acetazolamide
to HCAC.   (From Ref. 22, with permission.)

that the activity-linked group in the active site is a histidyl
residue [56,57,59]. In particular, the N-H protons of the histidines
that serve as ligands have been found to exchange slowly with the
solvent [59], in contradiction to the predictions of this hypothesis
[58].

The only *known* physiological function for carbonic anhydrase
is the catalysis of the reversible hydration of carbon dioxide
($H_2O + CO_2 = H^+ + HCO_3^-$). It should be noted, however, that physiological concentrations of Cl^- inhibit HCAB (but not HCAC) [54].
This observation has led to the suggestion that in vivo HBAB may be
utilized in reactions of intermediate metabolism rather than as a
carbonic anhydrase, although a recent detailed investigation has
failed to provide evidence for this hypothesis [60].

Primarily as the result of the work of Pocker and co-workers,
carbonic anhydrase is now known to catalyze a variety of hydration
and hydrolysis reactions which, although not necessarily of biological
significance, are of great mechanistic interest. They involve the
addition of hydroxide to a carbon-oxygen double bond or an analogue
thereof [32]. The rates of the hydration and dehydration reactions
catalyzed by carbonic anhydrase have inverse pH-rate profiles, with
the hydration velocity varying with pH as though it were proportional
to the basic form of an enzymatic group with a pK ~7, and as though
the dehydration velocity were proportional to the acidic form of the
same group. Interaction with monoanionic inhibitors shifts the
apparent pK_a of this group to higher pH values. As indicated above,
this group probably is the metal-bound water molecule, although the
ionization of a histidyl residue in the active site cannot be excluded
a priori [32]. The pH dependence of k_{cat} of both reactions parallel
these pH-rate profiles, while the K_m values for CO_2 and HCO_3^- are
essentially pH independent (Fig. 4) [63].

The available evidence suggests that zinc exerts its catalytic
function through the bound water molecule (see above). It is thought
that the zinc greatly lowers the pK_a of this H_2O molecule and affects
its ionization. The metal-bound hydroxide ion so generated is then
postulated to attack CO_2, which may be bound at a fifth site on the

metal. If this latter conjecture is correct, the metal has multiple
roles in the enzyme: It acts as a template to bring together the two
reactants, it behaves as a Lewis acid and produces the nucleophile
(OH^-), and it activates the electrophile (CO_2).

3.   STRUCTURAL ROLE OF ZINC: ASPARTATE TRANSCARBAMYLASE

Aspartate transcarbamylases are ubiquitous in living matter [64,65].
They catalyze the condensation of carbamylphosphate with L-aspartate
to give carbamyl-L-aspartate, the key precursor in the biosynthesis
of pyrimidines (Fig. 7). The enzyme from *E. coli* is reviewed here.
It has been studied intensively and requires zinc for structural
stability.

     E. coli aspartate transcarbamylase is oligomeric with a molecu-
lar weight of 310,000, as determined by sedimentation equilibrium
studies [66,67]. The holoenzyme exhibits marked homotropic and
heterotropic allosteric properties [68]. L-Aspartate binds coopera-
tively, and the activity of the enzyme is inhibited by several pyrimidine nucleotides (most notably CTP) and activated by the purine ATP.
Thus the enzyme is under feedback regulation, keeping a balance
between the concentration of the pyrimidine and purine nucleotides
in the organism.

     Treatment with mercurial reagents (e.g., p-mercuribenzoate or
neohydrin) dissociates the native enzyme into five subunits [66,69].

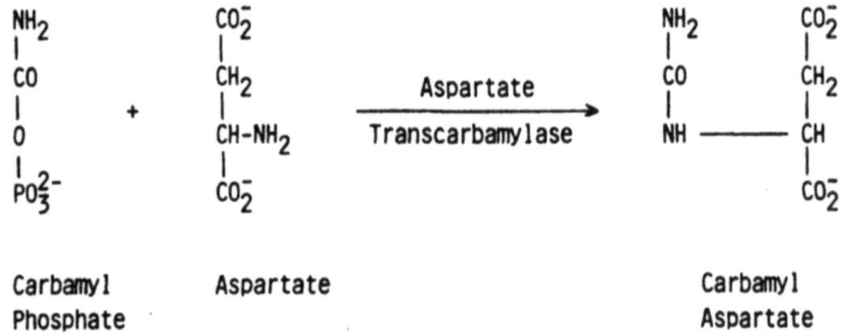

FIG. 7.   Reaction catalyzed by aspartate transcarbamylase.

Ion-exchange chromatography [69,70] or heat treatment separates the
subunits into two distinct types [71]. One, the catalytic or C subunit, has a molecular weight of 100,000 and is fully active but not
subject to heterotropic allosteric effects. The other, the regulatory or R subunit, with a molecular weight of 34,000, is enzymatically
inactive but binds the allosteric effector CTP [66,67]. Treatment of
an unfractionated mixture of C and R subunits with β-mercaptoethanol
allows reassociation to result in the native enzyme, with a subunit
composition of C_2R_3. The C subunit consists of three identical polypeptide chains, each with a molecular weight of 33,000 [67,72], and
the R subunit consists of two identical chains, each with a molecular
weight of 17,000 [72]. Denaturing agents [e.g., sodium dodecyl
sulfate (SDS) or guanidine hydrochloride] reversibly dissociate the
subunits into their constituent chains [67,72]. Hence the holoenzyme
consists of 12 polypeptide chains, 6 of which occur in two sets of
trimers (the C subunits), while the other 6 occur as three sets of
dimers (the R subunits). Differential scanning calorimetry shows
that the stability of the regulatory subunits against heat denaturation is increased by over 17°C when they are incorporated into the
native enzyme, whereas the catalytic subunits are much less affected
[73].

   E. coli aspartate transcarbamylase contains 6 g-atoms zinc per
mol [74,75]. Zinc is bound to the R subunits (one atom of zinc per
chain or two atoms per subunit) and is not required for the catalytic
activity of the enzyme. It is, however, essential for the maintenance
of the quaternary structure of the holoenzyme (see below).

   The primary structure of the R chain has been determined (Table
4), and that of the C chain is reported to be near to completion
(W. Konigberg, 1979; quoted in Ref. 76). The R chain consists of
152 residues with Met and Asn at the N and C termini, respectively
(Table 4). The chain contains four cysteinyl residues, all of which
are located in the C-terminal third of the molecule, in positions
109, 114, 137, and 140 (Table 4). Ala and Leu are the N- and C-
terminal residues of the C chain, respectively, and its sole Cys
residue is in position 46. A mutant form of E. coli aspartate

TABLE 4

Amino Acid Sequence of the R Chain of
E. coli Aspartate Transcarbamylase

	5	10	15
Met-Thr-His-Asn-Asp-Lys-Leu-Gln-Val-Ala-Glu-Ile-Lys-Arg-Gly-Thr-			

20
Val-Ile-Asn-His-Ile-Pro-Ala-Glu-Ile-

	30	35	40
-Gly-Phe-Lys-Leu-Leu-Ser-Leu-Phe-Lys-Leu-Thr-Glu-Thr-Gln-Asp-Arg-			

45   50
Ile-Thr-Ile-Gly-Leu-Asn-Leu-Pro-Ser-

	55	60	65
-Gly-Glu-Met-Gly-Arg-Lys-Asp-Leu-Ile-Lys-Ile-Glu-Asn-Thr-Phe-Leu-			

70   75
Ser-Glu-Asx-Glx-Val-Asx-Glx-Leu-Ala-

	80	85	90
-Leu-Tyr-Ala-Pro-Gln-Ala-Thr-Val-Asn-Arg-Ile-Asn-Asp-Tyr-Glu-Val-			

95   100
Val-Gly-Lys-Ser-Arg-Pro-Ser-Leu-Pro-

	105	110	115
-Glu-Arg-Asn-Ile-Asp-Val-Leu-Val-Cys-Pro-Asp-Ser-Asn-Cys-Ile-Ser-His-			

120   125
Ala-Glu-Pro-Val-Ser-Ser-Ser-Phe

	130	135	140
-Ala-Val-Arg-Arg-Ala-Asx-Asx-Ile-Ala-Leu-Lys-Cys-Lys-Tyr-Cys-Glu-Lys-			

145   150
Glu-Phe-Ser-His-Asn-Val-Val-Leu-Ala-Asn

Source: Ref. 72.

transcarbamylase lacking catalytic activity has been isolated [77,78]. This mutant is due to a single change in the C chain where Gly 125 replaces Asp.

X-ray diffraction studies have determined the tertiary structure of the enzyme in the presence and absence of CTP to a resolution of 2.8 Å [79]. Each C chain consists of a polar domain, which contains four helices (H1-H4) and a five-stranded parallel β sheet, and an equatorial domain, which is situated in the center of the molecule and composed of three helices (H6-H8) and a six-stranded parallel β sheet (Fig. 8); two helices (H5 and H9) join the two domains. The

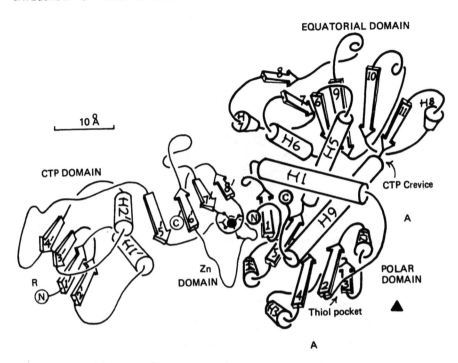

FIG. 8.   Schematic representation of a CR unit of aspartate transcarbamylase showing helices (cylinders) and sheets (arrows).   (From
Ref. 79, with permission.)

three chains in the C subunit are positioned in a triangular fashion
such that helix H2 of one chain is in contact with helix H3 and strand
S3 of the adjacent chain (Figs. 8 and 9), while interactions between
the two C subunits occur via the loops of the polypeptide chain connecting S7 to H7 and S9 to S10.   The active site of the C chain has
not yet been identified, but it is postulated to occur at one end of
the polar domain in a region that constitutes a secondary binding site
for CTP.   (In addition to acting as an allosteric effector, CTP, in
common with many other phosphates, is a competitive inhibitor of the
enzyme [80].)   If this interpretation is correct, the adjacent C
chains within the C subunit share the active site.

The zinc-binding site is located in the C-terminal region of
the R chain.   Zinc is bound by the four cysteinyl residues, in

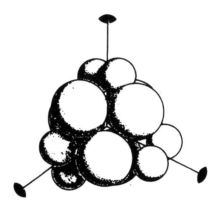

FIG. 9. Quaternary structure of aspartate transcarbamylase. The larger spheres represent the C subunits, and the smaller spheres the R subunits. (From Ref. 79, with permission.)

tetrahedral coordination geometry. The main CTP-binding domain is situated in the N-terminal region of the R domain and consists of two helices (H1' and H2') and a four-stranded β sheet (Fig. 8). The two chains of the R subunit are connected in the region of the CTP domain, such that the two β sheets in this domain form one continuous sheet and the helices H1' are in contact with the sheet. The zinc domain represents the major site of interaction between the R and C chains, with the polypeptide loops between the zinc-binding cysteinyl ligands being in close contact with the polar domain of the C chain. This interaction explains the importance of zinc for the association of the R and C subunits and the dissociative effect of mercurial reagents, which react with the cysteinyl ligands (see below).

X-ray diffraction studies have also defined the quaternary structure of the native enzyme [81,82]. The gross molecular structure has D_3 symmetry, with the three chains in the two different C subunits being nearly eclipsed (Fig. 9). Each C chain is associated with a C chain in the opposite subunit by interactions through a regulatory dimer; the two C chains connected through a given R subunit are 120° apart about the threefold axis (Fig. 9). In the

center of the molecule (between the two C subunits) is a large
(50 × 50 × 25 Å) aqueous cavity, bordered by the polar domains
(which include the active sites) of the C chain.

It has been postulated [82,83] that modulation of the access
of substrates to the central cavity mediates the allosteric properties of the enzyme. However, x-ray studies fail to indicate significant effects of the allosteric effector CTP on the conformation
of the molecule [79], and hence the basis of the allosteric properties of aspartate transcarbamylase is not apparent from these
studies.

Native E. coli aspartate transcarbamylase contains six atoms
of zinc per molecule [67,74,75]. When E. coli is cultured in zinc-
deficient media supplemented with Cd(II), the resultant enzyme is
fully active but contains 6 g-atoms of cadmium per mole [74]. Prolonged dialysis against chelating agents does not remove the metal,
and after 40 days of incubation no detectable exchange with $^{65}Zn(II)$
occurs [75]. Dissociation of the holoenzyme with mercurials results
in R subunits that do not contain zinc; instead, significant amounts
of mercury derived from the reagents are found [75], indicating that
Hg(II) displaces zinc during the dissociation reaction. However,
dissociation in the presence of excess Zn(II) results in R subunits
containing two atoms of zinc per subunit. In contrast, the C subunits do not contain any metal, and addition of zinc does not affect
their activity. Dialysis against chelating agents removes the zinc
from the isolated R subunit. The resultant metal-free subunits are
unstable, and their cysteinyl residues tend to dimerize [74]. They
also partly dissociate into R chains so that a monomer-dimer equilibrium is established, and they do not recombine with the C subunits.
The addition of Zn(II) stabilizes the dimeric structure of the R
subunits and restores their capacity of associating with the C subunits. Similarly, the addition of Cd(II), Co(II), Mn(II), Cu(II),
or Ni(II) restores the capacity of the R subunits to associate [75,
84]. The metal, however, is not required for the binding of CTP,
and the apo R subunits can bind this effector with an affinity comparable to that of the metalloderivative [69,74].

The essentiality of zinc for the stabilization of the quaternary structure of aspartate transcarbamylase has also been demonstrated in vivo [75]. Thus, when *E. coli* is grown in a zinc-
deficient medium, 70% of the enzyme synthesized is found as discrete
subunits.

The absorption spectra of the Zn(II)-, Cd(II)-, Hg(II)-, and
Ni(II)-substituted enzymes, and those of the metalloderivatives of
the isolated R subunits, indicate that the metals are bound to
cysteinyl ligands, since the charge-transfer bands in the UV region
are characteristic of the corresponding metal-mercaptide complexes
[75,84]. X-ray diffraction studies have confirmed this mode of
metal binding (see above). The substrate carbamylphosphate and the
substrate analogue succinate, but not CTP, perturb the absorption
and CD spectra of the zinc and cadmium enzymes [85]. Similarly, the
binding of the bisubstrate ligand N-(phosphonacetyl)-L-aspartate
perturbs the absorption spectrum of the nickel enzyme [84]. These
results indicate that the binding of substrates to the C subunit
affects the conformational state of the R subunit and that the metal
is not involved in binding the effector.

Chemical modifications have identified several amino acid
residues in the C chain that are necessary for catalytic activity.
Thus pyridoxal-5'-phosphate is an inhibitor of the enzyme, competitive with carbamylphosphate, and binds tightly (K_i = 1 µM) at a
single site per C chain [86]. This interaction involves the formation of a Schiff base, as judged by spectrophotometric measurements.
The pyridoxyl phosphate/enzyme complex can be reduced by sodium
borohydride to yield an inactive derivative that contains one pyri-
doxal phosphate per chain. Amino acid analyses reveal that the
reagent is incorporated as an N-ε-pyridoxyl-L-lysine [87]. The
modified residue has been identified as Lys 84, and x-ray diffraction studies place it at one end of the polar domain, in helix H3
[79]. The modification of lysyl residues by succinic anhydride also
results in loss of activity, although in this instance up to six
lysines per chain are modified [88].

The unreduced pyridoxyl phosphate/enzyme complex can be photo-oxidized to an inactive derivative in which two histidyl residues per C chain are oxidized specifically [86]. In this instance, the pyridoxyl phosphate acts as a photosensitizing agent and limits the oxidation to residues in the vicinity of the active site through its specific interaction with the enzyme. Bromosuccinate inactivates *E. coli* aspartate transcarbamylase with concomitant incorporation of one group per C chain, and this, too, is thought to be due to the modification of histidyl residues [65,89].

The sole cysteine in the C chain (at position 46) is unreactive toward most thiol reagents but may be modified specifically by potassium permanganate [90] and 2-chloromercuri-4-nitrophenol [81] with accompanying loss of activity; substrates protect against these modifications. However, this cysteine is not required for activity per se, for the introduction of a small group on the thiol, such as cyanide [91] or methyl [90], does not abolish activity. It is thought that bulky substituents on this cysteine sterically interfere with aspartate binding; the x-ray diffraction studies support this hypothesis, as Cys 46 is located close to Lys 84 in the polar domain.

Phenylglyoxal reacts with one arginine per C chain with concomitant loss of activity [92]. Substrates protect against this inactivation, and it appears that the modified residue is involved in the binding of carbamyl phosphate, although its position in the primary and tertiary structures has not been established.

Several of the eight tyrosyl residues per C chain can be modified selectively by tetranitromethane. Under mild conditions one tyrosine residue per C chain is modified [93,94]. The resultant enzyme retains 85% of the enzymatic activity and retains the homotropic and heterotropic effects seen with the native protein [94]. Further nitration successively leads to the loss of cooperativity in substrate binding and of enzymatic activity [95]; these two effects can be correlated with the modification of Tyr 213 and Tyr 160, respectively [95,96]. Carbamyl phosphate protects against the modification of the latter residue, and hence it is thought that Tyr 160 may be in the vicinity of the carbamyl phosphate binding site. CD difference spectra also suggest the involvement of tyrosine in the transmission of homotropic interactions, as the binding of carbamyl phosphate and of succinate (a substrate analogue) to the native enzyme is accompanied by the perturbation of tyrosyl residues [85].

The two tryptophans in the C chain can be replaced by 7-azatryptophan by including this reagent in the growth medium of *E. coli* [76]. The R chain does not contain tryptophan. The specific activity of the resulting protein is the same as that of native enzyme, but its allosteric properties are altered. The x-ray structure of aspartate transcarbamylase indicates that one of the tryptophyl residues of the native enzyme, Trp 257, is near the surface of the molecule pointing toward the exterior, whereas the other, Trp 199, is buried and close to the carbamyl phosphate-binding site. It is therefore postulated that the altered kinetic behavior of the azatryptophyl enzyme is due to the modification of the latter residue [76]. A role for tryptophan in the allosteric properties of the native enzyme has also been inferred from changes in the enzyme's UV and CD spectra that accompany the binding of substrate and effectors [85,97].

Finally, as indicated above, the replacement of Gly 125 by Asp in the C chain results in the complete loss of catalytic activity [77,78]. This residue is distant from the active site, but its modifications affect the reactivity of Lys 84, Cys 46, and Tyr 213. Hence it is postulated that this substitution causes a change in the tertiary structure of the enzyme, rendering it inactive [78].

In spite of the large number of amino acid residues which are known to be important for catalysis (see above), the exact role of any one of these in the reaction is unknown, and thus the details of the mechanism of the enzyme remains obscure.

4. REGULATORY AND CATALYTIC ROLES OF
ZINC: LEUCINE AMINOPEPTIDASE

Aminopeptidases have now been identified in all forms of life, and
a considerable number of them have been shown to be zinc metalloenzymes. They catalyze the specific hydrolysis of N-terminal amino
acid residues from proteins, peptides, and amino acid amides, and
generally require a free α-amino (or α-imino) group in the L configuration. Otherwise, their substrate specificity is broad, and they
can remove most amino acids from the N terminus of amide linkages
[98]. In general, the zinc-containing aminopeptidases isolated from
mammalian sources are oligomeric with molecular weights in excess of
200,000 and contain 2 g-atoms of Zn(II) per mol of subunit. In contrast, the microbial enzymes tend to be monomeric, with molecular
weights of approximately 40,000 containing 1-2 g-atoms Zn(II) per mol
enzyme. The leucine aminopeptidases (α-aminoacyl peptide hydrolase;
EC 3.4.11.1), isolated from mammalian tissues, are the best characterized, and hence the remainder of this discussion will be limited to
them.

Leucine aminopeptidase was first observed in extracts of porcine
intestinal mucosa by Linderstrom-Lang in 1929 [99]. Subsequently,
immunologically homologous enzymes were identified in porcine kidney,
bovine lens, and other cystosolic fractions of a variety of vertebrate species [100-102]. Particulate leucine aminopeptidase, isolated
from pig kidney, is seemingly a distinct enzyme and unrelated to
cystosolic leucine aminopeptidase [102].

The designation leucine aminopeptidase is actually a misnomer
because hydrolytic activity is also observed toward a large number
of other N-terminal amino acids, particularly those with hydrophobic
amino acids. However, the name leucine aminopeptidase has been retained to distinguish this enzyme from other aminopeptidases and to

indicate that leucyl amides are among those which are hydrolyzed
most rapidly. The order for the rate of hydrolysis of N-terminal
amino acids is Leu > Phe > Val > Ala > Gly > other [103]. The
presence of a D amino acid residue or of proline in the penultimate
position will retard hydrolysis. Leucine aminopeptidase also exhibits
esterase activity; L-leucine and L-tryptophan esters are the best substrates [104]. Leucine aminopeptidase from bovine lens has been characterized extensively and will be discussed below. It is hexameric
with a subunit molecular weight of 54,000 [105,106] and can be dissociated into six identical subunits by treatment with 7 M urea,
3.7 M guanidine chloride, or 0.17% SDS [105]. Electron microscopic
studies of the native enzyme in solution reveal that the six subunits
are arranged at the vertices of a distorted triangular prism [107,
108].

Limited tryptic digestion of bovine lens leucine aminopeptidase
results in the cleavage of one specific peptide bond per subunit
[109]. Despite the cleavage of this bond, the enzyme aggregate remains intact and retains all of its catalytic properties, including
activation by Mn(II) (see below). Dissociation of the trypsin-cleaved
enzyme into its constituent subunits reveals that this treatment
splits each subunit into two fragments, with molecular weights of
17,000 and 34,000, corresponding to its N- and C-terminal parts,
respectively. Trypsin does not further digest these fragments, and
they do not appear to be linked by a disulfide bond in the native
enzyme. Chymotrypsin, plasmin, and thrombin do not cleave the native
enzyme. The amino acid sequence of the N-terminal cyanogen bromide
fragment of the enzyme has been determined [110]. This fragment
contains 171 residues and has a molecular weight of 18,637 (Table 5).
The sequence indicates that the bond specifically cleaved by trypsin
is between Arg 137 and Lys 138. Circular dichroism studies have been
interpreted to denote that the native enzyme possesses little β
structure [111,112]. The secondary structure is reported to be 36%
α helix, 15% β sheet, and 49% random coil. The determination of the
three-dimensional structure of leucine aminopeptidase by x-ray

TABLE 5

Amino Acid Sequence of the N Terminal Cyanogen Bromide
Fragment of Bovine Lens Leucine Aminopeptidase

10
Thr-Lys-Gly-Leu-Val-Leu-Gly-Ile-Tyr-Ser-Lys-Glu-Lys-Glu-Glu-Asp-Glu-
20   30
Pro-Gln-Phe-Thr-Ser-Ala-Gly-Glu-Asn-Phe-Asn-Lys-Leu-
                                 40
Val-Ser-Gly-Lys-Leu-Arg-Glu-Ile-Leu-Asn-Ile-Ser-Gly-Pro-Pro-Leu-Lys-Ala-
50   60
Gly-Lys-Thr-Arg-Thr-Phe-Tyr-Gly-Leu-His-Glu-Asp-
                                 70
Phe-Pro-Ser-Val-Val-Val-Val-Gly-Leu-Gly-Lys-Lys-Thr-Ala-Gly-Ile-Asp-Glu-
80   90
Gln-Glu-Asn-Trp-His-Glu-Gly-Lys-Glu-Asn-Ile-Arg-
                                 100
Ala-Ala-Val-Ala-Ala-Gly-Cys-Arg-Gln-Ile-Gln-Asp-Leu-Glu-Ile-Pro-Ser-Val-
110   120
Glu-Val-Asp-Pro-Cys-Gly-Asp-Ala-Gln-Ala-Ala-Ala-
                                 130
Glu-Gly-Ala-Val-Leu-Gly-Leu-Tyr-Glu-Tyr-Asp-Asp-Leu-Lys-Gln-Lys-Arg-Lys-
140   150
Val-Val-Val-Ser-Ala-Lys-Leu-His-Gly-Ser-Glu-Asp-
                                 160
Gln-Glu-Ala-Trp-Gln-Arg-Gly-Val-Leu-Phe-Ala-Ser-Gly-Gln-Asn-Leu-Ala-Arg-
170
Arg-Leu-Met

Source: Ref. 110.

analysis is in progress [113]. The enzyme crystallizes in the hexa-
gonal space group $P6_322$ with unit cell dimensions a = 132 Å and C =
122 Å.

   The enzyme contains eight half-cysteinyl residues per subunit,
six of which are found as the free sulfhydryl group [114]. There are
indications that some of these sulfhydryl groups may act as zinc-
binding ligands [114]. Chemical modification has also implicated
two histidyl residues per subunit in metal binding [115].

   Chemical modifications of leucine aminopeptidase have given
little information concerning the groups or side chains (other than

the metal atoms) which are involved in catalytic activity. Oxidation
of tryptophan by N-bromosuccinimide decreases activity by 50%. On
the other hand, chemical modification of tyrosyl residues has little
effect on enzymatic activity.[116]. Thiol reagents specifically
modify a single cysteine group per subunit of the native enzyme but
without any effect on activity [114]. None of the residues noted
above appear to be essential for the hydrolytic activity of the
enzyme.

Crystalline bovine lens leucine aminopeptidase contains two
atoms of zinc per subunit [114,117]. Zinc can be removed by treatment with 1,10-phenanthroline to yield the enzymatically inactive
apoprotein. The electrophoretic and immunological properties of
the apoenzyme, which retains the native hexameric structure, are
identical with those of the holoprotein [102]. Hence zinc does not
play a structural role in this enzyme. Readdition of zinc to the
apoenzyme restores the enzymatic activity.

Prolonged incubation of zinc leucine aminopeptidase with $CoCl_2$
generates an active enzyme containing two atoms of cobalt per subunit
[118]. Cobalt and zinc compete for two independent binding sites
per subunit of the enzyme. At pH 7.5, the ratio of the association
constants for zinc and cobalt are 115 and 15.9 for sites 1 and 2,
respectively. The di-cobalt enzyme is 15 times as active as the
native zinc enzyme.

Addition of excess Mg(II) or Mn(II) to the native enzyme
enhances activity [114,119] due to the replacement of one of the
zinc atoms per subunit by the added metal (Fig. 10). However, addition of either Mg(II) or Mn(II) to the apoenzyme does not restore
enzymatic activity [114]. These results indicate that the binding
of zinc at one site per subunit is essential for catalytic activity.
Whereas cobalt can replace zinc at this site with retention of
activity, other metals are ineffective. On the other hand, a number
of metal ions [e.g., Zn(II), Co(II), Mg(II), Mn(II), Fe(II), Ni(II),
and Cu(II)] can bind at a second site per subunit and *regulate*
enzymatic activity. Hence the two metal atoms bound to this leucine
aminopeptidase function in distinct ways: one has a *catalytic*

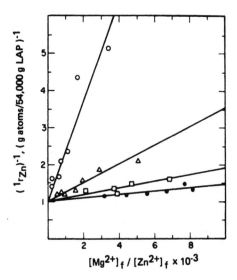

FIG. 10. Reciprocal plot of the number of zinc atoms per subunit
bound at the regulatory site of bovine lens aminopeptidase versus
the Mg/Zn ratio at pH 8.16 (●), 8.44 (□), 8.78 (△), and 9.14 (○).
The intercept of unity on the y axis indicates that Mg(II) replaces
zinc at only one site per subunit, while the increase in the slope
with increasing pH indicates that Mg(II) binds tighter to this site
at higher pH. (From Ref. 119, with permission.)

function and induces activity, whereas the other *regulates* the
activity induced by the zinc atom at the first site. The nature of
the metal occupying the activation site profoundly affects k_{cat}, but
K_m is only slightly affected. The order of activation for various
metals binding at this site is Mn(II) > Mg(II) > Fe(II) > Co(II) ~
Ni(II) > Zn(II) > Cu(II) [102]. However, zinc is bound much more
tightly at this activation site than any other metal [114,118,119];
for example, the ratio of the apparent association constants for
zinc and magnesium is estimated to be $2 \cdot 10^5$ (Fig. 10). Hence it is
doubtful whether under physiological conditions any cation other
than zinc acts as an activator.

It has also been known for a long time that porcine kidney
leucine aminopeptidase is activated by metal ions [100]. The enzyme
as isolated contains one zinc atom per subunit which is essential

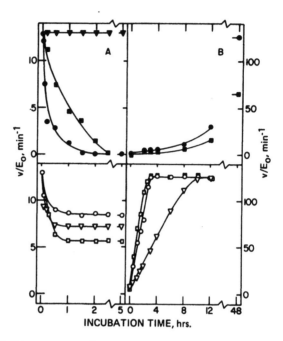

FIG. 11.   (A) Time course for the inhibition of porcine kidney
leucine amino peptidase, containing one zinc atom per subunit by:
(▼) Ca(II); (■) Cd(II); (●) Hg(II); (O) Ni(II); (▽) Cu(II); (□)
Zn(II).   (B) Reversal of this inhibition by added Mg(II).   (From
Ref. 120, with permission.)

for catalysis, and its activity is again regulated by the binding
of divalent metal ions at a second site per subunit [120].   However,
with this enzyme Mn(II) and Mg(II) activate, whereas Ni(II), Cu(II),
Zn(II), Hg(II) and Cd(II) inhibit (Fig. 11).   Since Zn(II) and Mg(II)
bind to the regulatory site of the porcine enzyme with comparable
affinities, it has been suggested that competition of these two
metals for the regulatory site may play a role in regulating the
activity of the enzyme under physiological conditions [120].

Anions can also exert an activating effect on bovine lens
aminopeptidase, but this phenomenon has been studied much less
extensively.   The following order of anion effectiveness has been
established: $F^- < SO_4^{2-} < Cl^- < N_3^- < Br^- < SCN^- < ClO_4^- < I^- < OH^-$
[121,122].

5. NONCATALYTIC AND CATALYTIC ROLES
OF ZINC: ALCOHOL DEHYDROGENASE

Alcohol dehydrogenases are NAD(H)-dependent enzymes that catalyze the interconversion of ethanol and other primary alcohols with the corresponding aldehydes; certain secondary alcohols and sterols are also substrates. Alcohol dehydrogenase activity has been detected in all organisms in which it has been sought, and the enzymes isolated from a wide variety of species have been shown to be zinc metalloenzymes [123]. The properties of the equine-liver enzyme are the focus of the following summary.

Equine-liver alcohol dehydrogenase is dimeric; each of the two identical subunits has a molecular weight of 40,000 and is composed of a single polypeptide chain containing 374 amino acids [124]. There are three major isoenzymes which are of genetic origin, due to the combination of two types of subunits (E, for ethanol active, and S, for steroid active), to result in the dimers EE, ES, and SS [125]. The activity of the EE isoenzyme toward ethanol is higher than that of the SS isoenzyme, whereas the converse is true for steroid substrates; the activity of the hybrid ES isoenzyme is intermediate between the two. In addition, there are at least nine minor nongenetic forms of the equine enzyme, usually attributed to secondary modification of the three major isoenzymes [126,127].

Each subunit of the enzyme contains two zinc atoms and binds one molecule of NAD(H) [128,129]; one of the zinc atoms is essential for activity, whereas the function of the other is unknown and is referred to as noncatalytic [130,131]. The integrity of the dimeric structure is independent of either zinc atom, since their complete removal seemingly does not affect it [131]. However, treatment with guanidinium chloride or urea dissociates the dimers of the native enzyme into monomers, with concomitant alterations of the tertiary structure and loss of activity [132,133].

The primary sequence of the E subunit (Table 6) has been determined [124]. The distribution of some of the amino acids in the sequence is markedly skewed. Thus six of the seven histidyl

TABLE 6

Amino Acid Sequence of the E Subunit of Equine-Liver Alcohol Dehydrogenase[a]

```
     1                 5                10                15                20
Acetyl-Ser-Thr-Ala-Gly-Lys-Val-Ile-Lys-Cys-Lys-Ala-Ala-Val-Leu-Trp-Glu-Glu-Lys-Lys-Pro-
                                                                                       Gln

    25                30                35                40
Phe-Ser-Ile-Glu-Glu-Val-Glu-Val-Ala-Pro-Pro-Lys-Ala-His-Glu-Val-Arg-Ile-Lys-Met-

    45                50                55                60
Val-Ala-Thr-Gly-Ile-Cys-Arg-Ser-Asp-Asp-His-Val-Val-Ser-Gly-Thr-Leu-Val-Thr-Pro-

    65                70                75                80
Leu-Pro-Val-Ile-Ala-Gly-His-Glu-Ala-Ala-Gly-Ile-Val-Glu-Ser-Ile-Gly-Glu-Gly-Val-

    85                90                95               100
Thr-Thr-Val-Arg-Pro-Gly-Asp-Lys-Val-Ile-Pro-Leu-Phe-Thr-Pro-Gln-Cys-Gly-Lys-Cys-
                                                  Ile

   105               110               115               120
Arg-Val-Cys-Lys-His-Pro-Glu-Gly-Asn-Phe-Cys-Leu-Lys-Asn-Asp-Leu-Ser-Met-Pro-Arg-
Ser                    Leu                              Ser

   125              ·130               135               140
Gly-Thr-Met-Gln-Asp-Gly-Thr-Ser-Arg-Phe-Thr-Cys-Arg-Gly-Lys-Pro-Ile-His-His-Phe-

   145               150               155               160
Leu-Gly-Thr-Ser-Thr-Phe-Ser-Gln-Tyr-Thr-Val-Val-Asp-Glu-Ile-Ser-Val-Ala-Lys-Ile-

   165               170               175               180
Asp-Ala-Ala-Ser-Pro-Leu-Glu-Lys-Val-Cys-Leu-Ile-Gly-Cys-Gly-Phe-Ser-Thr-Gly-Try-
```

```
        185         190         195         200
Gly-Ser-Ala-Val-Lys-Lys-Val-Ala-Lys-Val-Thr-Gln-Gly-Ser-Thr-Cys-Ala-Val-Phe-Gly-Leu-
        205         210         215         220
Gly-Gly-Val-Gly-Leu-Ser-Val-Ile-Met-Gly-Cys-Lys-Ala-Ala-Gly-Ala-Ala-Arg-Ile-Ile-
        225         230         235         240
Gly-Val-Asp-Ile-Asn-Lys-Asp-Lys-Phe-Ala-Lys-Ala-Lys-Glu-Val-Gly-Ala-Thr-Glu-Cys-
        245         250         255         260
Val-Asn-Pro-Gln-Asp-Tyr-Lys-Lys-Pro-Ile-Gln-Glu-Val-Leu-Thr-Glu-Met-Ser-Asn-Gly-
        265         270         275         280
Gly-Val-Asp-Phe-Ser-Phe-Glu-Val-Ile-Gly-Arg-Leu-Asp-Thr-Met-Val-Thr-Ala-Leu-Ser-
        285         290         295         300
Cys-Cys-Gln-Glu-Ala-Tyr-Gly-Val-Ser-Val-Ile-Val-Gly-Val-Pro-Pro-Asp-Ser-Gln-Asn-
        305         310         315         320
Leu-Ser-Met-Asn-Pro-Met-Leu-Leu-Leu-Ser-Gly-Arg-Thr-Trp-Lys-Gly-Ala-Ile-Phe-Gly-
        325         330         335         340
Gly-Phe-Lys-Ser-Lys-Asp-Ser-Val-Pro-Lys-Leu-Val-Ala-Asp-Phe-Met-Ala-Lys-Lys-Phe-
        345         350         355         360
Ala-Leu-Asp-Pro-Leu-Ile-Thr-His-Val-Leu-Pro-Phe-Glu-Lys-Ile-Asn-Glu-Gly-Phe-Asp-
        365         370
Leu-Leu-Arg-Ser-Gly-Glu-Ser-Ile-Arg-Thr-Ile-Leu-Thr-Phe
Lys
```

[a]The six known exchanges of the S subunit (at positions 17, 94, 101, 110, 115, and 336) are given below the corresponding E-chain residues.

Source: Refs. 124 and 134.

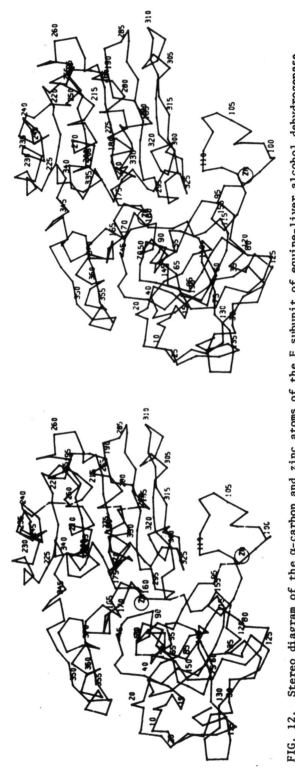

FIG. 12. Stereo diagram of the α-carbon and zinc atoms of the E subunit of equine-liver alcohol dehydrogenase. (From Ref. 123, with permission.)

residues are located in the N-terminal half of the molecule, and
none of the 14 cysteinyl residues occur in its C-terminal quarter
(Table 6). Peptide mapping of tryptic digests and partial sequence
studies of the EE and SS isoenzymes have revealed a high degree of
homology between the E and S subunits [134] since they differ in
only six positions (Table 6).

The x-ray crystallographic structure of equine-liver alcohol
dehydrogenase has been determined to a resolution of 2.4 Å [135].
The dimeric molecule is ellipsoid, with the ends being wider than
the central section (Fig. 12). The helical content is 28%, and the
β-pleated sheet content is 34%. Each subunit consists of two unequal
domains separated by a broad, deep cleft lined by hydrophobic residues
which constitute the active center. The larger domain consists of
231 residues and encompasses the catalytic area. The two zinc atoms
within each subunit are bound in this area, one (the "catalytic"
zinc) in the active site and the other (the "noncatalytic" zinc) 25 Å
away near the surface of the molecule. The other domain, composed of
143 residues, constitutes the coenzyme-binding site. The two subunits
of the dimeric molecule are joined at the coenzyme-binding areas so
that these regions form a central core in the dimeric enzyme surrounded by the active-site clefts on either side. Located at the
bottom of the active-site cleft, the catalytic zinc is coordinated
to the thiol groups of Cys 46 and Cys 174, the N_3 atom of His 67,
and the O atom of a water or a hydroxide molecule; the overall ligand
geometry around the metal is a distorted tetrahedron (Fig. 13). The
solvent molecule bound to the zinc atom is part of a system of hydrogen bonds that includes Ser 48 and His 51. The two catalytic metal
atoms of the dimer are 47 Å apart. In the center of a loop of the
peptide chain is the noncatalytic zinc, fully coordinated to the
thiol groups of Cys 97, Cys 100, Cys 103, and Cys 111 in a distorted
tetrahedral arrangement (Fig. 13); this region is probably not essential for the catalytic activity of the enzyme. Although it has been
postulated [128,136] that the noncatalytic zinc atom is required for
the structural stability of the enzyme, the tertiary structure provides no evidence for this conjecture.

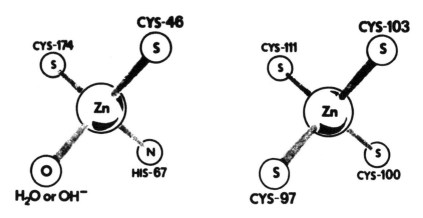

FIG. 13. Schematic representation of the ligands and coordination geometry of the catalytic (left) and noncatalytic (right) zinc atoms of equine liver alcohol dehydrogenase. (From Ref. 123, with permission.)

Since the formation of the enzyme-NADH complex induces major conformational changes, x-ray crystallographic studies of NAD(H) binding were confined initially to the study of the coenzyme analogue ADP-ribose [137]. The data showed that the adenine moiety is bound in a hydrophobic pocket at one end of the coenzyme-binding crevice, with the amino group pointing outwardly into the solution. The pyrophosphate group is salt linked to the guanidinium group of Arg 47, which serves as a general anion-binding site in alcohol dehydrogenase [135,138-140]. The adenosine ribose moiety is bound to one of the carboxyl oxygens of Asp 223, and the nicotinamide ribose is hydrogen-bonded to the carbonyl oxygens of Ile 269 and Gly 293. The nicotinamide group fits into a hydrophobic pocket, with the C_4 atom being 4.5 Å away from the catalytic zinc. More recently, technical problems in studying the binding of NADH to alcohol dehydrogenase have been overcome [141]. The crystallographic data indicate that NADH binds in a fashion similar to ADP-ribose, inducing a conformational change which involves rotation of the catalytic domain. In this manner, the active-site cleft is narrowed, and the substrate and the catalytic zinc atom are shielded from solution.

X-ray crystallographic analysis has also proven decisive in clarifying the mode of binding of competitive inhibitors and substrate analogues to the enzyme. The inhibitors 1,10-phenanthroline and imidazole bind to the catalytic zinc atoms, displacing the metal-bound water molecule; the binding of 1,10-phenanthroline, a bidentate agent, results in a pentacoordinate complex. Several alcohols and the substrate analogue dimethyl sulfoxide also bind directly to the catalytic zinc atom [142].

The functional groups implicated by the tertiary structure are those previously proposed on the basis of chemical modifications. Thus 2,3-butanedione and phenylglyoxal inactivate alcohol dehydrogenase through the modification of two arginyl residues [138]. The coenzyme no longer binds to the modified enzyme, and NAD(H) protects the enzyme against inactivation. These results implicate arginyl residues in coenzyme binding, and the tertiary structure confirms that an argininyl residue, Arg 47, is involved in the binding of the coenzyme (see above).

One of the 14 cysteinyl residues per subunit can be modified specifically by iodoacetate, with concomitant loss of activity [143, 144]. This residue has been identified as Cys 46, one of the ligands to the catalytic zinc atom. Interestingly, modification of Arg 47 by butanedione markedly reduces the rate of carboxymethylation of this cysteine residue [138,139]. When first observed, this effect was thought to be due to the juxtaposition of these residues, a conclusion confirmed by x-ray crystallography. The binding of the reagent to Arg 47 is thought to precede the modification of Cys 46 by iodoacetate, thus explaining the specific nature of this modification.

Bromoacetyl derivatives of NAD^+ inactivate alcohol dehydrogenase by selectively modifying another cysteinyl residue, and coenzyme protects against inactivation [145]. This residue has been identified as Cys 174, another ligand to the catalytic zinc atom [146]. These results suggest that this cysteine is close to the coenzyme-binding domain, as verified by studies of the tertiary structure.

The acetimdylation of lysyl residues enhances the enzymatic activity of equine-liver alcohol dehydrogenase by as much as 10-fold [147]. The group responsible for this effect is thought to be Lys 228 [148]. NAD^+ and NADH retard its modification, again indicating that it is located near the coenzyme-binding site. This suggestion has been corroborated by x-ray structural studies.

The presence of zinc in equine-liver alcohol dehydrogenase was first demonstrated by Vallee and Hoch in 1956 [149]. The differential reaction of the two zinc atoms in each subunit with 1,10-phenanthroline initially indicated that they might not play the same roles [136]. Since only one phenanthroline molecule binds to each subunit, completely inactivating the enzyme, it was concluded that only one of the two zinc atoms per subunit is involved in catalysis [136]. The x-ray studies have confirmed this deduction (see above).

Because a stable apoenzyme has not yet been prepared, metal restoration or substitution cannot be accomplished simply by metal addition. However, metal exchange can be achieved either by equilibrium dialysis [150,151] or removal and replacement of zinc from crystals [152]. The existence of both catalytic and noncatalytic metal atoms in the enzyme creates a significant problem in identifying the particular pair of metal atoms which is exchanged in the course of the procedure. Site-specific metal substitution in solution has been accomplished by monitoring the process with ^{65}Zn(II) [131,136,150,153,154]. Thus, when native equine-liver alcohol dehydrogenase (LADH) is dialyzed against 0.1 M sodium *acetate* buffer, pH 5.5, only two of the four zinc atoms exchange with ^{65}Zn(II), in a single first-order rate process, to yield [(LADH)^{65}Zn$_2$Zn$_2$] [150]. In marked contrast, when dialysis is carried out in 0.1 M sodium *phosphate* buffer, pH 5.5, *both* pairs of zinc atoms exchange with ^{65}Zn(II), each pair at a different first-order rate, resulting in [(LADH)^{65}Zn$_2$^{65}Zn$_2$]. The latter species is then redialyzed against stable Zn(II) in acetate buffer to give [(LADH)Zn$_2$^{65}Zn$_2$]. By this procedure, alcohol dehydrogenase specifically radiolabeled at the first (noncatalytic) or second (catalytic) metal-binding site is obtained. These derivatives allow precise monitoring of the rate,

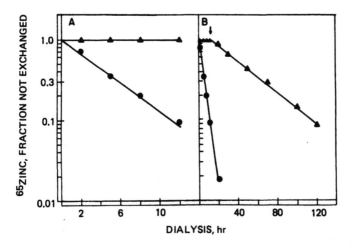

FIG. 14. Preparation, through controlled dialysis, of (A) $[(LADH)Co_2Zn_2]$ and (B) $[(LADH)Co_2Co_2]$. Aliquots of $[(LADH)^{65}Zn_2Zn_2]$ (•) and $[(LADH)Zn_2{}^{65}Zn_2]$ (▲) were dialyzed against 0.2 M $CoCl_2$, 0.1 M sodium acetate, pH 5.9. After 12 h the dialysate was replaced by 0.2 M $CoCl_2$, 0.1 M sodium acetate, pH 5.4. For details, see the text. (From Ref. 151, with permission.)

extent, and site specificity of subsequent replacement. Thus the dialysis of these two derivatives against Co(II) results in $[(LADH)Co_2Zn_2]$ and $[(LADH)Co_2{}^{65}Zn_2]$, respectively (Fig. 14), demonstrating that the Co(II) replaces only the Zn(II) at the first (noncatalytic) metal-binding site. The specific activity of this hybrid species is identical to that of the native enzyme, and addition of 1,10-phenanthroline results in circular dichroic spectra that are indistinguishable from those obtained with the native enzyme [150,151].

These observations clearly indicate that the metal atoms that are replaced first do correspond to the noncatalytic pair. The properties of the $[(LADH)Co_2Co_2]$ enzyme, obtained through the extensive dialysis of the hybrid species against Co(II) (Fig. 14), reinforce this conclusion. The specific activity of the fully substituted enzyme is lower than that of the native enzyme, and its reaction with 1,10-phenanthroline results in a distinctive absorption spectrum. Since it is known that 1,10-phenanthroline binds specifically

to the catalytic metal atoms (see above), these results further
demonstrate that the catalytic metal ions are the second pair to
exchange.

The absorption spectra of $[(LADH)Co_2Zn_2]$ and $[(LADH)Co_2Co_2]$
further differentiate the two pairs of metal atoms in the enzyme
[150,151]. Both $[(LADH)Co_2Zn_2]$ and $[(LADH)Co_2Co_2]$ are blue-green,
and their spectra exhibit absorption maxima at 340, 655, and 740 nm
(Fig. 15). However, the absorbance at 740 nm reflects only the
first pair of cobalt atoms (i.e., noncatalytic Co), whereas the
340- and 655-nm bands reflect both pairs of cobalt atoms (Fig. 15).
Moreover, both the wavelength and intensity of the absorption band
at 340 nm are consistent with charge transfer between cobalt and the
sulfur ligands of the two metal-binding sites.

Zinc in alcohol dehydrogenase can also be replaced by Cd(II)
to result in $[(LADH)Cd_2Zn_2]$ and $[(LADH)Cd_2Cd_2]$ [153]. The properties
of these species are in accord with the conclusions outlined above.

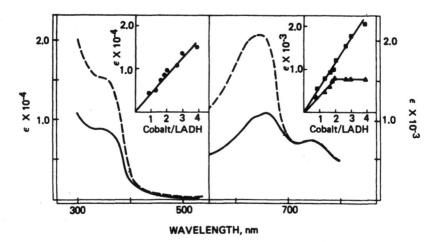

FIG. 15. Absorption spectra of $[(LADH)Co_2Zn_2]$ (——) and
$[(LADH)Co_2Co_2]$ (----). *Insets*: Absorption maxima of LADH versus
cobalt content: (left) 340 nm (●); (right) 655 nm (■) and 740 nm
(▲). (From Ref. 150, with permission.)

The fully substituted cadmium enzyme retains 14% of the enzymatic activity of the native enzyme. The UV spectra of the cadmium enzymes exhibit a band at 240 nm, with E_{240} of $1.6 \cdot 10^4$ M^{-1}/cm per noncatalytic cadmium and of $0.9 \cdot 10^4$ M^{-1}/cm per catalytic cadmium. This band is diagnostic of a cadmium thiolate charge-transfer transition.

The x-ray and the metal-replacement studies clearly show that the catalytic metal atoms are critically involved in the mechanism of the enzyme. In contrast, the function of the noncatalytic metal atoms is still obscure: they are remote from the active site, their replacement by Co(II) or Cd(II) does not affect activity, and there is no evidence that they are required for the structural stabilization of the enzyme. It was shown early, however, that their removal results in loss of enzymatic activity, although this seems to be an indirect consequence [154].

The kinetics of the reaction of alcohol dehydrogenase with primary alcohols and aldehydes follow a Theorell-Chance sequential mechanism, where the binding of the coenzyme is the first and dissociation the last (and rate-determining) step [155,156]. Transient kinetic studies have led to the suggestion that the enzyme shows coenzyme-induced half-site reactivity during catalysis, with negative cooperativity between two subunits [157,158]. Others maintain that the two subunits are kinetically equivalent [159-161].

The rate of dissociation of NADH from the enzyme is independent of pH, whereas that of NAD^+ is dependent on a pK_a of 7.6 [162-164]. The association rates for both NAD^+ and NADH are dependent on a pK_a of 9.2; the binding of NAD^+ reduces this pK_a to 7.6. The water molecule bound at the catalytic metal atom has been thought to be the group responsible for these pK_a values [129,165,166].

It has been proposed that the substrate can bind at the catalytic metal atom without displacing the water molecule, and that proton transfer between this water molecule and the substrate precedes hydride transfer during alcohol oxidation (Fig. 16) [166,167].

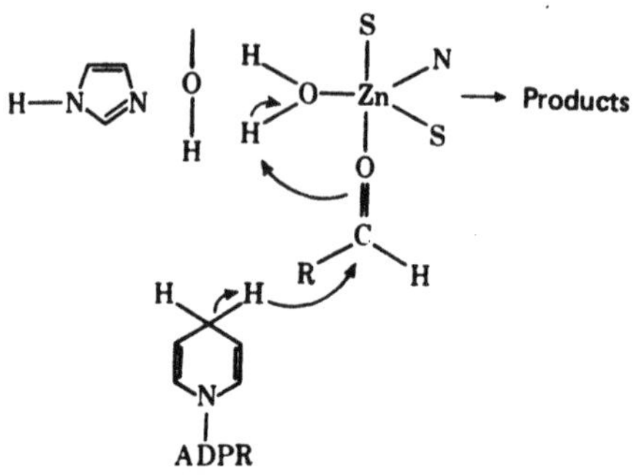

FIG. 16. Proposed mechanism for aldehyde reduction by alcohol
dehydrogenase via a five-coordinate intermediate. (From Ref. 166,
with permission.)

6. GENERAL CONCLUSIONS

The properties of the metalloenzymes reviewed above demonstrate that
zinc can play several distinct roles in enzymes. Its most common
function is to participate in catalysis; this function manifests by
the fact that the metal is indispensible for the enzymatic reaction.
Three of the four enzymes cited (i.e., carbonic anhydrase, leucine
aminopeptidase, and alcohol dehydrogenase) contain and are dependent
on catalytic zinc atoms. As indicated above, all evidence demonstrates that zinc is critically involved in their mechanism of action
but not in the stabilization of their tertiary or quartenary (for
leucine aminopeptidase) structure. A similar situation is found with
many other metalloenzymes. Indeed, about all of the enzymes listed
in Table 1 contain at least one catalytic zinc atom, with the exception of transcarbamylase (and possibly of superoxide dismutase),
although the exact mechanism of zinc in catalysis is not yet known.
Even less is known about the other roles of zinc in enzymes. A
structural role, whereby zinc acts by stabilizing the tertiary and/or
quartenary structure of the enzyme, has been demonstrated conclu-
sively only for two enzymes: aspartate transcarbamylase (reviewed
above) and B. subtilis α-amylase.

One of the two zinc atoms per subunit of equine and human
alcohol dehydrogenase has been thought to stabilize the tertiary
structure of the enzyme, although experimental evidence for this
assumption has never been found. There is, in fact, no known role
for this zinc atom, which has noncommitally been referred to as
noncatalytic (see Sec. 5). Interestingly, this noncatalytic zinc
atom is absent in yeast alcohol dehydrogenase, in spite of the fact
that the residues which serve as its ligands in the equine enzyme
are conserved in the yeast enzyme. *E. coli* alkaline phosphatase
similarly contains zinc atoms of unknown function, which by analogy
are best termed noncatalytical.

In exercising its regulatory function in, for example, bovine
and porcine leucine aminopeptidase (see Sec. 4) and fructose-1,6-
biphosphatase, zinc activates or inhibits but is not itself essential
for enzymatic activity.

Future investigation of metalloenzymes will probably uncover
yet other roles for zinc, which in terms of its biological functions
is thus far the most versatile of the transition and group IIB metals.

REFERENCES

1. J. Raulin, *Am. Sci. Natl. Bot. Biol. Veg.*, *11*, 93 (1869).

2. E. J. Underwood, in *Trace Elements in Human Animal Nutrition*, 3rd ed., Academic Press, New York, 1971, p. 208.

3. D. Keilin and T. Mann, *Biochem. J.*, *34*, 1163 (1940).

4. B. L. Vallee and H. Neurath, *J. Biol. Chem.*, *217*, 253 (1955).

5. B. L. Vallee, W. E. C. Wacker, A. F. Bartholomay, and E. D. Robin, *N. Engl. J. Med.*, *255*, 403 (1956).

6. E. J. Moynahan, *Lancet*, *2*, 399 (1974).

7. B. L. Vallee, E. A. Stein, W. N. Summerwell, and E. H. Fischer, *J. Biol. Chem.*, *234*, 2901 (1959).

8. B. L. Vallee and R. J. P. Williams, *Proc. Natl. Acad. Sci. USA*, *59*, 498 (1968).

9. M. F. Dunn, in *Structure and Bonding* (J. D. Dunitz, P. Hemmerich, R. H. Holm, J. A. Ibers, C. K. Jørgensen, J. B. Neilands, D. Reinen, and R. J. P. Williams, eds.), Springer-Verlag, Heidelberg, 1975, p. 103.

48 ALPHONSE GALDES AND BERT L. VALLEE

10. K. K. Kannan, M. Petef, K. Fridborg, H. Cid-Dresdner, and S. Lövgren, FEBS Lett., 73, 115 (1977).

11. M. J. Carter, Biol. Rev., 47, 465 (1972).

12. B. Andersson, P. O. Nyman, and L. Strid, Biochem. Biophys. Res. Commun., 48, 670 (1972).

13. L. E. Henderson, D. Henriksson, and P. O. Nyman, Biochem. Biophys. Res. Commun., 52, 138 (1973).

14. M. Sciaky, N. Limozin, D. Filippi-Foveau, J. M. Gulian, C. Dalmasso, and G. Laurent, C.R. Hebd. Seances Acad. Sci., Ser. D, 279, 1217 (1974).

15. L. E. Henderson, D. Henriksson, and P. O. Nyman, J. Biol. Chem., 251, 5457 (1976).

16. K. K. Kannan, B. Notstrand, K. Fridborg, S. Lövgren, A. Ohlsson, and M. Petef, Proc. Natl. Acad. Sci. USA, 72, 51 (1975).

17. K. K. Kannan, A. Liljas, I. Vaara, P.-C. Bergstén, S. Lövgren, B. Strandberg, U. Bengtsson, U. Carlbom, K. Fridborg, L. Järup, and M. Petef, Cold Spring Harbor Symp. Quant. Biol., 36, 221 (1971).

18. A. Liljas, K. K. Kannan, P.-C. Bergstén, I. Vaara, K. Fridborg, B. Strandberg, U. Carlbom, L. Järup, S. Lövgren, and M. Petef, Nature New Biol., 235, 131 (1972).

19. B. Notstrand, I. Vaara, and K. K. Kannan, in The Isozymes, 3rd Int. Conf. Isoenzymes, Vol. 1: Molecular Structure (C. L. Markert, ed.), Academic Press, New York, 1975.

20. J. E. Coleman, Biochemistry, 4, 2644 (1965).

21. S. Beychok, J. M. Armstrong, C. Lindblow, and J. T. Edsall, J. Biol. Chem., 241, 5150 (1966).

22. I. Vaara, in "The Molecular Structure of Human Carbonic Anhydrase, Form C and Inhibitor Complexes," Inaugural dissertation, UNIC-B22-2, Uppsala University, Sweden, 1974.

23. A. Yazgan and R. W. Henkens, Biochemistry, 11, 1314 (1972).

24. W. S. Craig and B. P. Gaber, J. Am. Chem. Soc., 99, 4130 (1977).

25. P. L. Whitney, P. O. Nyman, and B. G. Malmström, J. Biol. Chem., 242, 4212 (1967).

26. A. Nilsson and S. Lindskog, Eur. J. Biochem., 2, 309 (1967).

27. J. A. Verpoorte and C. Lindblow, J. Biol. Chem., 243, 5993 (1968).

28. S. L. Bradbury, J. Biol. Chem., 244, 2002 (1969).

29. S. L. Bradbury, J. Biol. Chem., 244, 2010 (1969).

30. P. L. Whitney, P. O. Nyman, and B. G. Malmström, J. Biol. Chem., 242, 4212 (1967).

31. P. L. Whitney, *Eur. J. Biochem.*, *16*, 126 (1970).

32. Y. Pocker and S. Sarkanen, *Adv. Enzymol.*, *47*, 149 (1978).

33. J. E. Coleman, *Nature (Lond.)*, *214*, 193 (1967).

34. R. Bauer, P. Limkilde, and J. T. Johansen, *Biochemistry*, *15*, 334 (1976).

35. S. Lindskog, *J. Biol. Chem.*, *238*, 945 (1963).

36. S. Lindskog and P. O. Nyman, *Biochim. Biophys. Acta*, *85*, 462 (1964).

37. J. E. Coleman, *J. Biol. Chem.*, *242*, 5212 (1967).

38. A. Y. Romans, M. E. Graichen, C. H. Lochmüller, and R. W. Henkens, *Bioinorg. Chem.*, *9*, 217 (1978).

39. B. Holmquist, T. A. Kaden, and B. L. Vallee, *Biochemistry*, *14*, 1454 (1975).

40. Y. Pocker and J. T. Stone, *J. Am. Chem. Soc.*, *87*, 5497 (1965).

41. Y. Pocker and J. T. Stone, *Biochemistry*, *7*, 2936 (1968).

42. S. Lindskog, *Biochemistry*, *5*, 2641 (1966).

43. A. B. Dennard and R. J. P. Williams, in *Transition Metal Chemistry*, Vol. 2 (R. L. Carlin, ed.), Marcel Dekker, New York, 1966, p. 115.

44. J. E. Coleman, in *Inorganic Biochemistry*, Vol. 1 (G. L. Eichhorn, ed.), Elsevier, New York, 1973, p. 488.

45. I. Bertini, G. Canti, C. Luchinat, and A. Scozzafava, *J. Am. Chem. Soc.*, *100*, 4873 (1978).

46. S. Lindskog, L. E. Henderson, K. K. Kannan, A. Liljas, P. O. Nyman, and B. Strandberg, in *The Enzymes*, 3rd ed., Vol. 5 (P. D. Boyer, ed.), Academic Press, New York 1971, p. 587.

47. A. Galdes and H. A. O. Hill, in *Specialist Periodical Report*, Vol. 1 (H. A. O. Hill, ed.), The Chemical Society, London, 1979, Chap. 8.

48. P. H. Haffner and J. E. Coleman, *J. Biol. Chem.*, *248*, 6630 (1973).

49. R. Bauer, P. Limkilde, and J. T. Johansen, *Carlsberg Res. Commun.*, *42*, 325 (1977).

50. T. H. Maren, *Physiol. Rev.*, *47*, 595 (1967).

51. K. Fridborg, K. K. Kannan, A. Liljas, J. Lundin, B. Strandberg, R. Strandberg, B. Tilander, and G. Wiren, *J. Mol. Biol.*, *25*, 505 (1967).

52. G. S. Brown, G. Navon, and R. G. Shulman, *Proc. Natl. Acad. Sci. USA*, *74*, 1794 (1977).

53. J. L. Sudmeier and S. J. Bell, *J. Am. Chem. Soc.*, *99*, 4499 (1977).

54. T. H. Maren, C. S. Rayburn, and N. E. Liddell, *Science, 191,* 469 (1976).

55. J. E. Coleman, *Ann. Rev. Pharmacol., 15,* 221 (1975).

56. I. D. Campbell, S. Lindskog, and A. I. White, *J. Mol. Biol., 90,* 469 (1974).

57. I. D. Campbell, S. Lindskog, and A. I. White, *J. Mol. Biol., 98,* 597 (1975).

58. R. K. Gupta and J. M. Pesando, *J. Biol. Chem., 250,* 2630 (1975).

59. I. D. Campbell, S. Lindskog, and A. I. White, *Biochim. Biophys. Acta, 484,* 443 (1977).

60. S. K. Chapman and T. H. Maren, *Biochim. Biophys. Acta, 527,* 272 (1978).

61. P.-C. Bergstén, I. Vaara, S. Lövgren, A. Liljas, K. K. Kannan, and U. Bengtsson, in *Proceedings of the Alfred Benzen Symposium IV* (M. Rorth and P. Astrup, eds.), Munksgaard, Copenhagen, 1971, p. 363.

62. P. W. Taylor, R. W. King, and A. S. V. Burgen, *Biochemistry, 9,* 3894 (1970).

63. Y. Pocker and D. W. Bjorkquist, *Biochemistry, 16,* 5698 (1977).

64. M. E. Jones, L. Spector, and P. Lipmann, *J. Am. Chem. Soc., 77,* 819 (1955).

65. G. R. Jacobson and G. R. Stark, in *The Enzymes,* 3rd ed., Vol. 9 (P. D. Boyer, ed.), Academic Press, New York, 1973, p. 226.

66. J. C. Gerhart and H. K. Schachman, *Biochemistry, 4,* 1054 (1965).

67. J. P. Rosenbusch and K. Weber, *J. Biol. Chem., 246,* 1644 (1971).

68. J. C. Gerhart and A. B. Pardee, *J. Biol. Chem., 237,* 891 (1962).

69. J. A. Cohlberg, V. P. Pigiet, and H. K. Schachman, *Biochemistry, 11,* 3396 (1972).

70. J. C. Gerhart and H. Holoubek, *J. Biol. Chem., 247,* 2886 (1967).

71. P. D. J. Weitzman, and I. B. Wilson, *J. Biol. Chem., 241,* 5481 (1966).

72. K. Weber, *Nature (Lond.), 218,* 1116 (1968).

73. L. P. Vickers, J. W. Donovan, and H. K. Schachman, *J. Biol. Chem., 253,* 8493 (1978).

74. J. P. Rosenbusch and K. Weber, *Proc. Natl. Acad. Sci. USA, 68,* 1019 (1971).

75. M. E. Nelbach, V. P. Pigiet, J. C. Gerhart, and H. K. Schachman, *Biochemistry, 11,* 315 (1972).

76. J. Foote, D. M. Ikeda, and E. R. Kantrowitz, *J. Biol. Chem., 255,* 5154 (1980).

77. K. A. Wall, J. E. Flatgaard, I. Gibbons, and H. K. Schachman, *J. Biol. Chem.*, *254*, 11910 (1979).

78. K. A. Wall and H. K. Schachman, *J. Biol. Chem.*, *254*, 11917 (1979).

79. H. L. Monaco, J. L. Crawford, and W. N. Lipscomb, *Proc. Natl. Acad. Sci. USA*, *75*, 5276 (1978).

80. R. W. Porter, M. O. Modebe, and G. R. Stark, *J. Biol. Chem.*, *244*, 1846 (1969).

81. D. C. Wiley, D. R. Evans, S. G. Warren, C. H. McMurray, B. F. P. Edwards, W. A. Franks, and W. N, Lipscomb, *Cold Spring Harbor Symp. Quant. Biol.*, *36*, 285 (1971).

82. S. G. Warren, B. F. P. Edwards, D. R. Evans, D. C. Wiley, and W. N. Lipscomb, *Proc. Natl. Acad. Sci. USA*, *70*, 1117 (1973).

83. D. R. Evans, S. G. Warren, B. F. P. Edwards, C. H. McMurray, P. H. Bethge, D. C. Wiley, and W. N. Lipscomb, *Science*, *179*, 683 (1973).

84. R. S. Johnson and H. K. Schachman, *Proc. Natl. Acad. Sci. USA*, *77*, 1995 (1980).

85. J. H. Griffin, J. P. Rosenbusch, K. K. Weber, and E. R. Blout, *J. Biol. Chem.*, *247*, 6482 (1972).

86. P. Greenwell, S. L. Jewett, and G. R. Stark, *J. Biol. Chem.*, *248*, 5994 (1973).

87. T. D. Kempe and G. R. Stark, *J. Biol. Chem.*, *250*, 6861 (1975).

88. E. A. Meighen, V. Pigiet, and H. K. Schachman, *Proc. Natl. Acad. Sci. USA*, *65*, 234 (1970).

89. D. S. Gregory and I. B. Wilson, *Biochemistry*, *10*, 154 (1971).

90. G. R. Jacobson and G. R. Stark, *J. Biol. Chem.*, *248*, 8003 (1973).

91. T. C. Vanaman and G. R. Stark, *J. Biol. Chem.*, *245*, 3565 (1970).

92. E. R. Kantrowitz and W. N. Lipscomb, *J. Biol. Chem.*, *251*, 2688 (1976).

93. M. W. Kirschner and H. K. Schachman, *Biochemistry*, *12*, 2987 (1973).

94. M. W. Kirschner and H. K. Schachman, *Biochemistry*, *12*, 2997 (1973).

95. S. M. Landfear, W. N. Lipscomb, and D. R. Evans, *J. Biol. Chem.*, *253*, 3988 (1978).

96. A. M. Lauritzen, S. M. Landfear, and W. N. Lipscomb, *J. Biol. Chem.*, *255*, 602 (1980).

97. K. D. Collins and G. R. Stark, *J. Biol. Chem.*, *244*, 1869 (1969).

98. R. J. Delange and E. L. Smith, in *The Enzymes*, 3rd ed., Vol. 3 (P. D. Boyer, ed.), Academic Press, New York, 1971, p. 81.

99. K. Linderstrom-Lang, *z. Physiol. Chem.*, *182*, 151 (1929).

100. D. H. Spackman, E. L. Smith, and D. M. Brown, *J. Biol. Chem.*, *212*, 255 (1955).

101. D. Glasser, M. John, and H. Hanson, *Hoppe-Seyler's Z. Physiol. Chem.*, *351*, 1337 (1970).

102. H. Hanson and M. Frohne, *Methods Enzymol.*, *45*, 504 (1976).

103. H. Hanson, D. Glasser, M. Ludewig, H. G. Mannsfeldt, M. John, and H. Nesvadba, *Hoppe-Seyler's Z. Physiol. Chem.*, *348*, 689 (1967).

104. R. Kleine and H. Hanson, *Acta Biol. Med. Ger.*, *9*, 606 (1962).

105. S. W. Melbye and F. H. Carpenter, *J. Biol. Chem.*, *246*, 2459 (1971).

106. F. H. Carpenter and K. T. Harrington, *J. Biol. Chem.*, *247*, 5580 (1972).

107. N. A. Kiselev, V. Ya. Stel'mashchuk, V. L. Tsuprun, M. Ludewig, and H. Hanson, *J. Mol. Biol.*, *115*, 33 (1977).

108. A. Taylor, F. H. Carpenter, and A. Wlodawer, *J. Ultrastruct. Res.*, *68*, 92 (1979).

109. L. Van Loon-Klaassen, H. T. Cuypers, and J. Bloemendal, *FEBS Lett.*, *107*, 366 (1979).

110. L. A. H. Van Loon-Klaassen, H. T. Cuypers, H. Van Westreenen, W. W. DeJong, and H. Bloemendal, *Biochem. Biophys. Res. Commun.*, *95*, 334 (1980).

111. M. Becker, W. Schalike, and D. Zirwer, *Stud. Biophys.*, *35*, 203 (1973).

112. W. Schalike, M. Becker, and D. Zirwer, *Stud. Biophys.*, *35*, 221 (1973).

113. F. Jurnak, A. Rich, L. Van Loon-Klaassen, H. Bloemendal, A. Taylor, and F. H. Carpenter, *J. Mol. Biol.*, *112*, 149 (1977).

114. F. H. Carpenter and J. M. Vahl, *J. Biol. Chem.*, *248*, 294 (1973).

115. M. Ludewig, M. Frohne, I. Marquardt, and H. Hanson, *Eur. J. Biochem.*, *54*, 155 (1975).

116. M. Frohne, R. Michael, S. Fittkau, and H. Hanson, *Hoppe-Seyler's Z. Physiol. Chem.*, *350*, 223 (1969).

117. U. Kettmann and H. Hanson, *FEBS Lett.*, *10*, 17 (1970).

118. G. A. Thompson and F. H. Carpenter, *J. Biol. Chem.*, *251*, 1618 (1976).

119. G. A. Thompson and F. H. Carpenter, *J. Biol. Chem.*, *251*, 53 (1976).

120. H. E. Van Wart and S. H. Lin, *Biochemistry*, *20*, 5682 (1981).

121. M. Ludewig, J. Lasch, U. Kettmann, M. Frohne, and H. Hanson, *Enzymologia*, *41*, 59 (1971).

122. J. Lasch, W. Kudernatsch, and H. Hanson, *Eur. J. Biochem.*, *34*, 53 (1973).

123. C. I. Brändén, H. Jörnvall, H. Eklund, and B. Furugren, in *The Enzymes*, 3rd ed., Vol. 2 (P. D. Boyer, ed.), Academic Press, New York, 1975, p. 103.

124. H. Jörnvall, *Eur. J. Biochem.*, *16*, 25 (1970).

125. R. Pietruszko, H. J. Ringold, T.-K. Li, B. L. Vallee, Å. Åkeson, and H. Theorell, *Nature (Lond.)*, *221*, 440 (1969).

126. H. Theorell, in *Pyridine Nucleotide-Dependent Dehydrogenases* (N. Sund, ed.), Springer-Verlag, New York, 1970, p. 121.

127. U. M. Lutstorf, P. M. Schurch, and J. P. Von Wartburg, *Eur. J. Biochem.*, *17*, 497 (1970).

128. Å. Åkeson, *Biochem. Biophys. Res. Commun.*, *17*, 211 (1964).

129. S. Taniguchi, H. Theorell, and Å. Åkeson, *Acta Chem. Scand.*, *21*, 1903 (1967).

130. B. L. Vallee and F. L. Hoch, *J. Biol. Chem.*, *225*, 185 (1957).

131. D. E. Drum, J. H. Harrison, T.-K. Li, J. L. Bethune, and B. L. Vallee, *Proc. Natl. Acad. Sci. USA*, *57*, 1434 (1967).

132. F. J. Castellino and R. Barker, *Biochemistry*, *7*, 2207 (1968).

133. J. A. Koepke, Å. Åkeson, and R. Pietruszko, *Enzyme*, *13*, 177 (1972).

134. H. Jörnvall, *Eur. J. Biochem.*, *16*, 41 (1970).

135. H. Eklund, B. Nordström, E. Zeppezauer, G. Söderlund, I. Ohlsson, T. Boiwe, B. O. Söderberg, O. Tapia, C. I. Brändén, and Å. Åkeson, *J. Mol. Biol.*, *102*, 27 (1976).

136. D. E. Drum and B. L. Vallee, *Biochemistry*, *9*, 4078 (1970).

137. M. A. Abdallah, J.-F. Biellmann, B. Nordström, and C.-I. Brändén, *Eur. J. Biochem.*, *50*, 475 (1975).

138. L. G. Lange, J. F. Riordan, and B. L. Vallee, *Biochemistry*, *13*, 4361 (1974).

139. L. G. Lange, J. F. Riordan, B. L. Vallee, and C. I. Brändén, *Biochemistry*, *14*, 3497 (1975).

140. H. Eklund, B. Nordström, E. Zeppezauer, G. Söderlund, I. Ohlsson, T. Boiwe, and C. I. Brändén, *FEBS Lett.*, *44*, 200 (1974).

141. H. Eklund and C. I. Brändén, *J. Biol. Chem.*, *254*, 3458 (1979).

142. B. V. Plapp, H. Eklund, and C. I. Brändén, *J. Mol. Biol.*, *122*, 23 (1978).

143. T.-K. Li and B. L. Vallee, *Biochem. Biophys. Res. Commun.*, *12*, 44 (1963).

144. T.-K. Li and B. L. Vallee, *Biochemistry*, *3*, 869 (1964).

145. C. Woenckhaus and R. Jeck, *Hoppe-Seyler's Z. Physiol. Chem.*, *352*, 1417 (1971).

146. H. Jörnvall, C. Woenckhaus, and G. Johnscher, *Eur. J. Biochem.*, *53*, 71 (1975).

147. B. V. Plapp, *J. Biol. Chem.*, *245*, 1727 (1970).

148. R. Dworschack, G. Tarr, and B. V. Plapp, *Biochemistry*, *14*, 200 (1975).

149. B. L. Vallee and F. L. Hoch, *Fed. Proc.*, *15*, 619 (1956).

150. A. J. Sytkowski and B. L. Vallee, *Proc. Natl. Acad. Sci. USA*, *73*, 344 (1976).

151. A. J. Sytkowski and B. L. Vallee, *Biochemistry*, *17*, 2850 (1978).

152. W. Maret, I. Andersson, H. Dietrich, H. Schneider-Bernlöhr, R. Einarsson, and M. Zeppezauer, *Eur. J. Biochem.*, *98*, 501 (1979).

153. A. J. Sytkowski and B. L. Vallee, *Biochemistry*, *18*, 4095 (1979).

154. D. E. Drum, T.-K. Li, and B. L. Vallee, *Biochemistry*, *8*, 3792 (1969).

155. H. Theorell and B. Chance, *Acta Chem. Scand.*, *5*, 1127 (1951).

156. H. Sand and H. Theorell, in *The Enzymes*, 2nd ed., Vol. 7 (P. D. Boyer, H. Lardy, and K. Myrback, eds.), Academic Press, New York, 1963, p. 25.

157. S. A. Bernhard, M. F. Dunn, P. L. Luisi, and P. Schack, *Biochemistry*, *9*, 185 (1970).

158. M. F. Dunn, S. A. Bernhard, D. Anderson, A. Copeland, R. G. Morris, and J. P. Roque, *Biochemistry*, *18*, 2346 (1979).

159. J. D. Shore and H. Gutfreund, *Biochemistry*, *9*, 4655 (1970).

160. J. Kvassman and G. Pettersson, *Eur. J. Biochem.*, *69*, 279 (1976).

161. R. J. Kordaland and S. M. Parsons, *Arch. Biochem. Biophys.*, *194*, 439 (1979).

162. K. Dalziel, *J. Biol. Chem.*, *238*, 2850 (1963).

163. M. C. DeTraglia, J. Schmidt, M. F. Dunn, and J. T. McFarland, *J. Biol. Chem.*, *252*, 3493 (1977).

164. J. Kvassman and G. Pettersson, *Eur. J. Biochem.*, *100*, 115 (1979).

165. S. Subramanian and P. D. Ross, *J. Biol. Chem.*, *254*, 7827 (1979).

166. J. Schmidt, J. Chen, J. DeTraglia, D. Minkel, and J. T. McFarland, *J. Am. Chem. Soc.*, *101*, 3634 (1979).

167. R. T. Dworschack and B. V. Plapp, *Biochemistry*, *16*, 2716 (1977).

Chapter 2

MODELS FOR ZN(II)-BINDING SITES IN ENZYMES

Robert S. Brown, Joan Huguet,[*] and Neville J. Curtis[†]
Department of Chemistry, University of Alberta
Edmonton, Alberta, Canada

1. INTRODUCTION

Since it is now accepted that Zn(II) is essential for all forms of
life [1,2] it is perhaps not surprising that at least 80 Zn(II)-con-
taining metalloenzymes have been identified fulfilling specific roles
in each of the six major biochemical categories as oxidoreductases,
transferases, hydrolases, lyases, isomerases, and ligases (see Chapter
1). Being a d^{10} transition element, the ubiquitous role of Zn(II)

[*]*Present affiliation:* Institute of Organic Chemistry, University of
Zürich, Zürich, Switzerland
[†]*Present affiliation:* Research School of Chemistry, Australian
National University, Canberra, Australia

55

probably results from a combination of factors. It is a reasonable
Lewis acid and undergoes rapid ligand exchange; does not readily form
hydroxo derivatives at physiological pH; accepts a wide variety of
O-, N-, and S-containing ligands; and importantly, does not partake
in metal redox reactions under normal conditions. The biological
roles of Zn(II) have been well reviewed [3-7] and seem dominated by
its ability in enzymes to assume distorted four- or five-coordinate
complexes. For the most part, Zn(II) alone is not a sufficiently
viable entity to catalyze so efficiently the myriad of reactions for
which nature has chosen it, but rather is held in the protein matrix
by at least three enzyme-based ligands in such a way as to allow only
one or two additional ligand sites to be available to solvent and
substrate. Of the five Zn(II) enzymes whose active-site structures
have been established by x-ray, nuclear magnetic resonance (NMR), or
metal-substitution techniques, a combination of histidine imidazole,
glutamate or aspartate COO^-, and cysteine S(H) are the enzyme-derived
ligands which hold the metal in a distorted four- or five-coordinate
geometry.

The study of small-molecule analogues of the metal-binding
sites in enzymes at first glance appears to offer an attractive
opportunity to model the metal-based chemistry without protein-based
complications inherently attendant on studying the enzymes themselves. Models may be designed to reproduce one or more of the
biological properties of the system to test the hypothesis that a
"sufficient condition for the action of an enzyme is a particular
conjunction of reacting atoms or groups" [8]. Should such prove to
be the case, this in no way should be taken to imply that the enzyme
must operate similarly, but rather offers some precedence for the
chemistry that nature can make use of in the biological system. It
should always be borne in mind that small molecules may interact
with the metal ions differently since in simple systems the stereochemistry of ligation may be dictated by the metal, whereas in the
enzyme, the metal stereochemistry is undoubtedly controlled by the
ligand geometry [6], an eventuality that can have profound effects

on the overall chemistry. For the most part the models conceived
for the enzymes to be discussed in this chapter have not yet evolved
to the point where they could be considered truly effective enzyme
mimics. Their construction is generally dictated by what can be
readily synthesized such that rarely (if ever) is a model presented
which includes all the known Zn(II)-binding groups present in the
enzyme. Nevertheless, as we shall see, substantial advances in our
understanding of possible catalytic roles of enzymes can be derived
from model studies.

2. CARBONIC ANHYDRASE

Carbonic anhydrase (CA) is a widely distributed enzyme found in both
plants and animals whose only known function is to facilitate the
interconversion of CO_2 and HCO_3^- as in Eq. (1) [3,4,6,9-14]. The

$$CO_2 + H_2O \rightleftharpoons H^+ + HCO_3^- \qquad (1)$$
$$H_2CO_3$$

complete structures of the human B and C isoenzymes have been determined by x-ray crystallographic analysis (Fig. 1) [15-17] and show
the essential Zn(II) to be bound to three histidine imidazole residues; a fourth ligand position is said to be occupied by H_2O or OH^-
(hydrogen bonded to the protein through a network of Thr 199 and
Glu 106 [18]), which may be important for the catalytic process.
The ability of the enzyme to hydrate CO_2 depends on its "basic"
form, generated by ionization of some active site group having a pK_a
value of ~7 under certain conditions. Although many candidates for
this group have been offered [19,20], including imidazolium [3,21,22],
the Glu 106 in an unusual environment [18], zinc-bound imidazole [23,
24], or zinc-bound H_2O [18,25-27], most recent work has centered on
the involvement of the latter group [28], suggested originally by
Davis [29].

  Detailed spectroscopic investigation of the reddish blue
Co(II)CA, which is also catalytically active [30-32], originally

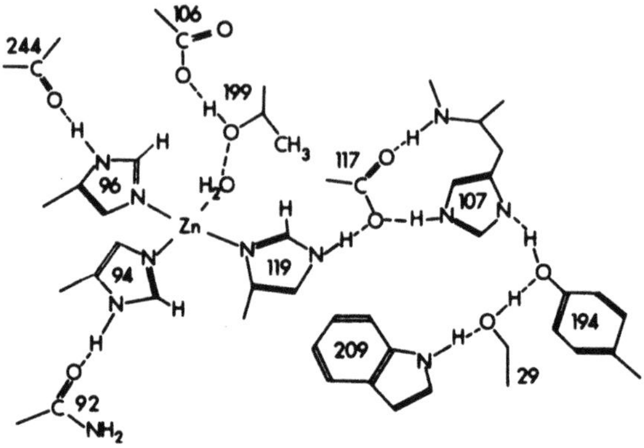

FIG. 1.   Portion of the active site of human CAB.   (Redrawn from Ref. 18.)

showed that the characteristic "basic" absorption spectrum and catalytic activity was again tied to some active-site residue having a
pK_a of ~7, but more recent evidence in the absence of buffers or
inhibitory anions indicate that the "pK_a" is somewhat more complicated than can be explained on the basis of a single ionizing group
[33].   In fact, formation of the high-pH form of Co(II)CA in the
strict absence of anions appears to have an apparent pK_a significantly lowered to 5.6 [34].   The absorption spectrum shows the
Co(II) environment to be distorted four- or five-coordinate, the
Co(II) therefore providing an excellent spectroscopic model [7] for
the situation in the native enzyme.

Model work on CA has centered on the design of chelating
ligands that approximate the distorted four(five)-coordinate active
site, or catalyze certain of the reactions mediated by the enzyme,
such as CO_2 hydration [10,25,31,35-37], aldehyde hydration [38,39],
and ester hydrolysis [40].   Of particular interest is the metal-
associated group responsible for the apparent pK_a in the enzyme.

In an attempt to demonstrate that a Zn-OH_2 with low coordination numbers could be both easily ionized and a potent nucleophile,
Wooley studied tetradentate ligands 1a-c [8,41,42].   Titration of
1a:Zn^{2+} at 0°C shows a single deprotonation according to Eq. (2)

$$1a, \quad R' = CH_3, \quad R = H$$
$$b, \quad R' = R = H$$
$$c, \quad R' = R = CH_3$$

$$1a:Zn^{2+} \underset{}{\overset{K_a}{\rightleftharpoons}} 1a:Zn^+OH + H^+ \tag{2}$$

having a pK_a of 9.17 [8], which is within experimental error of the point of inflection of a plot of its catalytic activity for CH_3CHO hydration against pH. The studies indicate that reducing the coordination number around Zn^{2+} from 6 to 5 also reduces the pK_a for $Zn^{2+}-OH_2$ from 10 or greater [8] and provides a Zn^+-OH which is nucleophilic enough to compare favorably with the enzymic rates for acetaldehyde hydration but not CO_2 hydration; further reducing the coordination number from five to four should have a commensurate effect on the pK_a [8].

While titration data for some imidazole-Zn(II) complexes were interpreted [43] to suggest that Zn-imidazole could undergo pyrrole N-H ionization at pH values near 7, these conclusions were criticized [44] on the basis of lack of precedence and the fact that they were drawn from titrations performed under nonequilibrium conditions. It appears that pyrrole N-H pK_a values in Zn(II) complexes should be closer to 13, and that $Zn^{2+}-OH_2$ should deprotonate at much lower values. NMR data for histamine-Zn(II) complexes are in agreement [45] in that ionization of $Zn^{2+}-OH_2$ commences at pH ~8.

In studying the reactivity of metal-coordinated nucleophiles toward ester carbonyls, Sargeson and co-workers [46] investigated the cleavage of p-nitrophenyl acetate by substitution-inert six-coordinate Co(III) complexes 2 and 3. Both are nucleophiles when deprotonated (pK_a of 2 = 6.4; pK_a of 3 = 10.0), as evidenced by the formation of acetylated reactants, the reactivity of deprotonated 3

$(NH_3)_5 \, Co \, OH_2^{+3}$

2

$(NH_3)_5 \, Co \, Im \, H^{+3}$

3

4

5

6

7

being comparable to that of free OH^-. The hydroxide derivative of 2 is 10^3 less nucleophilic than deprotonated 3 in water, but in dimethyl sulfoxide (DMSO), the two are of comparable activity. Although $(NH_3)_5M(III)OH^{2+}$ complexes [M(III) = Co, Rh, Ir] will attack CO_2 nucleophilically [47,48], they do so some 10^5-fold less efficiently than CA. Of course, parallels between the two are some-what tenuous since metal type, charge, and coordination are different.

Recently, Tabushi et al. reported [49] a β-cyclodextrin model containing pendant imidazoles (4) which in conjunction with Zn(II) shows modest but significant ability to hydrate CO_2 ($k_{cat_{4:Zn(II)}}$= 166 M^{-1}/s at pH 7.5 [49] versus k_{cat}/K_M for human CAB = 10^7 M^{-1}/s [25]). The reactivity of 4:Zn is suggested to arise from a combina-tion of ligation and the hydrophobic environment provided by the cyclodextrin cavity, although the exact mechanism of action remains to be shown [49].

It should be apparent that the models described above, while providing important clues for catalysis, do not approximate the known coordination site in the enzyme with respect to either numbers or types of Zn(II) ligands, features that should have profound influ-ences on both catalytic and spectral properties, particularly if a central role is played by the low coordination number of the metal. "Tripod" molecules containing three ligands anchored to a central pivotal point, such as 5 [50], 6 [51-53], 7 and 8 [54,55], 9 [55-58], and 10 [59-61], have been extensively investigated with respect to

their mode of metal binding. Co(II) absorption spectra of 5 and 6 having nitrogen as a central pivotal point are characteristic of trigonal bipyramidal species severely distorted toward tetrahedral [50-53]. Crystallographic data on the related Cu(II) complex 11 [62] confirms that the apical nitrogen acts as a fifth ligand with a Cu(II)-N_1 distance of 2.059 Å, roughly the same as the other Cu(II)-N distances (2.048-2.120 Å).

As a function of pH the Co(II)(ClO$_4$)$_2$ complex of 6 reveals some interesting analogies to the spectrum of Co(II)CA [53]. Two pKa

values are seen in a plot of the absorption of $\underset{\sim}{6}$:Co(II) as a function
of pH [53]. The first, at 7.4, results from deprotonation of the
apical N, while the second, at 8.8 ± 0.2, is said to be that of a
Co(II)-OH$_2$, the ionization of which generates a complex spectrum
which at pH 9.65 is remarkably similar in appearance to that of the
high-pH form of Co(II)CA, although the extinction coefficient is much
lower (ε_{640} $\underset{\sim}{6}$:Co(II) = 60 versus ε_{640} Co(II)CA = 310 M^{-1}/cm [33]).
Despite the spectral similarities between Co(CA) and Co(II) $\underset{\sim}{6}$, the
latter does not catalyze CO$_2$ hydration, suggesting that the enzyme
relies on additional structure requirements other than simply the
bound metal [63].

Further reducing the M^{2+} coordination numbers in complexes
containing three neutral nitrogen ligands proves to be a rather more
difficult task. Removal of the apical nitrogen does provide triden-
tate chelation in ligands $\underset{\sim}{7-9}$, as evidenced by NMR data derived from
the Zn(II) complexes, but the ligands are too small to enforce four
or five coordination [54] since they generally bind as octahedral
2:1 complexes. Alkyl substitution of the tris-imidazolecarbinol $\underset{\sim}{7}$
at the 4 and 5 positions does sterically encumber 2:1 complexation
as expected, but in the presence of Zn(II) the complexes are fairly
unstable toward OH⁻ and tend to lose H$_2$O to form what is reported
to be the fulvene-like $\underset{\sim}{12}$, as in Eq. (3) [55]. Problems of 2:1

complexation and complex instability are circumvented with the
phosphine analogue 10c [61], the longer P-C bonds extending the
imidazoles outward such that stable complexes with both zinc and
Co(II) are formed. While the analogues 10a and 10b show little
(if any) propensity to form four(five)-coordinate complexes, as
judged by their Co(II) absorption spectra, the 4,5-diisopropyl
groups in 10c encapsulate the bound M^{2+} sufficiently to

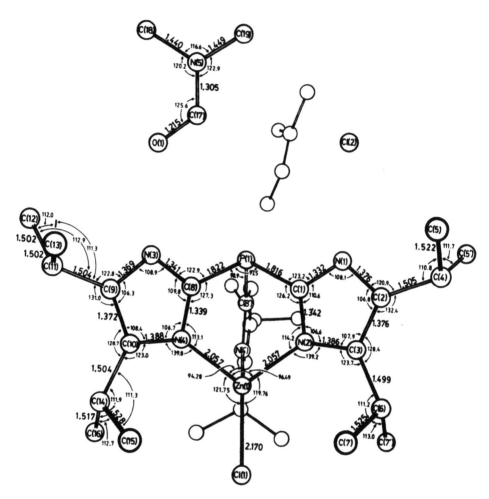

FIG. 2. X-ray crystal structure of [tris-2(4,5-diisopropylimidazolyl) phosphine]dichlorozinc(II). (Reprinted with permission from Ref. 64. © 1981 American Chemical Society.)

allow access of only one (or possibly two) additional ligands. The x-ray structure of 10c:ZnCl$_2$ portrayed in Fig. 2 distinctly shows the distorted tetrahedral ligation, the three N-Zn-N angles being in the order of 95°.

In solution 10c:Co(II) shows features in common with Co(II)CA, including anion-dependent visible absorption spectra [59] and a pH

dependence, with an intensely blue four(five)-coordinate species
having an apparent pK_a for formation of 5.6 in a medium of 80%
ethanol-water [60]. As its Zn(II) complex, 10c catalyzes $CO_2 \rightleftharpoons HCO_3^-$
interconversion, although its maximum k_{cat} of ~900 M^{-1}/s at "pH 6.4"
is modest relative to that for the enzyme $(k_{cat}/K_M = 10^7 M^{-1}$/s [25]).
Since the catalysis afforded by 10c:Zn(II) maximizes at "pH 6.4" and
is diminished at both higher and lower values, a deficiency in the
model is evident which is ascribable to competition of the ligand
and Zn(II) for H^+ and OH^-, respectively, as shown in Eq. (4) [60].

$$P\underset{Im^+H}{\overset{Im\cdots}{<}}{\overset{-Im\cdots}{\underset{-Im\cdots}{}}}:Zn^{++} \underset{H^+}{\overset{}{\rightleftharpoons}} P\overset{Im\cdots}{\underset{-Im\cdots}{<}}:Zn^{++} \overset{OH^-}{\rightleftharpoons} PIm_3 + Zn(OH)_2$$

(4)

Although it is not possible on the basis of existing data to attribute
the catalysis to any specific mechanism, it is important to consider
that monovalent anions specifically inhibit activity as they do CA
[65-67]. It is not unreasonable to suggest that anions bind to the
fourth Zn(II) site in the complex, thereby restricting access of
other potential reactants, such as H_2O, OH^-, or CO_2.

Quantitative titration data for 13·$2HNO_3$ in the presence of
1.0 equivalent of $Zn(ClO_4)_2$ or $Co(ClO_4)_2$ were performed in 80% ethanol-
H_2O, the data being shown in Fig. 3 [68]. It is clear that whereas
in the absence of added M^{2+}, 2 equivalents of OH^- are consumed by
the time the third pK_a for 13 has passed, in the presence of Zn(II),
3 equivalents of OH^- are consumed by "pH" 7, indicating an additional
ionization of some group associated with the complex. Similar titration of 13:Co(II) shows a complex to be fully formed by pH 7 (2 equivalents of OH^- are consumed) and a single additional ionization with
an apparent pK_a of ~8. More interestingly, as shown in Fig. 4, the
latter ionization gives rise to a blue species indicative of a four
(five)-coordinate species ($\varepsilon_{598nm} \approx 550 M^{-1}$/cm). Like Bertini et al.'s
pH study of the absorption spectrum of 6:Co(II) [53,63], it seems that
formal titration of a Co(II)-coordinated H_2O reduces the coordination
number of the complex, creating an environment akin to that in
Co(II)CA.

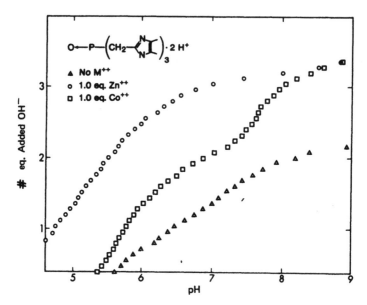

FIG. 3. Titration curves for $\underset{\sim}{13} \cdot 2H^+$ in the absence (△) and presence of 1 equivalent $Zn(ClO_4)_2$ (O) and $Co(ClO_4)_2$ (□) in a medium of 80% ethanol:H_2O. (Reprinted with permission from Ref. 68. © 1982 American Chemical Society.)

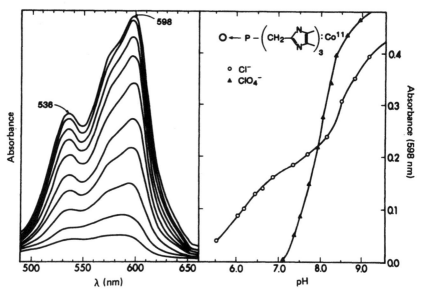

FIG. 4. Co(II) complex of the $\underset{\sim}{13}$ absorption spectrum as a function of pH in a medium of 80% ethanol:H_2O showing a spectrophotometric pK_a of ~8 in the presence of 0.2 M ClO_4^- (△). In the presence of 0.2 M Cl^- (O), an additional four(five)-coordinate species is generated which at higher pH gives way to the same spectrum as that obtained in the absence of Cl^-. (Reprinted with permission from Ref. 68. © 1982 American Chemical Society.)

Originally, Lindskog noted that monovalent anions perturbed the absorption spectrum of Co(II)CA, and that the effect could be reversed by an increase in pH such that the inhibitor anion is displaced from the metal to yield ultimately the spectrum characteristic of the high-pH form [30,67]. This has been discussed at length by Koenig in terms of competition of anion and OH$^-$ (H$_2$O) for a common metal site [34,69]. Spectroscopic titration studies of 13:Co(II) as a function of pH in the presence of various anions, such as halides, nitrate, acetate, and picolinate, reveal a similar phenomenon, as typified by the titration in the presence of Cl$^-$ (Fig. 4). It is clear that these anions bind to 13:Co(II) to form four(five)-coordinate complexes, as evidenced by the formation of characteristic absorption spectra, and that the effect can be reversed by increasing [OH$^-$] until at high pH only the spectrum of 13:Co(II)-OH$^+$ is seen. The overall process is best explained as in Eq. (5), in which formation of 14 is suppressed by the addition of anions to give 15, the

OP$\left(\text{CH}_2\text{-}\underset{\text{N}}{\overset{\text{N}}{\diagup\!\!\!\diagdown}}\right)_3$

13

$$13 : 2H^+ + Co(H_2O)_6 \xrightleftharpoons[]{OH^-} OP\text{-}(CH_2\,Im)_3 : Co(H_2O)_3 \qquad (5)$$

$$OP\text{-}(CH_2\,Im)_3 : Co\diagdown \underset{(OH_2)_{0,1}}{} \xrightleftharpoons[OH^-]{A^-} OP\text{-}(CH_2\,Im)_3\,Co \diagdown \underset{(OH_2)_{0,1}}{OH^-}$$

15   **14**

effect being reversed by yet higher [OH$^-$]. On the basis of these observations, it is tempting to ascribe a similar behavior to the enzyme, the well-known shift in either spectroscopic or activity related pK$_a$ to higher values in the presence of inhibitory anions being attributable to a mass-action effect involving a competition between A$^-$ and OH$^-$ for a common metal site [33,68].

Theoretical models for the active site of CA have been offered by several groups [70-76]. *Ab initio* pseudopotential SCF computations [70,71] indicate that while imidazole alone deprotonates easier than does H$_2$O, complexation of these to Zn(II) reduces the deprotonation energy of both such that Zn^{2+}-OH$_2$ ionizes easier than does

Zn^{2+}-Im-H. Thus the function of Zn(II) could be to reduce the pK_a of bound H_2O by electrostatic interaction, the OH⁻ retaining a significant negative character to nucleophilically attack a CO_2 which is somewhat polarized by virtue of its interaction with the metal as a distant ligand. The overall process is similar to the five-coordinate transition state (16) proposed for CO_2 hydration on the basis of x-ray [18] and electron spin resonance (ESR) data with Co(II)-CA [77]. The gas-phase reaction is computed to be an almost barrierless process with an activation energy of ~1 kcal/mol, and an overall exothermicity of ~13 kcal/mol [70,71].

Clemmenti's Monti Carlo calculation of the reaction of CO_2 and H_2O in the field of Zn(II) requires an activation energy of 9 kcal/mol, with the associated proton transfer occurring either simultaneously

$$(NH_3)_3 \, Zn \cdots \overset{\ominus}{O}\text{-}H \quad \cdots \quad C\text{=}O \quad \longrightarrow \quad (NH_3)_3 \, Zn \overset{H}{\underset{O}{\overset{O}{\ominus}}} C\text{=}O$$

16   17   2.05A

or subsequent to OH_2 attack on CO_2 [72]. According to the same calculations [72-74], the Zn^+-OH mechanism is disfavored since a much higher activation energy is required, contrasting the *ab initio* results noted above.

3. PEPTIDASES: CARBOXYPEPTIDASE AND THERMOLYSIN

At present, x-ray structures for the Zn(II)-binding sites of carboxypeptidase A (CPA), carboxypeptidase B (CPB) [78-80], and thermolysin [81, 82] indicate very similar ligation (Fig. 5) [83] in that for all three, the required Zn(II) is bound into the protein by two histidine imidazoles and a glutamate residue, with water residing at a fourth ligand position.* Co(II)CPA retains activity for certain ester and peptide

---

*A recent refinement of the CPA structure at 1.75 Å has revealed that in the native enzyme, the Zn(II) environment is five-coordinate, with four of the ligands being enzyme-based (two imidazole nitrogens and the two carboxylate oxygens of glutamate 72) and the fifth being H_2O. The Gly-Tyr/enzyme complex shows the H_2O ligand to be replaced by the

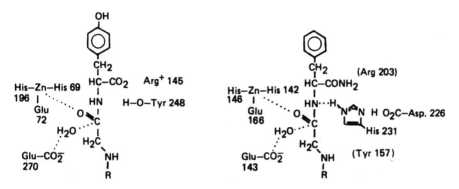

FIG. 5. X-ray crystal structure of carboxypeptidase A and thermolysin, showing the binding modes of substrates or analogues into the active sites. (Reprinted with permission from Ref. 83. © 1975 American Chemical Society.)

substrates and shows characteristic absorption bands at 555 (ε = 160 M^{-1}/cm) and 572 nm (ε = 160 M^{-1}/cm) as well as a weak near-infrared band at 940 nm (ε = 25 M^{-1}/cm) not unlike the situation in Co(II)CA [85].

Mechanistically, two main proposals for activity of CPA have been offered and will be considered only briefly since more elaborate discussions appear in this volume and elsewhere [86-88]. We will assume on the basis of structural similarities between CP and thermolysin that similar mechanisms are operative, although this remains to be proven. In the first, originally proposed by Lipscomb [79a], the peptide binds to the enzyme to displace $Zn-OH_2$, the metal function being to polarize the C=O, thereby rendering it more susceptible to attack by Glu 270 or a general base-delivered H_2O. The former requires that an anhydride acyl enzyme intermediate be formed along the reaction pathway [Eq. (6)]. The second possibility, which is generally overlooked in discussions of CP mechanisms, considers Zn^+-OH attack similar to that involved in discussions of CA activity [89].

---

peptide carbonyl oxygen and amide nitrogen. The amide N statistically occupies a second position near Glu 270. Consequently, the coordination number of Zn(II) may vary from five to six in CPA-substrate complexes [84].

$$(6)$$

Breakdown of the tetrahedral intermediate in peptide hydrolysis probably requires assistance from a phenolic OH of Tyr 248 which is appropriately positioned for this function. Modification studies show that Tyr 248 is not required for ester hydrolysis, and may not even be required for peptides [85,86]. Mechanistic analyses have been discussed at length [79-88] and a unifying mechanism for both esters and peptides has been presented by Breslow and Wernick [90].

Modeling CPA action on peptides requires not only construction of the Zn(II) binding site, but appropriate positioning of a carboxylate anion, and phenolic OH aligned correctly in a single small molecule, a formidable synthetic task. Since the phenolic hydroxyl is not required for esterase activity, model complexity can be reduced to include only the binding site and remote carboxylate, and several revealing investigations have appeared. Conceptually, it is simplest to consider the effects of the individual catalytic groups prior to analyzing their cooperative interactions.

Several studies have shown that M^{2+} enhances the attack of nucleophiles (notably OH^-) on associated esters or amides. Breslow showed that Ni^{2+} coordinated to 18 enhanced OH^- attack by 400-fold

relative to the situation where no metal was present [91], and favored a mechanism whereby external OH^- attacked the Ni^{2+} complex. The kinetically equivalent delivery of $M^{2+}-OH^-$ is difficult to distinguish from the latter event, but can sometimes be accomplished on the basis of effects of added nucleophiles and ^{18}O labeling experiments. Pertinent to the latter is the work of Sargeson and co-workers on ligand exchange inert Co(III) complexes 19 and 20 [92-94]. Simple C=O activation by coordination enhances external OH^- attack on the amide in 19 by a factor of 10^4 over the noncomplexed case, and ^{18}O tracer studies showed the major pathway to yield a single ^{18}O incorporated into the formate product, indicative of retention of the Co(III)-O bond [93]. In the case of glycinamide hydrolysis from 20, an increase of 10^{10} in rate is seen over uncatalyzed hydrolysis. The studies show that while the C=O unit is not initially polarized by the metal, a correctly positioned Co(III)-OH^- can be a potent nucleophile even though such coordination significantly reduces its basicity (10^9-fold for OH^- in 20 [94]).

Nucleophilic activation by M^{2+} coordination is not limited to H_2O. Zn(II) mediated trans esterification of N-(β-hydroxyethyl)ethylenediamine (21) by p-nitrophenylpicolinate (22) involves a dual role for the metal [95]: reduction of the pK_a of the coordinated hydroxyethyl group to ~8.4, and a template effect leading to a ternary complex (23) as in Eq. (7), which was shown to be the kinetically competent species [95]. The isolated trans-esterified product of 21 could

$$21 \ (R=H) \quad\quad 22 \quad\quad\quad\quad\quad 23 \quad\quad\quad 21\left(R= \overset{O}{\underset{\parallel}{C}}\right) \quad\quad (7)$$

only have been formed by an intracomplex reaction. Similar observations have been made for Zn(II):24 complex reactions with 22 [96].

24 : Zn(11)   25   26

Wells and Bruice reported a detailed kinetic study of the decomposition of Co(II) and Ni(II) complexes of ester 25 [97]. These 1:1 complexes hydrolyze 10^3- to 10^5-fold faster than hydroxide-

mediated hydrolysis alone. Complexation with other divalent ions
such as Ca(II), Mn(II), and Mg(II) is too weak to be observed, while
Zn(II) forms 2:1 complexes which are not particularly reactive. All
experimental results can be accommodated by the process shown in
Scheme 1, where attack of a metal-bound OH⁻ gives rise to preequilib-
rium formation of a tetrahedral intermediate, the decomposition of
which is acid mediated [97].

Scheme 1   (charges omitted for simplicity)

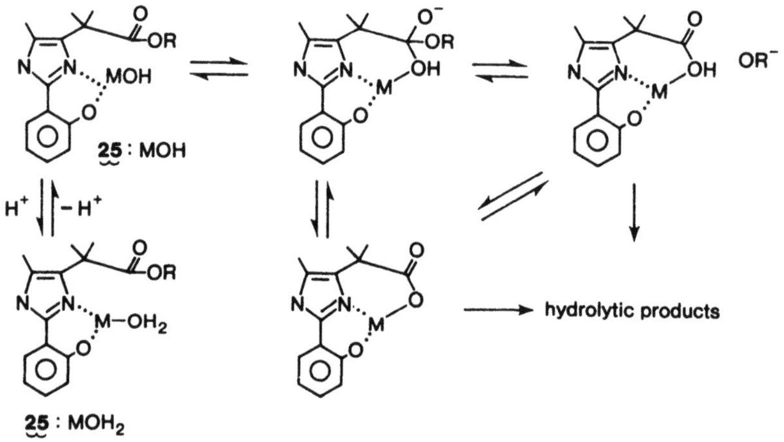

Should the CPA mechanism involve formation of an anhydride
intermediate, its decomposition must be at least as fast as the
turnover number for the enzyme. No such intermediate has been
detected for peptide hydrolysis, and only one for CPA-mediated
hydrolysis of the ester O-trans-p-cinnamoyl-L-β-phenyl lactate at
low temperature [88]. At subzero temperatures, hydrolysis of the
intermediate is rate limiting, while at higher temperatures its
formation is, which accounts for failures to observe buildup of
acyl enzyme intermediates under normal conditions. In a study of
Zn(II)-promoted hydrolysis of a model for such an intermediate,
Breslow's group [98] studied anhydride 26 as a function of pH. In
the absence of Zn(II), hydrolysis of 26 is pH independent between pH
1 and 7.50, while under saturating conditions of Zn(II) the cleavage
becomes first order in OH⁻ above pH 5, the extrapolated rate constant
of 3.0 ± 0.5 s⁻¹ at pH 7.50 being well within the range of enzymic

k_{cat} values for several esters (0.5-230 s^{-1} [99-101]). Since Zn(II)
does not catalyze the addition of a more powerful nucleophile, NH$_2$OH,
the mechanism probably involves delivery of a metal-coordinated OH$^-$
[98].

Although from the above it appears that excellent catalysis of
both coordinated esters and anhydrides by M^{2+}-OH$^-$ does occur, aside
from Sargeson's Co(III) complexes [92-94] and Breslow's 2-phenanthro-
line carboxamide-Ni(II) studies [91], catalysis of amide hydrolysis
by divalent metal ions appears to be quite sluggish, with the important exception of 27 [102]. For the Cu(II) complex titration data

$$\text{27} : M^{++} \qquad \text{28} \qquad \text{29} \tag{8}$$

are consistent with a metal-OH$_2$ pK$_a$ of 7.6, the decomposition rate
being ~10^5- and 10^7-fold faster in the presence of Zn(II) and Cu(II),
respectively, than that of the uncomplexed lactam at pH 7.6. A sig-
moidal pH-rate profile is analyzed in terms of a two-step process in
which formation of a tetrahedral intermediate is rate limiting at
low pH, and its decomposition is at high pH [102]. It is important
to note that this remarkable catalysis is manifested when the metal
is positioned above the π face of the cleaving amide, allowing favorable delivery of M^{2+}-OH$^-$, whereas in related systems where the metal
lies in the amide plane, only limited acceleration [91] or even
inhibition is seen.

Of relevance to the involvement of Glu$^-$ 270 in CPA is whether
a neighboring COO(H) acting as a nucleophile or general base will be
able to compete with delivery of M^{2+}-OH$^-$ in ester and amide hydrolysis.
Decomposition of both 30a and 30b is accelerated by Ni(II), the rate
of OH$^-$ attack being increased by 9300 and 3100 times, respectively
[103]. The data indicate the decomposition of 30b:Ni(II) to be increased by a factor of 2 when the COO$^-$ group is ionized and suggests
a "semicooperative" simultaneous catalysis by M^{2+} and COO$^-$ likely by
a general-base involvement of the latter on the metal-polarized ester.

30 a, X = H
    b, X = COO(H)

31 a, X = COO(H)
    b, X = H

32

Fife and Squillacote [104] later investigated metal-promoted cleavage of **31** and **32**. Although at pH 3.5, there is at least a 10^4-fold rate enhancement of uncomplexed **31a** over **31b** due to nucleophilic involvement of the carboxylate, in the presence of either Ni(II) or Co(II), **31a** shows a twofold rate *reduction* at all pH values, indicating an inhibitory effect of M^{2+} in this system. The observation that **32**, in the presence of Cu(II), Co(II), or Zn(II) (all of which bind to the phenanthroline nitrogens), hydrolyzes 20- to 40-fold slower than the free amide at pH 2.75, and at pH 5.35 shows no hydrolysis whatsoever, probably indicates that the coordinated C=O will not permit a proton transfer to the leaving group which is required for amide hydrolysis.

In an additional comparative study of M^{2+} effects on the hydrolysis of amide **33** and ester **34**, the same authors [105] observed

33

34

35

large rate enhancements for the ester (10^2- to 10^4-fold at 100 times excess of M^{2+}) but not for the amide. Since the accelerating effect for **34** is first order in $[OH^-]$, an M^{2+}-OH^- attack is indicated at pH values above 6, but below that value this pathway cannot compete with the already efficient nucleophilic role of COO^-, which shows little or no metal-ion catalysis [105]. This is probably due to a competition between M^{2+} and H^+ for binding to the quinoline N such that no complex is formed at low pH and hence the preferred pathway involves COO^- attack within the zwitterionic species **35**. On the basis of these observations, it may be that CPA cleaves ester substrates by dual mechanisms, a nucleophilic role for COO^- being favored at low pH values, and M^{2+}-OH$^-$ at high pH. It is important to note, however, that since 34 does not bind M^{2+} efficiently at low pH, the relevant question of metal catalysis of COO$^-$ involvement cannot be tested with this system.

Ester 36, on the other hand, introduces an additional carboxylate to strengthen M^{2+} binding. In the absence of M^{2+}, 36 hydrolysis

shows a bell-shaped pH-rate profile between pH 2 and 7, indicating involvement of a zwitterionic species akin to 35 (X = COO$^-$). A plateau between pH 7 and 9 indicates nucleophilic attack of the glutarate on the unprotonated quinoline ester. Interestingly, pH-rate profiles for 37 in the presence of saturating M^{2+} show a first-order dependence on [OH$^-$] throughout the region where neighboring COOH is ionizing, clearly showing that the rapid reaction is independent of the state of ionization so that intramolecular general-base attack of carboxylate is not capable of competing with M^{2+}-OH$^-$ attack [106]. As well, pH-rate profiles for 36 show a marked dependence on saturating M^{2+}, the plots being linear in [OH$^-$] at high and low pH, with a small plateau in between. The OH$^-$ catalyzed attack is extremely large at high pH (5.5-7.5) ranging from $2 \cdot 10^6$- to $4 \cdot 10^7$-fold with saturating Ni(II) and Cu(II), again favoring delivery of M^{2+}-OH$^-$ to the ester. In these systems, upper limits for catalysis of COO$^-$ nucleophilic attack by M^{2+} are 10^2-10^3, but this still cannot compete with the metal hydroxide reaction. For significant nucleophilic attack to dominate, the ester C=O bond and COO$^-$ would require an excellent steric fit which cannot yet be demonstrated in model systems but could be accomplished in the enzyme. Although most mechanisms postulated for CPA involve a Lewis acid polarization of the C=O by Zn(II), it might be anticipated that this should not be a dominant factor for nucleophilic involvement of COO$^-$, because the rate-limiting step in hydrolysis is the breakdown of the tetrahedral

intermediate and not the attack if the leaving group pK_a greatly
exceeds that of COO^-. As the studies above show, strongly chelated
M^{2+} does not enhance intramolecular COO^- nucleophilic attack on
esters or amides since this process cannot compete with the efficient
metal-promoted OH^- attack [104-106]. Fife and Przystas [107] investigated the metal-ion effects on the hydrolysis of 2-pyridylmethyl-
hydrogen phthalate, a reactant that does not bind M^{2+} strongly, and
incorporates a relatively poor leaving group (2-pyridylmethoxide).
Divalent ions such as Ni(II), Zn(II), and Cu(II) do exert large
catalytic effects even though saturation is not observed when $[M^{2+}]$
is 100-fold that of the reactant. The metal-ion-catalyzed reactions
importantly are pH independent between values of 4.74 and 7.15, thus
indicating that OH^- is not a reactant. It appears, then, that the
metal catalysis is associated with COO^- participation, and that the
ion exerts its large catalytic effect in the breakdown of the tetrahedral intermediate by stabilizing the leaving-group oxygen in the
transition state [107]. This provides an important example of situations in which leaving-group stabilization by M^{2+} is capable of
generating sizable rate enhancements in carboxyl nucleophilic reactions. Such could also be important in CPA reactions with the types
of ester substrates commonly employed.

A preliminary report of other CPA models involving macrocyclic
paracyclophanes with pendant imidazole and oximes [108] has appeared,
this species showing a 15-fold rate enhancement for decomposition of
p-nitrophenyl hexadecanoate in the presence of saturating Cu(II).

Other than the report by Groves and Dias [102], significant
catalysis of amide hydrolysis by divalent M^{2+} has not been reported,
probably reflecting the inability of M^{2+} to stabilize tetrahedral
intermediate breakdown in most model systems [104-106]. In CPA, a
remote Tyr 248 hydroxyl has been suggested as a candidate for this
process [78-80]. Breslow and McClure studied the cooperative interaction of COO^- and phenolic OH in some maleamic acid derivatives
(38a, b) [109]. In aqueous media, both 38a and 38b undergo rapid
cyclization to produce the anhydride; since no evidence for phenol

assistance is observed, the rate-limiting step is probably nucleophilic attack. In CH_3CN containing 1 M H_2O as a solvent modeling
the enzyme interior, both compounds undergo cyclization if the
medium contains a 10:1 ratio of HOAc/OAc⁻. Their behavior diverges,
however, if the ratio is 1:10 or 1:50, the rates for both compounds
being significantly retarded but to different extents. It appears
that in the latter medium, protonation of the leaving amide becomes
a problem such that 38b containing the phenolic OH hydrolyzes 66
times as fast as 38a. Of note is that under the latter conditions,
no evidence for the appearance of anhydride is seen, a k_{H_2O}/k_{D_2O}
ratio of 1.5-2.3 indicating that the mechanism switches from nucleophilic to general base [109].

It is finally pertinent in terms of models for CPA to consider
briefly theoretical studies for this system [89,110,111]. Hayes and
Kollman [110] modeled several important active-site residues by fractional point charges and studied the effect of these on an amide
model of N-methylacetamide. Not surprisingly, placing (+) adjacent
to the amide C=O increases polarization, with the O becoming more
(-) at the expense of the other groups. Enforced rehybridization
from $sp^2 \rightarrow sp^3$ for the C=O and amide N is endothermic and exothermic,
respectively, the catalytic effect being interpreted in terms of an
overall lowering of the transition state energy for cleavage [110].
Lipscomb's approximate molecular orbital calculations indicate that
nucleophilic attack on formamide in the field of an electrophile
placed close to the C=O is markedly stabilized. The resulting tetrahedral intermediate is stabilized by M^{2+}-alkoxy interaction which
strengthens bonding between the tetrahedral carbon and its four substituents in the adduct [89]. Calculationally, the Zn^+-OH mechanism
is disfavored but it was modeled using $(NH_3)_2Be(OH)_2$ as the electrophile, which may be an inappropriate choice. Interaction between
this species and formamide is slightly attractive at large distances,
but repulsive when the distance between the carbonyl C and attacking
OH⁻ is <2.7 Å [89]. Morokuma and co-workers' recent *ab initio* calculations [111] indicate that proton transfer from $Zn-OH_2$ to Glu⁻ 270

leads to a more stable Zn^{2+}-OH—HOCO Glu, the role of the Zn(II)
being to stabilize OH$^-$ by charge transfer. Model building shows a
Gly-Tyr dipeptide can fit into the enzymic cavity as a fifth ligand
to Zn(II) but requires the sissile C=O to be complexed through its
π face. Nucleophilic attack by Zn^+-OH can now occur, the incipient
(-) being accommodated by Zn(II) in a Lewis acid role. The possibility of π-face attack by M^+-OH as being important is an intriguing
one which receives experimental support from the chemical model of
Groves and Dias [102].

4. ALCOHOL DEHYDROGENASE

Alcohol dehydrogenases (ADHs) from liver and yeast sources are Zn(II)
enzymes catalyzing the interconversion of aldehydes or ketones and
their corresponding alcohols and requiring NADH or NAD^+ cofactors as
in Eq. (9), the 4-hydrogen of the dihydronicotinamide being specifically transferred to the carbonyl carbon [112-114]. X-ray structure

$$\tag{9}$$

studies of the horse-liver enzyme [115-117] show the active site to
consist of a Zn(II) coordinated to two cysteine S(H) residues and a
histidine imidazole; water completes a somewhat distorted tetrahedral
coordination. The x-ray structure of a DMSO-inhibited ternary complex is shown in Fig. 6 [117,153]. Zn(II) can be replaced by Cd(II)
or Co(II) to give active species for which the absorption spectrum
of the Co(II) enzyme indicates a distorted four(five)-coordinate
environment for the catalytic metal [118,119]. A Zn(II)-coordinated
C=O is involved in the reduction of the ternary complex of ADH, NADH,
and 39, which shows spectroscopic similarities to several Lewis acid

$(CH_3)_2N$— ⟨⟩ —CH=CH—CHO

39

H H O
NH$_2$

40 R = nPr
    R = CH_2Ph

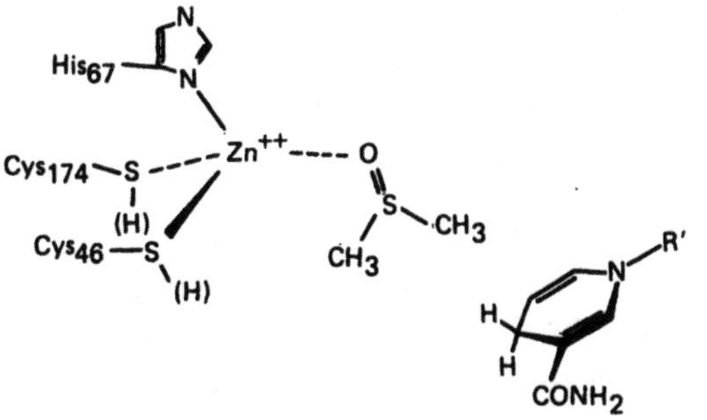

FIG. 6. Active-site structure of a DMSO-inhibited ternary complex of HLADH depicting the Zn^{2+} binding site. (Redrawn from Ref. 153.)

complexes of 39 in aprotic solvents [120,121]. The role of Zn(II) is proposed to be one of a Lewis acid which polarizes the C=O to activate it toward reduction. Proposed mechanisms have been well reviewed [3,6,9,122,123].

For the most part, model studies have centered on whether a C=O group held in proximity to a chelated M^{2+} [Zn(II) or Mg(II)] can be activated toward bimolecular reduction by N-propyl or N-benzyl derivatives of 40. An early example showed 41 to be reduced by 40 (R = nPr) in CH_3CN in the presence of $ZnCl_2$ [124]; metal-promoted reduction must be substantial since none is observed in its absence. Further work showed a direct transfer of the 4-hydrogen from 40 to the aldehyde group in 41, a process that was unaffected by radical

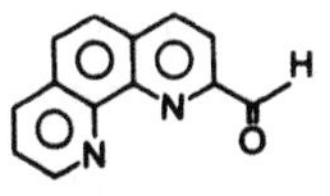

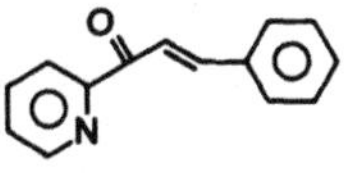

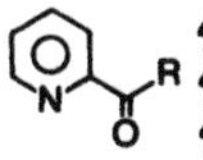

quenching agents [125]. However, disparity between kinetic and product isotope partitioning effects suggested an intermediate to be formed along the reaction pathway. The same study [125] showed that BH_4^- reduction of Zn(II) complexes of 2- and 4-pyridine carbox-aldehyde was enhanced by $7 \cdot 10^5$- and 100-fold, respectively, relative to the free aldehyde. Hence exhalted effects are seen only if the aldehyde C=O is close enough to the M^{2+} that its reduction is

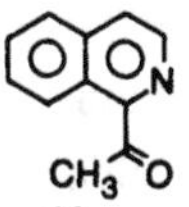

ROBERT S. BROWN ET AL.

favorably influenced. Bidentate complexation of C=O and N has been
observed in 2:1 complexes of 42 [126], octahedral or tetrahedral
species being formed depending on the M^{2+} counterion. In dry CH_3CN,
conjugate reduction of 42 by 40 in the presence of Zn(II) or Mg(II)
yields the corresponding saturated ketone [126]. It is now clear
that a wide variety of carbonyls such as 43-49 can undergo M^{2+}-
promoted reduction by dihydronicotinamides in dry CH_3CN [127-129]
or CH_3OH [130]. Hydroxylic solvents undergo an acid-catalyzed
addition across the 5,6-double bond of 40 to yield a nonreducing
adduct 51 [131], which complicates kinetic analyses in aqueous media.
This can be prevented by placing substituents at the 5,6-position,
as in 52 [132] or 53 [133,134].

Strictly speaking, coordination of the reducing carbonyl in 47
[134], in 3- and 4-cinnamoyl pyridines [126] or 3- and 4-pyridine
carboxaldehydes [130] does not appear to be a necessity for reduction, although M^{2+} effects are largest when this is allowed. Other
ketones, such as 50 or hexachloroacetone, have their rates of reduction by 40 retarded by addition of Zn(II) or Mg(II) [135,136], and
still others such as 44 or 45, show reduction rates to be maximized
as a function of increasing $[M^{2+}]$ and then diminish on further additions [128,137,138]. For the reduction of 45 by 40 in CH_3CN, Hughes
and Prince reported a detailed mechanistic analysis and concluded
that the reactive species was a $40:M^{2+}$ complex which then binds a
more weakly coordinating aldehyde [137]. This is evidenced by observations that the absorption spectrum of 40 shifts markedly in the

presence of ZnX_2 salts, which can exist in that medium as 1:1 singly
ionized electrolytes. However, the absorption spectrum of 2-pyridine
carboxaldehyde shows no shift in the presence of metal ions, indicating only a very weak association. [This apparently contrasts the
situation in CH_3OH, where addition of $Zn(OAc)_2$ is reported to cause
spectral shifts in 45 [130], although the salt is probably ionized
in CH_3OH but not CH_3CN.]

   The kinetic analysis [137] shows reduction to be first order
in 45 since plots of k_{obs} against [45] are linear. Importantly,
however, for constant concentrations of both 45 and 40, k_{obs} versus
increasing $[M^{2+}]$ shows a maximizing effect and then dimunition, the
process being shown in Eq. (10). Here the decrease in k_{obs} as $[MX^+]$

$$Mx^+ + NBDN \xrightarrow{k_2} [\text{MX: NBDN}]^+ \tag{10}$$

$$[\text{MX: NBDN}]^+ + PCA \xrightarrow{k_3} [\text{MX: NBDN:PCA}]^+$$

NBDN = 40 (R = CH_2Ph) $\downarrow k$

PCA = 45   Products

is increased beyond a certain point is attributable to a nonproduc-
tive association of MX^+ and 45, which effectively removes the latter
from being able to bind with the already formed dihydronicotinamide:MX^+
complex.

   ^{13}C and ^1H NMR shifts of the dihydropyridine resonances indicate
the site of M^{2+} binding to be the amide unit [138]. Analogous experiments as a function of [Zn(II)] show the NMR resonances of 45 to indicate that complexation can occur at N, but little or no change in the
aldehydic group is seen, suggesting a *transoid* orientation away from
the metal [138]. A picture emerges which is of direct relevance to
any model study in which M^{2+} binding by the aldehyde C=O moiety is
weaker than that of the dihydronicotinamide reducing agent. Structures 54 and 55 illustrate that although the C=O is not polarized by
the metal, it could be positioned close enough to the 4-position of
complexed 40 to allow H delivery, the incipient alcohol(ate) being
finally bound by the M^{2+} [138,139]. This scheme has a direct bearing
on observations of Ohno et al. that a dihydronicotinamide with a

chiral N substituent can induce large enantiomeric excesses in its
metal-promoted reductions of certain α-keto esters [140,141]. Ternary
complexes that associate both reductant and reducing agent together
on the metal provide a readily acceptable rationale [140,141] for the
observations, particularly if the M^{2+} is bound by the chiral amide
unit of the reducing agent.

Of course, bearing in mind Hughes and Prince's observations
[137-139], one might question the relevance of most of the previously
mentioned models to the situation in the enzyme, since it is already
clear that the enzymic metal is not associated with the NADH reducing
agent.

Much discussion has centered on whether dihydronicotinamides
in model systems or enzyme reduce substrates by direct transfer of
H^- or by single electron transfer (SET) from the dihydronicotinamide
to substrate followed by H abstraction (for a compendium of the
literature, see Ref. 126). It is clear that 40 can and does partake
in reductive processes by mechanisms which can only involve radical
intermediates, as is evidenced by its reduction of tertiary nitro
compounds [142] and electron transfer to 2,4-dinitrobenzonitrile
[143]. However, to date, no unambiguous evidence exists for SET
pathways mediated by ADH. For the most part evidence for SET reductions by 40 has rested on an observed disparity between kinetic (H/D)
isotope effects and product isotope partitioning ratios, which is
reasonably interpreted as arising from an intermediate along the
reaction pathway [124,125,144,145]. This was originally thought to
be the radical formed from SET between 40 and the substrate, but
recent work clearly shows it to arise from a competitive reaction
generating a covalent addition product as in Eq. (11) [146]. Such
addition products have been observed for several other aldehydes and
ketones [147,148], indicating that the foregoing isotope effect dis-

$$(11)$$

parity alone cannot be taken as valid evidence for a radical intermediate along the pathway for H transfer [146]. Recent work has
demonstrated the viability of intermediates of a radical nature in
the reduction of substrates by 40 [148,149], but since this was done
in the absence of metal, the exact mechanism by which C=O reduction
by 40 is promoted by Zn(II) still remains to be determined before
any statement can be made concerning the enzyme. It might be tempting to speculate that should SET processes be viable in the enzyme,
the purpose of Zn(II) would be to lower the reduction potential of
the carbonyl by stabilizing the incipient ketyl.[*] The ubiquitous
occurrence of cysteine S(H) in the active-site region of ADHs possibly serves to initiate a radical process, although exactly how remains
to be established.

To this point it is clear that models for ADHs have concentrated
exclusively on the reduction aspects and have as yet not approximated
the known metal-binding ligands. Since it is now clearly established
that the catalytic Zn(II) in both mammalian liver and yeast ADHs are
ligated to a histidine imidazole and two cysteine S(H) residues, it
is important to investigate the effect that such ligation embues on
the catalytic and spectroscopic features of models. An early report
from Williams's group [150] compared the spectra of monomeric Co(II)
complexes of thioglycollate and cysteine with those of Co(II)-substi-
tuted Zn(II) enzymes. Complexes 55a and 55b are four- or five-
coordinate, as judged from the appearance of absorption bands from

---

[*]For evidence against the SET process in ADH see S.-K. Chung and
S.-U. Park, *J. Org. Chem.*, *47*, 3197 (1982).

500 to 700 nm ($\varepsilon_{max} \geq 100$ M^{-1}/cm) and near 1000 nm ($\varepsilon_{max} \approx 10$ M^{-1}/cm). The main characteristic is the appearance of bands at ≤ 350 nm (ε_{max} ~3000 M^{-1}/cm) which are due to RS$^-$ → Co(II) charge transfer. In proteins containing no sulfhydryl ligands to the metal, there are no such charge-transfer bands between 250 and 500 nm unless phenolate is bound as a ligand. The effect is clearly demonstrated in the absorption spectra of Co(II) horse-liver ADH in that both the structural and catalytic metal (the former ligated to four cysteines and the latter to two) show charge-transfer bands at ~340 nm [ε >1000 M^{-1}/cm per Co(II)] [118,151]. A spectroscopic model for the structural metal has been offered in 56, a tetrahedral environment being enforced by the wide bite of the two dithiosquarates [152].

56   57 a, x = H   58
             b, x = CRO

A preliminary report has appeared of the synthesis and physical properties of some N, S, S ligands such as 57 and 58 which appear to be reasonable models for the catalytic metal binding site in ADH [153]. Chelation of divalent ions such as Cu(II), Ni(II), Co(II), and Zn(II) is very strong, as would be expected with two thiol ligands, and quantitative titration experiments show both thiols to be deprotonated when metal bound. It might be anticipated that 57b would be a reasonable model for the substrate-bound active site of the enzyme; however, studies of the reduction process with 40 have not yet appeared.

5. ALKALINE PHOSPHATASE

Alkaline phosphatases (APs) are widely distributed Zn(II)- and Mg(II)-containing enzymes, which as the name implies show activity toward phosphate monoesters at alkaline pH [154]. At present it is clear

that AP binds four Zn(II) and two Mg(II) ions into its dimeric structure. Two of the metals are tightly bound and occupy sites 32 Å
apart [155] which are of low symmetry and coordination number, based
on the d-d transitions of the Co(II) derivative [156,157]. These
are termed the "catalytic" metals, while the next pair of metals
which bind to the enzyme are termed "structural" [158,159]; from the
Co(II) absorption spectra of the structural metals, their binding
sites are of much higher symmetry and are probably octahedral,
although of unknown constitution. Two remaining sites have a higher
affinity for Mg(II) than Zn(II) and probably serve a structural or
stabilizing role.

X-ray analysis of the enzyme, although in progress for some
time [155,160], has not yet identified the active-site coordinating
ligands. ^{13}C NMR spectra of the enzyme prepared with histidine
enriched with ^{13}C at the γ position are consistent with at least
three imidazole ligands per monomer unit being coordinated to the
catalytic Zn(II) [161,162]. ESR spectra of Cu(II)$_2$AP are consistent
in that at least three magnetically equivalent N's are ligands to
the Cu(II) ions [163].

The mechanism of cleavage of phosphate monoesters by AP apparently involves formation of a phosphoryl enzyme, the phosphorylated
group being identified as a serine OH [154]. Since the k_{cat} term
for the enzyme is nearly invariant with structure, the kinetics
require that some step after phosphorylation be rate determining.
This could be either cleavage of the Ser O-PO$_3^{2-}$ group or phosphate
debinding, depending on pH, as shown in Eq. (12) [154]. Recently,

$$E + ROP \rightleftharpoons E \cdot ROP \xrightarrow{k_2} E-P \xrightarrow{k_3} E \cdot Pi \rightleftharpoons E + Pi \quad (12)$$

$$5 \times 10^3 \text{ S}^{-1} \quad \text{in acid 0.7 S}^{-1} \quad 20\text{---}70 \text{ S}^{-1}$$
$$\text{in base } 10^2 \text{ S}^{-1}$$

Knowles and co-workers have demonstrated a complete retention of
stereochemistry of a chiral phosphate in the enzyme-mediated trans
phosphorylation to acceptor alcohols, which is consistent with a
double displacement on P, the first being by Ser OH and the second
by the acceptor alcohol [164]. Mechanistically, the role of metal

might be a positioning or charge-neutralizing one in the Ser attack
step, and to then activate H_2O (similar to that process in CA) for
the subsequent dephosphorylation step [154].

Model systems for AP have not been studied as extensively as
those for the preceding three Zn(II) enzymes, presumably because the
active-site region has not yet yielded to x-ray analysis, and has
only recently been suggested on the basis of ^{13}C NMR data [161,162].
However, an early report from Sigman's group showed that an appropriately located OH in 59 could be trans-phosphorylated by ATP to
form 60 [165].

The reaction is proposed to involve a ternary complex in which
the function of the Zn(II) is twofold: to position the metal-associated
ATP appropriately for transfer of the terminal phosphoryl residue, and
to activate the proximal hydroxyl by reducing its pK_a. The latter is
evidenced by a kinetic pK_a of ~7.5 in the hydrolysis of p-nitrophenylpicolinate by 59 [165]. Apparently, once 60 forms it is stable under
the reaction conditions such that the closely located Zn(II) does not
rapidly promote its hydrolysis as might be expected if a Zn^+-OH were
able to be generated.

Such stability toward Zn(II)-catalyzed hydrolysis is also noted
for 2-pyridylmethylphosphate (61) [166].

Later work by Murakami [167] on the metal-promoted hydrolysis
of 8-quinolinyl phosphate (62) indicated that Cu(II) was effective
but Ni(II) was not, and in particular the effect increased markedly
as a function of pH, indicating involvement of OH^- as an external
or metal-associated nucleophile. Since 3-pyridylphosphate shows no

activation by either metal, a primary requirement for catalysis
appears to be at least bidentate chelation by phosphate and the
heterocyclic N. If the role of M^{2+} were simply one of charge
neutralization, the additional binding group would not necessarily
be required, although certainly larger formation constants result
from bidentate chelation. A second possible role for M^{2+} in these
systems appears to be its ability to stabilize the leaving quinolinol
anion in the transition state (63) by binding to the ester-type
oxygen [167]. In this example it is not clear whether the phosphorus
unit leaves as metaphosphate [167] or phosphate, although inorganic
phosphate is certainly liberated in the proposed enzymic mechanism
[154].

Metal-ion activation of phosphate transfer by bidentate coordination of the phosphate group to M^{3+} has been proposed to account
for the rapid rate of hydrolysis of 64 over 65 [168]. For the Co(III)

systems, although polarization or charge neutralization of the
phosphate is undoubtedly greater for bidentate than for monodentate
chelation, the dramatic difference in hydrolysis rates can scarcely
be explained in this way. It is proposed that the effect has a
steric origin resulting from a constraint of the O-P-O angle in the
metallo cycle, similar to the effects that cause a cyclic five-
membered phosphate to hydrolyze some 10^{6}- to 10^{8}-fold faster than
its acyclic counterparts [169,170].

Somewhat later, complementary observations were made in studies
of 66 [171] and 67, whose structure was confirmed by x-ray crystallography. Hydrolysis of 66 followed a rate law $k_{obs} = 7 \cdot 10^{-5} \, s^{-1}$ +
5.1 M^{-1}/s [OH^{-}], and ^{18}O tracer studies indicated that ester cleavage
from, and chelate opening on the metal occur through different pathways: through a chelated five-coordinate phosphorane intermediate

and the conjugate base hydrolysis mechanism, respectively. Ester hydrolysis is accelerated 10^9-fold relative to the uncoordinated ester in basic solution. In this example, attack by external hydroxide is especially favorable since it relieves the strain at the four-membered chelate.[*] A new question, perhaps more germaine to the proposed involvement of M^+-OH in the enzyme, is whether internal nucleophiles can be delivered by the metal to a coordinated phosphate.

Sargeson extended his Co(III) work to include studies on 68 to see whether a coordinated amido group could internally cleave p-nitrophenylphosphate [172]. Hydrolysis of 68 is first order in [OH⁻], and NMR analysis indicates the products to be p-nitrophenol(ate) and p-nitrophenylphosphate, these being formed by competing pathways, as shown in Eq. (13). ^{18}O tracer studies show no incorporation into p-nitrophenylphosphate, indicating that it was cleaved away from the complex intact. Based on a pK_a of the coordinated NH_3 of ~17, the accelerating effect of the internal NH_2 is ~10^8 for cleavage of

---

[*]Recent work by Lindoy and Sargeson et al. has established that the p-nitrophenylphosphate complex with the constitution [Co(L)₂(O₃POR)₂ (en)₂CO] (SO₃CF₃)₂ with L = 1,2-ethanediamine (en) or 1,3-propanediamine (tn) has shown that the earlier formulation [171] of 66 is incorrect and is, in fact, a dimeric species having two bridging phosphate ester groups coordinated to the two Co(III) centers to form a nonplanar eight-membered ring. It was previously proposed [171] that the reactivity of the chelated ester in base was attributable to strain within a four-membered chelate, but it now appears that the mechanism involves opening of the eight-membered dimer ring, capture of OH⁻ and subsequent intramolecular attack of the bound OH⁻ ion on the remaining bridging phosphate ester [172b].

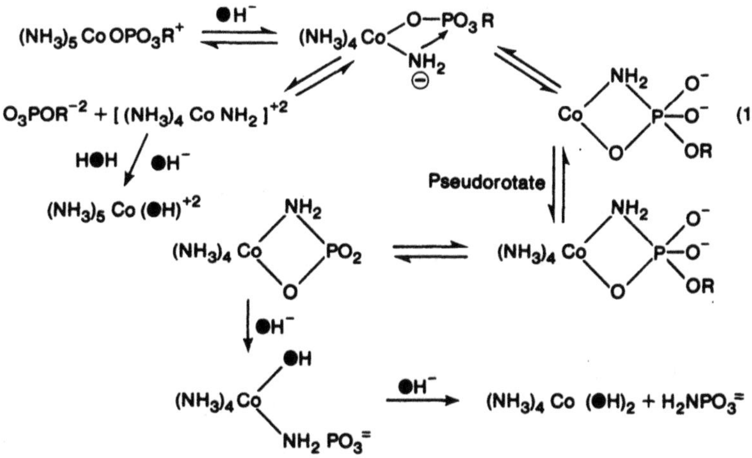

$$\begin{array}{c}
(NH_3)_5\,CoOPO_3R^+ \;\overset{\bullet H^-}{\rightleftharpoons}\; (NH_3)_4\,Co\overset{O-PO_3R}{\underset{\underset{\ominus}{NH_2}}{\diagup}}\\
\end{array}$$

$$O_3POR^{-2} + [\,(NH_3)_4\,Co\,NH_2\,]^{+2}$$

$$(NH_3)_5\,Co\,(\bullet H)^{+2}$$

(13)

Pseudorotate

$$(NH_3)_4\,Co\,(\bullet H)_2 + H_2NPO_3^=$$

p-nitrophenoxide, and is due largely to a proximity effect of
coordinated nucleophile and substrate, and to rapid decomposition
of the aminophosphorane.

6. SUPEROXIDE DISMUTASE

We consider last the metal-binding sites and function of superoxide
dismutase (SOD), the biological aspects of which have recently been
well reviewed [173]. These widely distributed enzymes operate using
several metals, such as Mn(II), or Fe(II), or both Cu(II) and Zn(II),
and facilitate the disproportion of O_2^- into H_2O_2 and O_2. The absence
of SODs produces severe consequences to the organism since a non-
trivial product of univalent reduction of O_2 to H_2O in biological
systems (O_2^-) is extremely toxic and therefore specific defense must
be made against it.

The x-ray structure of a Cu(II)-Zn(II) SOD from bovine erythrocytes [174,175] shows the enzyme to be a dimer with two active sites,
each containing a Cu(II) and Zn(II) bridged by a single imidazolate,
as in 69. In this case, the environment of the Zn(II) is an irregular tetrahedron, its four ligands being comprised of two histidine
imidazoles, one aspartate COO⁻ and the bridging imidazolate. The

69

Cu(II) is bound by four imidazole units in a nearly square planar
arrangement, with an axially oriented water molecule exposed to the
protein exterior. Apparently, the Cu(II) is the catalytic center
[176] and the Zn(II) has an auxillary role, since it can be replaced
by a variety of divalent metals such as Hg(II) and Co(II) without
loss of activity, while only Cu(II) can serve as the active metal
[177-180]. Recent ^{113}Cd NMR experiments with bovine Cd(II)Cu(II)SOD
show that on reduction it is the copper-imidazolate bond that breaks,
leaving the Cu(I)-O_2 moiety coordinated to the protein by the three
remaining histidines [176].

Since free aqueous Cu(II) ions are about 4.5 times as effective
as SOD in catalyzing superoxide dismutation, no special additional
role needs to be invoked for the enzyme. However, under conditions
found in the cell, there is no significant concentration of free
Cu(II), so that the role of the enzyme need only be one of holding
the metal in readiness for this exclusive task and preventing it from
undergoing additional deleterious processes. Hence models for SOD
have for the most part centered on the construction of binuclear
chelating agents capable of binding two metals with a bridging
imidazolate.

Recently, Lippard and co-workers reported preliminary studies
with complexes 70 and 71 [181,182]. The advantage of the binucleating

70   **71**

macrocycle in $\underline{70}$ appears to be one of enhanced stability over $\underline{71}$
since the former is stable from pH 6 to 10, while the latter is only
stable from pH 8.5 to 9.5. ESR spectra are remarkably similar to
those in bovine SOD, in which Zn(II) has been replaced by Cu(II).
A variety of other binuclear Cu(II) complexes have been reported
[183-186], although as yet no reports of a Zn(II)-Cu(II) complex
have appeared.

7. CONCLUSION

The account above has outlined results obtained on the basis of
studies of several small-molecule mimics for the active sites of
Zn(II) metalloenzymes. It is clear that consistent efforts by
researchers involved in this type of work has brought our understanding of possible modes of catalysis by these enzymes to a
reasonably sophisticated level. It cannot be overemphasized that
the great bulk of the modeling rests on a more or less complete
understanding of at least the active-site structure of the native
enzymes, as delineated by x-ray crystallography. It may be, however, that the relative positioning of the active-site residues is
subject to change once further refinement of the x-ray crystal data
is completed [63,64]. Nevertheless, one might conclude that at least
for CA, CPA, and ADH, model studies have established some precedence
for several of the spectroscopic and catalytic features of these
enzymes, although one could not claim with confidence that our
understanding is complete.

    With respect to CA, although it now seems possible to account
for many of the spectroscopic features of the Co(II) enzyme on the
basis of model studies, the phenomenal catalytic prowess of this
entity has not yet been approached. Further studies of chelating
"tripod" ligands which systematically vary the imidazole-M(II)
geometry will undoubtedly yield insights into how much structural
variations change the catalytic properties of the metal.

At present, Zn(II) peptidase and esterase models have demonstrated those features which promote large rate enhancements for
M(II)-catalyzed intramolecular hydrolysis of chelates. However, it
should be borne in mind that the great bulk of these studies are of
systems that are not catalytic in the true sense since turnover
capabilities have not yet been incorporated into the models. Clearly,
the next generation of models will need to incorporate a Zn(II)-
binding site and appropriately positioned COO(H) moiety into a
single molecule capable of catalyzing intermolecularly the hydrolysis
of esters and amides. This is a substantially more complex problem
since it requires some provision for substrate binding to the metal
chelate prior to the catalytic event, and then release of the hydrolyzed product to enable turnover.

Although it is presently demonstrated that Zn(II) can promote
bimolecular reduction of chelated aldehydes or ketones by N-alkyldi-
hydronicotinamides, presumably by acting as a Lewis acid, subsequent
model studies of ADH must address the function of the cysteine S(H)
groups in the enzyme active site. Since ADHs from both mammalian
and yeast sources contain these cysteines, it is reasonable to assume
that their presence influences the catalytic event in an as yet un-
determined way.

Modeling the action of alkaline phosphatase has not yet evolved
to the state of that for the foregoing three Zn(II) metalloenzymes,
although in the future we can expect a rapid increase in activity in
this area. Clearly, as more structural information becomes available
for this and the remaining Zn(II) enzymes, we can look forward to
intensified effort.

REFERENCES

1. E. J. Hewitt and T. A. Smith, *Plant Mineral Nutrition*, English
   Universities Press, London, 1975.
2. E. J. Underwood, *Trace Elements in Human and Animal Nutrition*,
   4th ed., Academic Press, New York, 1977.

3. M. F. Dunn, *Struct. Bonding (Berl.)*, *23*, 61 (1975).

4. A. Galdes and H. A. O. Hill, in *Inorganic Biochemistry* (H. A. O. Hill, ed.), Specialist Periodical Report, Chemical Society of London, 1979, p. 317ff.

5. A. M. Cheh and J. B. Neilands, *Struct. Bonding (Berl.)*, *29*, 123 (1976).

6. R. H. Prince, *Adv. Inorg. Radiochem.*, *22*, 349 (1979).

7. S. Lindskog, *Struct. Bonding (Berl.)*, *8*, 153 (1970).

8. P. Wooley, *Nature (Lond.)*, *258*, 677 (1975).

9. B. T. Golding and G. J. Leigh, in *Inorganic Biochemistry* (H. A. O. Hill, ed.), Specialist Periodical Report, Chemical Society of London, 1979, pp. 35-62.

10. Y. Pocker and S. Sarkanen, *Adv. Enzymol.*, *47*, 149-274 (1978).

11. P. Wyeth and R. H. Prince, *Inorg. Perspect. Biol. Med.*, *1*, 37 (1977).

12. J. E. Coleman, in *Inorganic Biochemistry*, Vol. 1 (G. L. Eichorn, ed.), Elsevier, Amsterdam, 1973, pp. 488-548.

13. S. Lindskog, L. E. Henderson, K. K. Kannan, A. Liljas, P. O. Nyman, and B. Strandberg, in *The Enzymes*, 3rd ed., Vol. 5 (P. D. Boyer, ed.), Academic Press, New York, 1971, pp. 587-665.

14. H. F. Bundy, *Comp. Biochem. Physiol.*, *57B*, 1 (1977).

15. A. Liljas, K. K. Kannan, P.-C. Bergstén, I. Vaara, K. Fridborg, B. Strandberg, U. Carlbom, L. Järup, S. Lövgren, and M. Petef, *Nature (Lond.)*, *235*, 131 (1972).

16. B. Notstrand, I. Vaara, and K. K. Kannan, in *The Isozymes*, Vol. 1 (C. L. Markert, ed.), Academic Press, New York, 1975, pp. 575-599.

17. K. K. Kannan, B. Notstrand, K. Fridborg, S. Lövgren, A. Ohlsson, and M. Petef, *Proc. Natl. Acad. Sci., USA*, *72*, 51 (1975).

18. K. K. Kannan, M. Petef, K. Fridborg, H. Cid-Dresdner, and S. Lövgren, *FEBS Lett.*, *73*, 115 (1977).

19. I. D. Campbell, S. Lindskog, and A. I. White, *Biochim. Biophys. Acta*, *484*, 443 (1977).

20. See J. Huguet, Ph.D. dissertation, University of Alberta, 1980, for a compendium of mechanisms.

21. J. H. Wang, *Proc. Natl. Acad. Sci. USA*, *66*, 874 (1970).

22. Y. Pocker and D. R. Storm, *Biochemistry*, *7*, 1202 (1968).

23. J. M. Pesando, *Biochemistry*, *14*, 681 (1975).

24. R. K. Gupta and J. M. Pesando, *J. Biol. Chem.*, *250*, 2630 (1975).

25. R. G. Khalifah, *J. Biol. Chem.*, *246*, 2561 (1971).

26. E. T. Kaiser and K.-W. Lo, *J. Am. Chem. Soc.*, *91*, 4912 (1969).

27. J. E. Coleman, *J. Biol. Chem.*, *242*, 5212 (1967).

28. J. E. Coleman, in *Biophysics and Physiology of CO₂* (C. Bauer, G. Gros, and H. Bartels, eds.), Springer-Verlag, Berlin, 1980, pp. 133-150.

29. R. P. Davis, *J. Am. Chem. Soc.*, *81*, 5674 (1959).

30. S. Lindskog, *J. Biol. Chem.*, *238*, 945 (1963).

31. Y. Pocker and D. W. Bjorkquist, *Biochemistry*, *16*, 5698 (1977).

32. S. Lindskog and P. O. Nyman, *Biochim. Biophys. Acta*, *85*, 462 (1964).

33. I. Bertini, C. Luchinat, and A. Scozzafava, *Inorg. Chim. Acta*, *46*, 85 (1980).

34. G. S. Jacob, R. D. Brown, and S. Koenig, *Biochemistry*, *19*, 3755 (1980).

35. H. DeVoe and G. B. Kistiakowski, *J. Am. Chem. Soc.*, *83*, 274 (1960).

36. J. C. Kernohan, *Biochim. Biophys. Acta*, *81*, 246 (1964).

37. C. K. Tu and D. N. Silverman, *J. Am. Chem. Soc.*, *97*, 5935 (1975), and references therein.

38. Y. Pocker and J. E. Meany, *J. Am. Chem. Soc.*, *87*, 1809 (1965).

39. Y. Pocker and J. E. Meany, *Biochemistry*, *4*, 2535 (1965).

40. Y. Pocker, L. C. Bjorkquist, and D. W. Bjorkquist, *Biochemistry*, *16*, 3967 (1977), and references therein.

41. P. Wooley, *J. Chem. Soc. Perkin Trans. II*, 318 (1977).

42. P. Wooley, in *Biophysics and Physiology of CO₂* (C. Bauer, G. Gros, and H. Bartels, eds.), Springer-Verlag, Berlin, 1980, pp. 216-225.

43. D. W. Appleton and B. Sarkar, *Proc. Natl. Acad. Sci. USA*, *71*, 1686 (1974).

44. R. B. Martin, *Proc. Natl. Acad. Sci. USA*, *71*, 4346 (1974).

45. I. Sóvágó, T. Kiss, and A. Gergely, *J. Chem. Soc., Dalton*, 964 (1978).

46. J. M. Harrowfield, V. A. Norris, and A. M. Sargeson, *J. Am. Chem. Soc.*, *98*, 7282 (1976).

47. E. Chaffee, T. P. Dasgupta, and G. M. Harris, *J. Am. Chem. Soc.*, *95*, 4169 (1973).

48. D. A. Palmer and G. M. Harris, *Inorg. Chem.*, *13*, 965 (1974).

49. I. Tabushi, Y. Kuroda, and A. Mochizuki, *J. Am. Chem. Soc.*, *102*, 1152 (1980).

50. L. K. Thompson, B. S. Ramaswamy, and E. A. Seymour, *Can. J. Chem.*, *55*, 878 (1977).

51. F. Mani, *Inorg. Nucl. Chem. Lett.*, *17*, 45 (1981).

52. F. Mani and G. Scapacci, *Inorg. Chim. Acta*, *38*, 151 (1980).

53. I. Bertini, G. Canti, C. Luchinat, and F. Mani, *Inorg. Chim. Acta*, *46*(B1), L91 (1980).

54. C. C. Tang, D. Davalian, P. Huang, and R. Breslow, *J. Am. Chem. Soc.*, *100*, 3918 (1978).

55. R. S. Brown and J. Huguet, *Can. J. Chem.*, *58*, 889 (1980).

56. R. K. Boggess and S. J. Boberg, *J. Inorg. Nucl. Chem.*, *42*, 21 (1980).

57. W. R. McWhinnie, G. C. Kulasingam, and J. C. Draper, *J. Chem. Soc. A*, 1199 (1966).

58. J. C. Lancaster and W. R. McWhinnie, *Inorg. Chim. Acta*, *5*, 515 (1975).

59. J. Huguet and R. S. Brown, *J. Am. Chem. Soc.*, *102*, 7571 (1980).

60. R. S. Brown, N. J. Curtis, and J. Huguet, *J. Am. Chem. Soc.*, *103*, 6953 (1981).

61. N. J. Curtis and R. S. Brown, *J. Org. Chem.*, *45*, 4038 (1980).

62. L. K. Thompson, R. G. Ball, and J. Trotter, *Can. J. Chem.*, *58*, 1566 (1980).

63. I. Bertini, G. Canti, C. Luchinat, and F. Mani, *Inorg. Chem.*, *20*, 1670 (1981).

64. R. J. Read and M. N. G. James, *J. Am. Chem. Soc.*, *103*, 6947 (1981).

65. B. H. Gibbons and J. T. Edsall, *J. Biol. Chem.*, *238*, 3502 (1963).

66. J. C. Kernohan, *Biochim. Biophys. Acta*, *96*, 304 (1965).

67. S. Lindskog, *Biochemistry*, *5*, 2641 (1966).

68. R. S. Brown, N. J. Curtis, S. Kusuma, and D. Salmon, *J. Am. Chem. Soc.*, *104*, 3188 (1982).

69. S. H. Koenig, R. D. Brown, and G. S. Jacob, in *Biophysics and Physiology of CO_2* (C. Bauer, G. Gros, and H. Bartels, eds.), Springer-Verlag, Berlin, 1980, pp. 238-253.

70. D. Demoulin, A. Pullman, and B. Sarkar, *J. Am. Chem. Soc.*, *99*, 8498 (1977).

71. A. Pullman and P. Demoulin, *Int. J. Quantum Chem.*, *16*, 641 (1979).

72. D. Demoulin and A. Pullman, in *Catalysis in Chemistry and Biochemistry: Theory and Experiment* (B. Pullman, ed.), D. Reidel, Dordrecht, Holland, 1979, pp. 51-66.

73. E. Clementi, G. Corongiu, B. Jönsson, and S. Romano, *Gazz. Chim. Ital.*, *109*, 669 (1979), and references therein.

74. E. Clementi, G. Corongiu, B. Jönsson, and S. Romano, *J. Chem. Phys.*, *72*, 260 (1980); *FEBS Lett.*, *100*, 313 (1979).

75. R. P. Sheridan and L. C. Allen, *J. Am. Chem. Soc.*, *103*, 1544 (1981).

76. W. A. Sokalski, private communication in form of preprint.

77. P. E. Haffner and J. E. Coleman, *J. Biol. Chem.*, *250*, 996 (1975).

78. W. N. Lipscomb, *Tetrahedron*, *30*, 1725 (1974), and references therein.

79. (a) W. N. Lipscomb, *Acc. Chem. Res.*, *3*, 81 (1970). (b) F. A. Quiocho and W. N. Lipscomb, *Adv. Protein Chem.*, *25*, 1 (1971).

80. M. F. Schmidt and J. R. Herriott, *J. Mol. Biol.*, *103*, 175 (1976).

81. W. R. Kester and B. W. Matthews, *J. Biol. Chem.*, *252*, 7704 (1977).

82. B. W. Matthews, J. N. Jansonius, and P. M. Colement, *Nature (Lond.)*, *238*, 37 (1972).

83. M. K. Pangburn and A. Walsh, *Biochemistry*, *14*, 4050 (1975).

84. D. C. Rees, M. Lewis, R. B. Honzatko, W. N. Lipscomb, and K. D. Hardman, *Proc. Natl. Acad. Sci. USA*, *78*, 3408 (1981).

85. S. A. Latt and B. L. Vallee, *Fed. Proc.*, *28*, 534 (1968).

86. J. T. Johnson and B. L. Vallee, *Biochemistry*, *14*, 649 (1975).

87. E. T. Kaiser and B. L. Kaiser, *Acc. Chem. Res.*, *5*, 219 (1972).

88. M. W. Makinen, K. Yamamura, and E. T. Kaiser, *Proc. Natl. Acad. Sci. USA*, *73*, 3882 (1976).

89. S. Scheiner and W. N. Lipscomb, *J. Am. Chem. Soc.*, *99*, 3466 (1977).

90. R. Breslow and D. L. Wernick, *Proc. Natl. Acad. Sci. USA*, *74*, 1303 (1977).

91. R. Breslow, R. Fairweather, and J. Keana, *J. Am. Chem. Soc.*, *89*, 2135 (1967).

92. D. A. Buckingham, C. E. Davis, D. M. Foster, and A. M. Sargeson, *J. Am. Chem. Soc.*, *92*, 5571 (1970).

93. D. A. Buckingham, D. M. Foster, and A. M. Sargeson, *J. Am. Chem. Soc.*, *96*, 1726 (1974); *92*, 6151 (1970).

94. D. A. Buckingham, F. R. Keene, and A. M. Sargeson, *J. Am. Chem. Soc.*, *96*, 4981 (1974).

95. D. S. Sigman and C. T. Jorgensen, *J. Am. Chem. Soc.*, *94*, 1724 (1972).

96. T. Eiki, S. Kawada, K. Matsushima, M. Mori, and W. Tagaki, *Chem. Lett.*, 997 (1980).

97. M. A. Wells and T. C. Bruice, *J. Am. Chem. Soc., 99*, 5341 (1977).

98. R. Breslow, D. E. McClure, R. S. Brown, and J. Eisenach, *J. Am. Chem. Soc., 97*, 194 (1975).

99. G. Tomalin, B. L. Kaiser, and E. T. Kaiser, *J. Am. Chem. Soc., 92*, 6046 (1970).

100. P. L. Hall, B. L. Kaiser, and E. T. Kaiser, *J. Am. Chem. Soc., 91*, 485 (1969).

101. E. T. Kaiser and F. W. Carson, *J. Am. Chem. Soc., 86*, 2922 (1966).

102. J. T. Groves and R. M. Dias, *J. Am. Chem. Soc., 101*, 1033 (1979).

103. R. Breslow and C. McAllister, *J. Am. Chem. Soc., 93*, 7096 (1971).

104. T. H. Fife and V. L. Squillacote, *J. Am. Chem. Soc., 99*, 3762 (1977).

105. T. H. Fife and V. L. Squillacote, *J. Am. Chem. Soc., 100*, 4787 (1978).

106. T. H. Fife and V. L. Squillacote, *J. Am. Chem. Soc., 101*, 3017 (1979).

107. T. H. Fife and T. J. Przystas, *J. Am. Chem. Soc., 102*, 7297 (1980).

108. Y. Murakami, Y. Aoyama, M. Kida, and J.-I. Kikuchi, *J. Chem. Soc., Chem. Commun.*, 494 (1978).

109. R. Breslow and D. E. McClure, *J. Am. Chem. Soc., 98*, 258 (1976).

110. D. M. Hayes and P. A. Kollman, *J. Am. Chem. Soc., 98*, 3335, 7811 (1976).

111. S. Nakagawa, H. Umeyama, K. Kitaura, and K. Morokuma, *Chem. Pharm. Bull., 29*, 1 (1981).

112. C.-I. Bränden, H. Jörnvall, H. Eklund, and B. Furugren, in *The Enzymes*, 3rd ed., Vol. 11 (P. D. Boyer, ed.), Academic Press, New York, 1975, p. 103.

113. T.-K. Li, *Adv. Enzymol., 45*, 427 (1977).

114. J. B. Jones and J. F. Beck, in *Applications of Biochemical Systems in Organic Chemistry* (J. B. Jones, C. J. Sih, and D. Perlman, eds.), Wiley, New York, 1976, p. 107.

115. C.-I. Bränden, H. Eklund, B. Nordström, T. Boiwe, G. Söderlund, E. Zeppezauer, I. Ohlsson, and Å. Åkeson, *Proc. Natl. Acad. Sci. USA, 70*, 2439 (1973).

116. H. Eklund, B. Nordström, E. Zeppezauer, G. Söderlund, I. Ohlsson, T. Boiwe, B.-O. Söderberg, O. Tapia, and C.-I. Bränden, *J. Mol. Biol., 102*, 27 (1976).

117. H. Eklund and C.-I. Bränden, *J. Biol. Chem.*, *254*, 3458 (1979).

118. A. J. Sytkowski and B. L. Vallee, *Proc. Natl. Acad. Sci. USA*, *73*, 344 (1976).

119. D. E. Drum and B. L. Vallee, *Biochim. Biophys. Res. Commun.*, *41*, 33 (1970).

120. R. G. Morris, G. Saliman, and M. F. Dunn, *Biochemistry*, *19*, 725 (1980), and references therein.

121. C. T. Angelis, M. F. Dunn, D. C. Muchmore, and R. M. Wing, *Biochemistry*, *16*, 2922 (1977).

122. K. M. Welsh, D. J. Creighton, and J. P. Klinman, *Biochemistry*, *19*, 2005 (1980), and references therein.

123. M. Hughes and R. H. Prince, *Bioorg. Chem.*, *6*, 137 (1977).

124. D. J. Creighton and D. S. Sigman, *J. Am. Chem. Soc.*, *93*, 6314 (1971).

125. D. J. Creighton, J. Hajdu, and D. S. Sigman, *J. Am. Chem. Soc.*, *98*, 4619 (1976).

126. R. A. Gase and U. K. Pandit, *J. Am. Chem. Soc.*, *101*, 7059 (1979).

127. R. A. Gase, G. Boxham, and U. K. Pandit, *Tetrahedron Lett.*, 2889 (1975).

128. A. Ohno, S. Yasui, R. A. Gase, S. Oka, and U. K. Pandit, *Bioorg. Chem.*, *9*, 199 (1980).

129. A. Ohno, S. Yasui, and S. Oka, *Bull. Chem. Soc. Jpn.*, *53*, 2651 (1980).

130. M. Shirai, T. Chishina, and M. Tanaka, *Bull. Chem. Soc. Jpn.*, *48*, 1079 (1975).

131. S. L. Johnson and P. T. Tuazon, *Biochemistry*, *16*, 1175 (1977).

132. S. Shinkai and T. C. Bruice, *Biochemistry*, *12*, 1750 (1973).

133. S. Shinkai, H. Hamada, T. Ide, and O. Manabe, *Chem. Lett.*, 685 (1978).

134. A. Ohno, S. Yasui, and S. Oka, *Bull. Chem. Soc. Jpn.*, *53*, 3244 (1980).

135. A. Ohno, S. Yasui, K. Nakamura, and S. Oka, *Bull. Chem. Soc. Jpn.*, *51*, 290 (1978).

136. D. C. Dittmer, A. Lombardo, F. H. Batzold, and C. S. Greene, *J. Org. Chem.*, *41*, 2976 (1976).

137. M. Hughes and R. H. Prince, *J. Inorg. Nucl. Chem. Lett.*, *40*, 703 (1978).

138. M. Hughes and R. H. Prince, *J. Inorg. Nucl. Chem.*, *40*, 713 (1978).

139. M. Hughes and R. H. Prince, *J. Inorg. Nucl. Chem.*, *40*, 719 (1978).

140. Y. Ohnishi, M. Kagami, and A. Ohno, *J. Am. Chem. Soc.*, *97*, 4766 (1975).

141. A. Ohno, T. Kimura, S. G. Kim, H. Yamamoto, and S. Oka, *Bioorg. Chem.*, *6*, 21 (1977).

142. N. Ono, R. Tamura, and A. Kaji, *J. Am. Chem. Soc.*, *102*, 2851 (1980).

143. D. D. Tanner, G. Diaz, and R. S. Brown, unpublished results.

144. J. J. Stephens and D. M. Chipman, *J. Am. Chem. Soc.*, *93*, 6694 (1971).

145. S. Shinkai, T. Ide, H. Hamada, O. Manabe, and T. Kunitake, *J. Chem. Soc., Chem. Commun.*, 848 (1977).

146. D. M. Chipman, R. Yaniv, and P. van Eikeren, *J. Am. Chem. Soc.*, *102*, 3244 (1980).

147. W. Tagaki, H. Sakai, Y. Yano, K. Ozeki, and Y. Shimazu, *Tetrahedron Lett.*, 2541 (1976).

148. P. van Eikeren, D. L. Grier, and J. Eliason, *J. Am. Chem. Soc.*, *101*, 7406 (1979).

149. P. van Eikeren, D. L. Grier, P. Kenney, and R. Tomakian, *J. Am. Chem. Soc.*, *101*, 7402 (1979).

150. K. Garbett, G. W. Partridge, and R. J. P. Williams, *Bioinorg. Chem.*, *1*, 309 (1972).

151. A. J. Sytkowski and B. L. Vallee, *Biochemistry*, *17*, 2850 (1978).

152. D. Coucouvanis, D. G. Holah, and F. J. Hollander, *Inorg. Chem. Lett.*, *14*, 2657 (1975).

153. N. J. Curtis and R. S. Brown, *Can. J. Chem.*, *59*, 65 (1981).

154. J. E. Coleman and J. F. Chlebowski, in *Advances in Inorganic Biochemistry*, Vol. 1 (G. L. Eichorn and L. G. Marzilli, eds.), Elsevier/North-Holland, Amsterdam, 1979, pp. 1-66, and references therein.

155. J. R. Knox and H. W. Wyckoff, *J. Mol. Biol.*, *74*, 533 (1973).

156. M. L. Applebury and J. E. Coleman, *J. Biol. Chem.*, *244*, 709 (1969).

157. J. S. Taylor, C. Y. Lau, M. L. Applebury, and J. E. Coleman, *J. Biol. Chem.*, *248*, 6216 (1973).

158. R. A. Anderson, W. F. Bosron, F. S. Kennedy, and B. L. Vallee, *Proc. Natl. Acad. Sci. USA*, *72*, 2989 (1975).

159. W. F. Bosron, R. A. Anderson, M. C. Falk, F. S. Kennedy, and B. L. Vallee, *Biochemistry*, *16*, 610 (1977).

160. W. D. Carlson, Ph.D. dissertation, Yale University, 1976.

161. J. D. Otvos and D. T. Browne, *Biochemistry, 19*, 4011 (1980).

162. J. D. Otvos and I. M. Armitage, *Biochemistry, 19*, 4021 (1980).

163. J. S. Taylor and J. E. Coleman, *Proc. Natl. Acad. Sci. USA, 69*, 859 (1972).

164. S. R. Jones, K. A. Kindeman, and J. R. Knowles, *Nature (Lond.), 275*, 564 (1978).

165. D. S. Sigman, G. M. Wahl, and D. J. Creighton, *Biochemistry, 11*, 2236 (1972).

166. Y. Murakami and M. Tagaki, *J. Am. Chem. Soc., 91*, 5130 (1969).

167. Y. Murakami and J. Sunamoto, *Bull. Chem. Soc. Jpn., 44*, 1827 (1971).

168. F. J. Farrell, W. A. Kjellstrom, and T. G. Spiro, *Science, 164*, 320 (1969).

169. F. H. Westheimer, *Acc. Chem. Res., 1*, 70 (1968).

170. J. G. Verkade, *Bioinorg. Chem., 3*, 165 (1974).

171. B. Anderson, R. M. Milburn, J. M. Harrowfield, G. B. Robertson, and A. M. Sargeson, *J. Am. Chem. Soc., 99*, 2652 (1977).

172. (a) J. M. Harrowfield, D. R. Jones, L. F. Lindoy, and A. M. Sargeson, *J. Am. Chem. Soc., 102*, 7733 (1980). (b) D. R. Jones, L. F. Lindoy, R. M. Milburn, A. M. Sargeson, and M. R. Snow, private communication in form of a preprint submitted to *J. Amer. Chem. Soc.* in manuscript form.

173. (a) I. Fridovich, in *Advances in Inorganic Biochemistry*, Vol. 1 (G. L. Eichorn and L. G. Marzilli, eds.), Elsevier/North-Holland, Amsterdam, 1979, pp. 67-90. (b) J. A. Fee, *Met. Ions Biol. Syst., 13*, 259-298 (1981).

174. J. S. Richardson, K. A. Thomas, B. H. Rubin, and D. C. Richardson, *Proc. Natl. Acad. Sci. USA, 72*, 1349 (1975).

175. J. S. Richardson, D. C. Richardson, K. A. Thomas, E. W. Silverton, and D. R. Davies, *J. Mol. Biol., 102*, 221 (1976).

176. D. M. Bailey, P. D. Ellis, and J. A. Fee, *Biochemistry, 19*, 591 (1980).

177. K. M. Beem, W. E. Rich, and K. V. Rajogopalan, *J. Biol. Chem., 249*, 7298 (1974).

178. H. J. Forman and I. Fridovich, *J. Biol. Chem., 248*, 2645 (1973).

179. J. A. Fee, *J. Biol. Chem., 248*, 4229 (1973).

180. L. Calabrese, G. Rotilio, and B. Mondovi, *Biochim. Biophys. Acta, 263*, 827 (1972).

181. P. K. Coughlin, S. J. Lippard, A. E. Martin, and L. Bulkowski, *J. Am. Chem. Soc., 102*, 7616 (1980).

182. C.-L. O'Young, J. C. Dewan, H. R. Lilienthal, and S. J. Lippard, *J. Am. Chem. Soc.*, *100*, 7291 (1980).

183. W. Mori, A. Nakahara, and Y. Nakao, *Inorg. Chim. Acta*, *37*, L507 (1979).

184. H. M. J. Hendriks and J. Reedijk, *Inorg. Chim. Acta*, *37*, L509 (1979).

185. M. J. Haddad, E. N. Duesler, and D. N. Hendrickson, *Inorg. Chem.*, *18*, 141 (1979).

186. M. G. B. Drew, C. Cairns, A. Lavery, and S. M. Nelson, *J. Chem. Soc.*, *Chem. Commun.*, 1122 (1980).

Chapter 3

AN INSIGHT ON THE ACTIVE SITE OF ZINC ENZYMES
THROUGH METAL SUBSTITUTION

Ivano Bertini[*] and Claudio Luchinat
Institute of General and Inorganic Chemistry
Faculty of Pharmacy
University of Florence
Florence, Italy

[*]Present affiliation: Institute of General and Inorganic Chemistry,
Faculty of Mathematical, Physical and Natural Sciences, University
of Florence, Florence, Italy

1. INTRODUCTION

As an essential metal for most living organisms, Zn(II) is widely
distributed in many biochemical processes. As far as catalytic
processes are concerned, Zn(II) is well represented in all six
enzyme classes defined by the Enzyme Commission of the International
Union on Biochemistry: oxidoreductases, transferases, hydrolases,
lyases, isomerases, and ligases (see Chapter 1). The number of zinc-
containing enzymes is now around 100, but in many cases knowledge is
limited at present to the report of a tentative number of metal atoms
per enzyme unit.

    Metal substitution usually plays a primary role in the charac-
terization of zinc enzymes since Zn(II) is a d^{10} ion and, as such,
is diamagnetic, colorless, and hence not suitable for studies through
electronic or electron paramagnetic resonance (EPR) spectroscopy.
Although diamagnetic, it cannot be used as a nuclear magnetic resonance (NMR) probe since its NMR active isotope, ^{67}Zn, is a quadrupolar
nucleus (I = 5/2) of low natural abundance and very low sensitivity.
Therefore, the Zn(II) is substituted with other probes; some of them,
such as Co(II), Cu(II), Mn(II), and Ni(II), are paramagnetic; others,
despite being diamagnetic, display a d-d absorption spectrum [e.g.,
Co(III)] or are used as a direct NMR probe [e.g., Cd(II)]. The aim
of this chapter is to show the kind of information that can be
obtained on the structure and catalytic mechanisms by use of the
investigative tools appropriate for the various metal probes.

    We thought it would be interesting to compare all the physico-
chemical properties for each metal-substituted derivative. This

allows us to gain reliability on the conclusions obtained through
metal substitution. In order to appreciate the progresses, or some-
time the puzzles, that arise through this kind of approach, a section
is needed in which some information on the zinc enzymes discussed
here is gathered. At the end some general remarks are presented by
considering the information obtained through the various metal substitutions for some of the most representative enzymes.

2. ZINC(II) IN METALLOENZYMES

Metals in metalloenzymes often take part in the catalytic processes,
either directly binding substrates and products, or activating a
residue in the active site, or both. They may also have the task of
keeping the conformation of the active site so as to satisfy the
steric requirements of substrates or, more generally, to maintain
the quaternary structure of the whole protein. In the same enzyme
these different types of functions may coexist, associated with
different metal sites; as will be summarized in this section, Zn(II)
is well represented in all these roles.

2.1. Oxidoreductases

Among the various Zn(II)-containing metalloenzymes in this first
enzyme class, alcohol dehydrogenase [1,2] and superoxide dismutase
[3,4] have been investigated most extensively. For both of them
high-resolution x-ray structures are available [1,5]: the data relevant to the Zn(II) sites are summarized in Table 1.
　　　Alcohol dehydrogenase is an 80,000 molecular weight dimer consisting of two identical subunits, each of them containing two zinc
atoms. The enzyme catalyzes the oxidation of alcohols to aldehydes
with nicotinamide dinucleotide.

$$RR'CHOH + NAD^+ \rightleftharpoons RR'C=O + NADH + H^+ \qquad\qquad (1)$$

TABLE 1

Representative Metal-Substituted Zinc Enzymes

Enzyme	Donor set	Substituted metals[a]	References[b]
Alcohol dehydrogenase	S_4^c; S_2NO^c	$Co^{II}(70)$, $Cu^{II}(1)$, $Cu^I(8)$, $Cd^{II}(30)$, $Ni^{II}(12)$	122, 135, 193
Superoxide dismutase	N_3O^c	$Co^{II}(100)$, $Cu^{II}(50)$, $Cd^{II}(100)$	56, 201, 202
Aspartate transcarbamylase	S_4^c	$Mn^{II}(100)$, $Ni^{II}(100)$, $Cd^{II}(130)$	162, 192, 203
Transcarboxylase	?	$Co^{II}(100)$, $Cu^{II}(0)$	20
RNA polymerase	?	$Co^{II}(100)$	60
Carboxypeptidase	$N_2O_3^c$	$Mn^{II}(28)$, $Fe^{II}(29)$, $Co^{II}(215)$, $Ni^{II}(50)$, $Cu^{II}(0)$, $Cd^{II}(0)$, $Hg^{II}(0)$, $Co^{III}(0)$	204, 205
Thermolysin	$N_2O_2^c$	$Mn^{II}(60)$, $Co^{II}(100)$, $Cu^{II}(100)$, $Cd^{II}(105)$, $Hg^{II}(100)$	206, 207
Alkaline phosphatase	N_4^d; NO_5; O_6	$Mn^{II}(0)$, $Co^{II}(20)$, $Ni^{II}(0)$, $Cu^{II}(0)$, $Cd^{II}(0)$, $Hg^{II}(0)$	208, 209

Enzyme	Coordination	Metal-substituted derivatives	Ref.
β-Lactamase II	N_3S^e	$Mn^{II}(3), Co^{II}(11), Ni^{II}(0), Cu^{II}(0), Cd^{II}(11), Hg^{II}(4)$	210
Carbonic anhydrase	N_3O^c	$Mn^{II}(4), Co^{II}(56), Ni^{II}(5), Cu^{II}(1), Cd^{II}(4), Hg^{II}(0), Co^{III}(0)$	211
Aldolase	$N_3O^f + ?$	$Mn^{II}(15), Fe^{II}(67), Co^{II}(85), Ni^{II}(11), Cu^{II}(0), Cd^{II}(0) Hg^{II}(0)$	212
Phosphomannose isomerase	?	$Mn^{II}(50), Co^{II}(90), Ni^{II}(0), Cu^{II}(130)$	213
Pyruvate carboxylase	?	$Co^{II}(100)$	47

[a] Percent activities relative to the native zinc enzyme in parentheses.
[b] Referred to the activities of the metal-substituted derivatives.
[c] From x-ray data; see Sec. 2.
[d] Refs. 116-118.
[e] Refs. 75, 77, and 78.
[f] Ref. 161.

One of the two zinc atoms of each subunit is located in the middle
of a channel about 20 Å deep and is responsible for the catalytic
activity, while the other is near the surface of the molecule [1].
The former (Table 1) is coordinated to two sulfur atoms from Cys 46
and Cys 174, to a nitrogen atom from His 67, and probably to a water
molecule which completes a nearly tetrahedral coordination polyhedron
[1]. Although deeply buried in the protein, the Zn(II) ion is, in
fact, accessible to solvent and solute molecules; the x-ray structure
has shown that NAD^+ binds near the metal, blocking one of the two
entries of the channel [6]. The other Zn(II) ion is coordinated to
four sulfur atoms from Cys 97, 100, 103, and 111 in a pseudotetrahedral arrangement and is apparently not accessible to solvent [1].
The chemical role of the catalytic zinc ion is to polarize the carbonyl oxygen [7] through either direct coordination (substituting
the coordinated water or adding in a fifth position), or hydrogen
bonding to the coordinated water molecule. The catalytic activity
and inhibitor binding are pH dependent. At least two pK_a values
have been identified in the range 7.6-11.2, one of which could be
related to the deprotonation of the coordinated water [8-11].

Superoxide dismutase is also a dimeric enzyme, of molecular
weight 32,000 [3,4]. Each of the two subunits contains a Zn(II) and
a Cu(II) ion, bridged by an imidazolate ion from His 61 [5]. Its
biological role is to prevent accumulation of O_2^- in tissues by catalyzing the reaction

$$2O_2^- + 2H^+ \rightleftharpoons H_2O_2 + O_2 \qquad (2)$$

As in many other oxidoreductases, the Cu(II) ion is responsible for
the enzymatic activity, being reduced to Cu(I) and reoxidized during
the catalytic cycle.

The Zn(II) ion is coordinated to three nitrogens from His 61,
69, and 78, and to an oxygen atom from Asp 81, in a nearly tetrahedral arrangement; the Cu(II) ion is bound to four nitrogens from
His 61, 44, 46, and 118 in a grossly distorted square-planar arrangement [5]. According to NMR data [12] a water molecule completes its
coordination polyhedron. Although not directly involved in the

catalytic process, the Zn(II) ion is possibly important in maintaining a distorted coordination around Cu(II) through the strain imposed by His 61. The latter is protonated [13] at low pH, with consequent breaking of the bridge and possible loss of the Zn(II) ion, and during the catalytic reduction of Cu(II). However, the enzyme depleted of zinc still displays high catalytic activity [14].

2.2. Transferases

In this enzyme class aspartate transcarbamylase is one step ahead in the investigation, its x-ray structure having recently been solved to 3.0-Å resolution [15]. The enzyme is a 300,000 molecular weight dodecamer composed of two catalytic subunits of molecular weight 100,000, each containing three identical polypeptides and three regulatory subunits each composed of two identical regulatory chains [16]. The enzyme catalyzes the synthesis of carbamyl-1-aspartate from aspartate and carbamyl phosphate:

$$NH_2CO\text{-}OPO_3^{2-} + {}^-OOCCH(NH_3^+)CH_2COO^- \rightleftharpoons {}^-OOCCH(NHCONH_2)CH_2COO^- + H_2PO_4^-$$
$$(3)$$

which is the first step in pyrimidine biosynthesis.

There are six Zn(II) ions in the native enzyme, one for each of the regulatory chains, which are bound to four sulfurs from Cys 109, 114, 137, and 140 [17] (Table 1); their function seems to be mainly structural, since the various subunits lose their ability to reconstitute the intact enzyme in the absence of added divalent cations.

Bacterial transcarboxylase is also a complex enzyme [18] whose molecular weight is reported to be around 800,000. It is constituted of 18 subunits, six each of three different peptides. Two of them have molecular weight 60,000 (2.5 S_H and 2.5 S_E); the third is around 12,000 (1.3 S_E) [19]. Its biological function is the reversible carboxyl transfer from methylmalonyl coenzyme A to pyruvate.

$$CH_3CH(COO^-)COSCoA + CH_3COCOO^- \rightleftharpoons CH_3CH_2COSCoA + {}^-OOCCH_2COCOO^-$$
$$(4)$$

The metal ions (12 per integer enzyme) are probably located pairwise
in the six subunits 2.5 S_E. At variance with the other zinc enzymes,
transcarboxylase contains different amounts of Zn(II), Co(II), and
Cu(II), depending on the growth medium of the bacteria. The activity
is dependent only on the total Zn + Co, suggesting that the incorporation of copper yields an inactive enzyme [20].

   RNA polymerase is the most representative among the recently
discovered zinc-containing nucleic acid polymerases [21], whose
general function is to catalyze the growth of nucleic acid chains.
The enzyme, isolated from *Escherichia coli*, has a molecular weight
of 500,000 and contains two zinc ions per molecule [22].

2.3. Hydrolases

Of the zinc-containing hydrolases carboxypeptidase A is by far the
most thoroughly investigated [23]. The protein is a monomer of
molecular weight 34,500 containing a single zinc atom per molecule;
its biological role is the hydrolysis of the C-terminal peptide bond

$$\cdots\text{-CH-CO-NH-CHCOO}^- + H_2O \rightleftharpoons \cdots\text{-CH-COO}^- + {}^+H_3N\text{-CH-COO}^- \quad (5)$$
$$\quad\ \ R \qquad\quad R' \qquad\qquad\qquad\quad\ R \qquad\qquad\quad R'$$

of proteins, polypeptides, and N-acyl amino acids.

   The x-ray structure of carboxypeptidase A has recently been
refined at 1.75 Å [24]. The zinc ligands are two N-1 nitrogens from
His 69 and His 196, two oxygens from the Glu 72 residue, and a water
molecule (Table 1). The bidentate behavior of Glu 72, which was not
apparent in the earlier x-ray data [25], could be also the result of
disordered monodentate behavior. This would be consistent with the
rather long metal oxygen bonds. The zinc is believed to coordinate
the carbonyl oxygen of the substrate [26], activating the carbon for
nucleophilic attack by water or hydroxide; the latter could also be
produced by deprotonation of the coordinated water molecule.

   Thermolysin is an endopeptidase of molecular weight 34,600
with a single zinc atom per molecule [27]; it also contains four
calcium ions, the presence of which imparts a high thermal stability

to the molecule. From the x-ray data [28] the zinc ion appears to
be coordinated to two N-3 nitrogens from His 142 and His 146, to an
oxygen from Glu 166, and to a water molecule. The similarity with
carboxypeptidase is very strict [29], except that there is not yet
evidence of a bidentate or disordered behavior of the Glu residue.

Alkaline phosphatases are all believed to contain zinc in
variable amounts [30]. They catalyze the hydrolysis of phosphate
monoesters according to the reaction

$$R\text{-}O\text{-}PO_3^{2-} + H_2O \rightleftharpoons R\text{-}OH + HPO_4^{2-} \tag{6}$$

The most extensively characterized is an 80,000 molecular weight
enzyme from $E.\ coli$ [31,32] for which a 7.7-Å x-ray structure is also
available [33]. The enzyme is a dimer containing three nonequivalent
metal-binding sites in each subunit. Two of them are occupied by zinc
(catalytic and structural) and the third by magnesium (regulatory).
In a very recent x-ray structure at 6 Å of resolution, obtained on
a crystal form different from the previous one, the metal-binding
sites have been localized [34]. In each subunit two metals are
located very near each other (4.9 Å), while the third is about 30 Å
away from this couple. At least two out of four zinc atoms per
molecule are essential to reconstitute a partially active enzyme
[32].

β-Lactamases are synthesized by bacteria which develop resistance to β-lactam antibiotics [35]. They catalyze the hydrolysis
of the β-lactam ring in penicillins and cephalosporins:

$$\text{(structure with RCONH, S, N, COO}^-, \text{O)} \ + \ H_2O \ \rightleftharpoons \ \text{(structure with RCONH, S, }^-\text{OOC, H}_2^+\text{N, COO}^-\text{)} \tag{7}$$

β-Lactamase II from $Bacillus\ cereus$ appears to be a zinc enzyme,
molecular weight 23,000, with two nonequivalent zinc sites per
molecule [36]. No crystal structure has been reported so far.

2.4. Lyases

Carbonic anhydrase is perhaps the most thoroughly investigated zinc
enzyme [37-40], if not the most studied enzyme of all. It was the
first zinc enzyme discovered, more than 40 years ago [41], but it
still poses some puzzling questions about the details of the mechanism
of action [39,40]. Its biological role is to enhance dramatically the
rate of establishment of the simple equilibrium reaction

$$CO_2 + H_2O \rightleftharpoons HCO_3^- + H^+ \qquad\qquad (8)$$

at the level of several different tissues. The erythrocytic enzyme
in mammalians is a monomer of molecular weight 30,000 and contains
an essential single zinc ion per molecule. From the x-ray structure
at 2.0 Å at pH 8.7 [42], the metal appears to be coordinated to two
N-3 nitrogens from His 94 and His 96, to the N-1 nitrogen of His 119,
and to a solvent molecule (Table 1). The precise role of the Zn(II)
ion in the catalytic process is still rather controversial [39,40].
This is a meaningful example of how metal substitution allows us to ⁻
propose well-defined models [40].

Class II aldolases, which can be isolated from yeast or bacteria
such as *Aspergillus niger* or *Candida utilis*, are dimeric enzymes of
molecular weight around 80,000 containing a zinc ion per subunit [43].
Their general role is the cleavage of fructose-1,6-diphosphate to
yield dihydroxyacetone monophosphate and glyceraldehyde-3-phosphate,

$$^{2-}O_3POCH_2CO[CH(OH)]_3CH_2OPO_3^{2-} \rightleftharpoons {}^{2-}O_3POCH_2COCH_2OH +$$
$$^{2-}O_3POCH_2CH(OH)CHO \qquad (9)$$

an intermediate step in the metabolism of carbohydrates. The reaction is accelerated by the presence of potassium ions. No x-ray data
are available for this class of enzymes.

2.5. Isomerases and Ligases

Zinc enzymes in these two classes are less numerous and the role of
zinc has not been characterized in great detail. Phosphomannose

isomerase is the first and probably the only isomerase that can be
surely classified as a zinc enzyme [44]. The protein isolated from
yeast has a molecular weight of 45,000 and contains a single Zn(II)
ion per molecule. Its role is the isomerization of mannose-6-
phosphate to fructose-6-phosphate. Although direct involvement of
zinc in the catalyzed reaction has not been demonstrated, it is
interesting to note that the rate of nonenzymic isomerization of
mannose-6-phosphate is dramatically increased by the presence of
Zn^{2+} [45].

Pyruvate carboxylases are biotin-containing metalloenzymes
that catalyze the reaction of pyruvate with HCO_3^- to give oxalacetate:

$$\text{Pyruvate} + \text{ATP} + HCO_3^- \rightleftharpoons \text{oxalacetate} + \text{ADP} + P_i \qquad (10)$$

All of them contain 1 mol of divalent metal ion per mol biotin,
which can be Mn(II), Mg(II), or Zn(II) [46]. The latter has been
found in the 600,000 molecular weight enzyme isolated from *Saccharo-
myces cerevisiae* [47]. The enzyme seems to be a tetramer, thus containing four biotin molecules and four zinc ions.

3. ZINC SUBSTITUTION AND GENERAL REMARKS ON THE METAL DERIVATIVES

The removal of zinc and its replacement with other divalent metal
ions is achieved in different ways for the enzymes listed in Table 1.
The simplest procedure for abstracting the native metal is based on
dialysis against solutions of metal-complexing agents, such as 1,10-
phenanthroline, 2,6-dipicolinic acid, or ethylenediaminetetraacetic
acid (EDTA), sometimes performed at low pH to facilitate protonation
of the residues bound to the metal ion [48-54]. When there is more
than one metal ion bound to the native protein, the completely depleted enzyme is usually obtained [52,53]. Reconstitution of different metal derivatives is performed by dialysis against the appropriate
metal salt or by stoichiometric additions; in some cases selective
binding to only one of the nonequivalent metal sites can be obtained
[55]. The zinc ion in superoxide dismutase can also be partially

exchanged with cobalt by direct dialysis against an excess of cobalt
salts, without appreciably removing the copper ion [56]. This is
not the case with alcohol dehydrogenase, where dialysis against
metal(II) salts yields the completely exchanged metal derivative
[57]. A procedure has been recently developed to deplete selectively
liver alcohol dehydrogenase of the catalytic zinc ion by treating
crystal suspensions of the enzyme with chelating agents [58]. In
aspartate transcarbamylase as a whole, zinc does not exchange with
exogenous ^{65}Zn and cannot be removed by chelating agents [59]. Only
the dissociation of native enzyme into subunits apparently causes
loss of zinc; upon readdition of zinc or other divalent metal ions
the holoenzyme can be easily reconstituted [59].

Co(II) has been partially inserted in pyruvate carboxylase
from *S. cerevisiae* [47], and fully in RNA polymerase from *E. coli*
[60], by growing the bacteria in a Co(II)-rich medium. The same
procedure has been applied to transcarboxylase from propionic bacteria; the latter enzyme already contains besides zinc, variable
amounts of cobalt and copper [20]. Everytime the replacement of
zinc by chemical methods fails, in vivo substitution is a promising
technique to be developed to enlarge the number of metal-substituted
derivatives.

Such artificial metalloproteins are then investigated through
spectroscopic techniques. The type of information that can be obtained varies a great deal with the kind of metal ion chosen.
Whereas, for instance, Co(II) is extremely suitable for electronic
spectroscopy [61,62], Cu(II) derivatives are often prepared with
the aim of investigating their electron spin resonance (ESR) spectra
[63]; Mn(II) is usually employed as a good nuclear relaxing agent
for NMR spectroscopy [64], and cadmium is substituted for zinc as
an NMR active nucleus by itself [65]. It is, however, a good idea
to regard the apoenzyme as a chelating ligand, a good characterization of which requires detailed investigation of more than one of
its metal complexes through more than one of the spectroscopic techniques available. The various data should complement each other
and help to draw a sound picture of the native zinc complex.

In general, the structural information obtained on the Co(II) derivatives can be directly transferred to the native enzyme with some confidence since the Co(II) and Zn(II) enzymes display close catalytic activity (Table 1) and probably close structural properties. The information on nonactive protein derivatives or more generally on derivatives with metal ions that differ substantially from Zn(II) from the point of view of coordination chemistry may still be instructive concerning the properties of the protein ligands and of the active cavity. What is learned upon interactions with substrates in the latter case should be transferred on the native enzyme only with some caution.

4.   COBALT(II) DERIVATIVES

Co(II) is a d^7 ion and, as such, is expected to be colored and paramagnetic, either low spin (S = 1/2) or high spin (S = 3/2). The latter is the state encountered in the present enzyme systems, with the single exception of a dicyanide derivative [66]. Magnetic susceptibility measurements have been performed in a few cases, yielding typical values for high-spin Co(II) complexes [67-70]. In principle, magnetic susceptibility is sensitive to the coordination number; however, except for very accurate measurements [69,70], the error in the measurement hardly allows us to discriminate confidently between four and five or five and six coordination. In the case of $Co(II)_2$-$Cu(II)_2$ superoxide dismutase, variable temperature magnetic susceptibility data have permitted investigation of the magnetic coupling between Co(II) and native Cu(II) [71]. The electronic spectra are generally well shaped and sensitive to the coordination number and are treated in the following section. The information obtained through NMR spectroscopy has recently gained power and a section is also devoted to this method. Finally, the results of EPR and magnetic circular dichroism (MCD) measurements, which are harder to evaluate, are also discussed.

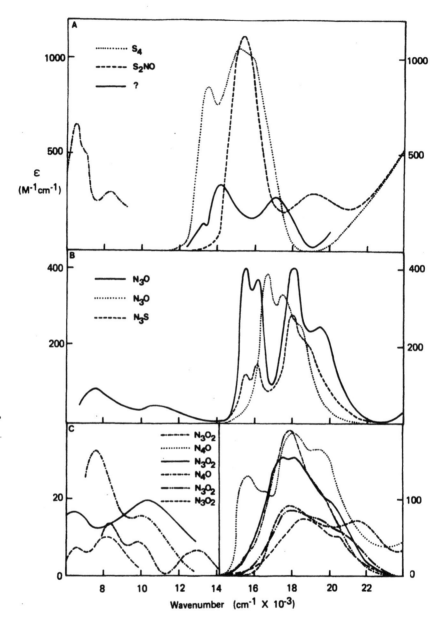

FIG. 1. Electronic absorption spectra of Co(II)-substituted metalloenzymes. (A) Co(II)$_2$ liver alcohol dehydrogenase, catalytic site
(----) [58,72]; Co(II)$_2$ liver alcohol dehydrogenase, noncatalytic
site (••••) [58,72], the infrared region (-•-•) being referred to
the Co(II)$_4$ derivative [57]; Co(II)$_2$ RNA polymerase (see the text)
(——) [60]. (B) Co(II) human carbonic anhydrase B, pH 9.9 (——)
[76]; Co(II)$_2$-Cu(II)$_2$ superoxide dismutase, after subtraction of the
spectrum of the native Zn(II)$_2$-Cu(II)$_2$ enzyme (••••) [71]; Co(II)

4.1. Electronic Spectra

The electronic spectra of several Co(II)-substituted enzymes are
reported in Fig. 1 together with the protein ligands' donor set.
The presence of absorption bands in the visible region for Co(II)
chromophores is due to electronic transitions from the ground level
originated by the ^{4}F term to those arising from the ^{4}P term; the
transitions observed in the near-IR region are those within the ^{4}F
levels [61].

From simple intensity considerations none of the spectra shown
in Fig. 1 can be due to octahedral Co(II) complexes, which should
display molar absorbances not greater than 20-30 M^{-1} cm^{-1}; five and
four coordinated cobalt complexes have intensities ranging from 50
to 800 M^{-1} cm^{-1}, with five coordinate in the lower and four coordinate
in the higher range, although without a sharp borderline [61,69,70].
The band energies should be qualitatively indicative of the ligand
field strength, which in turn depends on the nature and number of
donor atoms. The spectra of Co(II) alcohol dehydrogenase [58,72]
(Fig. 1A) are typically instructive: their high intensity strongly
suggests pseudotetrahedral coordination, whereas the low energy of
the F-P transitions is accounted for by the presence of "soft" sulfur
atoms in the donor sets [1,73].

The spectra of the other cobalt derivatives may be further
grouped into two classes. In the first class the following derivatives can be placed: Co(II)$_2$-Cu(II)$_2$ superoxide dismutase [56,71,74],
β-lactamase II [75], and the active or high-pH form of carbonic
anhydrase [76] (Fig. 1B). The large splitting observed in the four
F-P transitions of the latter two is an indication of large spin-

---

Fig. 1 (continued)
β-lactamase (----) [75].   (C) Co(II)$_2$ alkaline phosphatase, catalytic
site (••••) [82]; Co(II) human carbonic anhydrase B, pH 5.5 (-•-•)
[76]; Co(II) carboxypeptidase A (——) [50]; Co(II) bovine carbonic
anhydrase B-Au(CN)$_2^-$ derivative (--•--•) [88]; Co(II) thermolysin
(—••—••) [80]; Co(II)$_2$ yeast aldolase (----) [79]. The molar
absorbance is referred to the concentration of Co(II) ion. The
proposed donor sets are also shown.

orbit coupling, which may be associated with large distortions from regular symmetry. Co(II) in superoxide dismutase has definitely a pseudotetrahedral N_3O donor set [5]; in the case of carbonic anhydrase it is known from the crystal structure that there are three nitrogen donors from the protein and one oxygen from the solvent [42]. Three protein nitrogens have also been proposed to be ligands in β-lactamase II on the ground of ^{1}H histidine titration on the native enzyme [77] and the Co(II) derivative [78]; the presence of a charge-transfer band at 348 nm (not shown) was taken as an indication of a thiol group coordinated as the fourth ligand [75]. All of the derivatives noted above have been assigned four-coordinated pseudotetrahedral geometry and are characterized by intense absorptions. Regarding Co(II)$_2$ RNA polymerase (Fig. 1A), it should be pointed out that the reported molar absorbance which is calculated per mole of Co(II) ion can be twice as intense if the other Co(II) ion does not absorb in the same range, the two metal binding sites not being equivalent [22]. Therefore, from intensity considerations, tetracoordination has been suggested [60].

Carboxypeptidase [50], aldolase [79], thermolysin [80], and the low-pH form of carbonic anhydrase [76] also give rise to spectra of lower intensity (Fig. 1C) accompanied by a marked shift in the center of gravity of the F-P transitions. Although the interpretation of the spectral properties is still controversial, both the decrease in intensity and the shift to higher energy are, in our opinion, more consistent with a five- than a four-coordinate structure. In most cases this assignment appears consistent with the data obtained from other spectroscopic techniques (see Sec. 4.2). Also the spectrum of Co(II)$_2$ alkaline phosphatase [81,82] (Fig. 1C) is of relatively low intensity and therefore probably five-coordinated, although both four and five coordination have been suggested for the chromophore. Such a spectrum refers to Co(II) bound to the two equivalent catalytic sites [32,83].

There are at least three cases in which an octahedral environment around the Co(II) ion may be suggested from the electronic spectra or, better, from the *lack* of detectable bands in the visible

region. The first example is alkaline phosphatase, as far as the
structural and regulatory sites are concerned, since they do not
substantially contribute to the overall intensity of the Co(II)$_6$
derivative [83]. From the same types of considerations, six coordination has been proposed for Co(II) bound to the second site of
β-lactamase II [75]. By decreasing the pH of Co(II)$_2$ aldolase solutions, the electronic spectrum disappears reversibly while the two
cobalt ions are still bound to the protein, as shown by atomic
absorption after prolonged dialysis [79]. Although not explicitly
stated in the original paper, this may be an indication of an increase
in the coordination number up to six.

In the case of human carbonic anhydrase B the shape and intensity of the spectra for the two pH forms may be consistent with a pH-
dependent equilibrium between four (high pH) and five (low pH) coordination [84]. In the bovine enzyme the pH dependence is qualitatively similar, but the low-pH form still displays a molar absorbance
around 300 M^{-1} cm^{-1} [76], indicating that even at this pH it remains
largely tetrahedral. The general features of the acid-base equilibrium of Co(II) carbonic anhydrase have been mimicked by a model cobalt
complex, which in aqueous solution undergoes an equilibrium of the
type $CoLOH_2^{2+} \rightleftharpoons CoLOH^+ + H^+$, L being tris(3,5-dimethyl-1-pyrazolyl)
methyl amine [85]. Recent NMR data have shed further light on the
pH-dependent equilibrium of carbonic anhydrase (see Secs. 4.2 and
9.3).

Besides helping in guessing the stereochemistry and coordination number from the electronic spectra, the Co(II)-substituted
enzymes are also employed extensively to monitor the interaction
with substrates and inhibitors. Thorough studies have been performed
on Co(II) carbonic anhydrase [40], allowing us to obtain a deep insight into the mode of binding of anions, its pH dependence, and the
resulting stereochemistries. Several anionic inhibitors [e.g.,
CH_3COO^-, NCS^-, $Au(CN)_2^-$] give rise to five coordination, thus maintaining a solvent molecule in the coordination sphere [86,87]. For
the $Au(CN)_2^-$ derivative [88] (Fig. 1C), five coordination is also supported by the x-ray structure [89]. The resulting spectrum, arising

from an N_4O donor set, is fairly similar to the others reported in
Fig. 1C. The same coordination has been suggested for several bidentate ligands, the donor groups being the three histidine nitrogens
and the inhibitor itself [90,91]. Five-coordinate inhibitor derivatives of Co(II) carbonic anhydrase in general show a weak high
energy F-F transition in the region $12-14 \cdot 10^3$ cm^{-1}, which has been
suggested as being diagnostic of five coordination [86,87]. Other
inhibitors, such as CN$^-$ and the sulfonamides, give rise to four-
coordinated pseudotetrahedral geometries [87]. In other cases there
are equilibria between four- and five-coordinated species [92].

No change in coordination number seems to occur in any of the
other cobalt enzymes upon binding of ligands [50,79,80,93]; the
number of inhibitors, or species capable to interact with the metal
ion, is, however, much more limited. No interaction is detected
when Co(II) is substituted in superoxide dismutase [94] or in the
noncatalytic site of alcohol dehydrogenase [58,72]. In both cases
Co(II) is probably not accessible to solvent. A slight change in
the absorption maxima of $Co(II)_2Cu(II)_2$ superoxide dismutase is,
however, detected upon reduction of Cu(II) to Cu(I) [71], since the
latter ion is bridged to cobalt by a histidinato anion [5].

In $Co(II)_2$ RNA polymerase, ATP perturbs only the high-energy
peak, leaving unaltered the rest of the spectrum, while the template
analogue $d(pT)_{10}$ (oligodeoxythymidylic acid with 10 acid residues)
alters the entire spectrum [60]. This has been taken as evidence
that the ATP substrate binds only one of the two nonequivalent
cobalt chromophores [60].

4.2. Nuclear Magnetic Resonance

Paramagnetic metal ions perturb the NMR parameters of nearby nuclei;
this property has been widely used to detect the interaction of
solvent and solute molecules with the (paramagnetic) metal sites
of proteins. In particular, measurement of the proton relaxation

rates of water can yield information on the hydration state of the
metal.

Quantitative information on the spatial relationship between
the metal and the exchanging solvent nuclei can be obtained only if
a good estimate of the electronic relaxation time, τ_s, is available.
The latter can in principle be obtained through a set of T_1 measurements at different magnetic fields. It has recently been pointed out
that the Co(II) electronic relaxation times are also strongly dependent on the geometry [95] and, precisely, that they increase with
decreasing coordination number [96]; therefore, ^{1}H NMR is potentially
able to describe fully the cobalt chromophore in a metalloenzyme.

A thorough field dependence of the ^{1}H relaxation rates has
been performed on both bovine and human B carbonic anhydrases [84,
97,98], allowing us to prove definitely that exchangeable solvent
protons are in the coordination sphere of the metal in both the low-
and the high-pH forms; τ_s values typical of pseudotetrahedral Co(II)
($\sim 10^{-11}$ s) are obtained for both forms of the bovine enzyme, as well
as for the high-pH form of the human B derivative, while the low-pH
form shows τ_s values typical ($\sim 10^{-12}$ s) of five-coordinated species.
The chromophore probably contains two water molecules (see the electronic spectrum in Fig. 1C). Inhibitor derivatives that are assigned
as being five-coordinated on the basis of the electronic spectra
actually do not remove water from coordination and display typically
short τ_s values [98].

Water in the cobalt coordination sphere has also been detected
for carboxypeptidase A, with τ_s values indicative of five coordination [96], in accordance with electronic and magnetic susceptibility
measurements [69,70]. Furthermore, more than one water molecule
seems to be bound since even after addition of the inhibitor β-
phenylpropionate, the coordination number is unaltered and exchangeable water protons are still detected. It has been proposed that
the chromophore is constituted by two histidine nitrogens, one
oxygen from Glu 270, and two water molecules; only one of them is
replaced by the carboxylic oxygen of the inhibitor. Whereas the

Glu 270 residue appears to act as monodentate in the cobalt derivative
in an aqueous 1 M NaCl solution, it has been suggested to act as bidentate in the native enzyme in the solid state [24], (for a note
added in proof see Ref. 215).

Careful ^{1}H NMR studies have also been performed on Co(II)-
substituted liver alcohol dehydrogenase [99]; it has been shown that
water cannot be detected in any of the two pairs of sites (i.e.,
catalytic and structural). While cobalt in the structural site is
likely not to have coordinated water, that in the catalytic site is
probably coordinated by a water molecule. The sulfur donor atoms
in the donor set, which give rise to covalent bonds and have large
spin-orbit coupling constants, are believed to be responsible for
the exceptionally low τ_s values and for the consequent failure in
detecting the coordinated water. This conclusion is supported by
the data obtained on the copper derivative (see Sec. 5). It should
be noted that the foregoing properties of the sulfur ligands have
been proposed as being responsible for the small $A_{//}$ values of blue
copper proteins [100].

Much more limited data are available for the other cobalt-
substituted enzymes. T_1^{-1} values measured at 16 MHz on reduced
Co(II)$_2$-Cu(I)$_2$ and Zn(II)$_2$-Cu(I)$_2$ superoxide dismutase have shown
no detectable effect of Co(II) on water protons [101], consistent
with the inaccessibility of the site to solvent molecules.

The titration of apoalkaline phosphatase with cobalt chloride
at pH 7.0 has been followed through water proton relaxation measurements [102]. The relaxation-rate enhancement measured up to the
addition of 4 equivalents of metal ion is even smaller than that
observed for the aquo ion; although the authors conclude that there
is water bound to the metal, we feel that in their experimental
conditions it is not even clear which of the six sites are populated
[83] and to what extent.

Co(II)-enriched transcarboxylase has been investigated at 25
and 100 MHz [103]. There is an indication of metal-bound water;
the authors propose two water molecules bound per metal in the free
enzyme and one in the adducts with pyruvate and oxalate.

The nuclear relaxing properties of Co(II) have also been used to investigate the interaction of inhibitors and substrates with the metal center. A considerable amount of work has been done on cobalt carbonic anhydrase, where ^{1}H, ^{13}C, ^{19}F, and ^{31}P relaxation studies have allowed to establish that inhibitors [86,88,90,104-108], including the widely used $H_2PO_4^-/HPO_4^{2-}$ buffer system [109], do bind the metal; in the case of acetate a second, metal-independent binding site has also been proposed [104,106]. Kinetic information on ligand dissociation has often been obtained. A comparison of the kinetic behavior of acetate and oxalate led to proposing a bidentate behavior for the latter ligand [90]. Finally, the products of the catalyzed hydrolysis are found to bind the metal [104,105], including $H^{13}CO_3^-$ [110,111]; no direct information could be obtained, however, on $^{13}CO_2$, since in the presence of the active cobalt enzyme such a species is always in rapid equilibrium with $H^{13}CO_3^-$ and the information is averaged out.

The interaction of ethanol, acetaldehyde, and isobutylamide with Co(II)-substituted alcohol dehydrogenase has been investigated through ^{1}H NMR spectroscopy [112]. Although the assignment of the partially substituted cobalt as "catalytic" has been shown to be incorrect [58,72], the overall finding that only weak cobalt-proton interactions can be detected can be meaningful, leading to very short τ_s values and very long distances (14-17 Å) from the metal.

Co(II) has also been used recently as a "short-distance" paramagnetic probe [113], to complement relaxation studies on Mn(II) aldolase (see Sec. 6). The substrate acetol phosphate was found not to be directly bound to the metal ion, although the Co-^{31}P distance is still consistent with binding of the substrate in the active site. It should be noted that in the absence of a reliable value for τ_s, any quantitative estimate is rather risky.

Pyruvate has been proposed, through ^{13}C NMR, to form a second-sphere complex with Co(II) in transcarboxylase [103]; the reduction in the number of coordinated water molecules observed upon binding of pyruvate has been attributed to partial occlusion of the metal site.

^{31}P NMR of inorganic phosphate in the presence of $Co(II)_2$ alkaline phosphatase has shown that the first mole of phosphate added is NMR undetectable [114,115]. Any excess gives rise to the resonance of free phosphate, unaffected by the presence of the para-magnetic centers. Hence a long-lived Co(II)-phosphate complex must be formed whose ^{31}P resonance is broadened and possibly shifted beyond detection. The fact that only one mole is bound per enzyme dimer has been taken as evidence of strong negative cooperativity [115], although other possibilities have been recently proposed [116-118] (see Sec. 7).

The relatively low relaxing capability of Co(II) has found recent application in the detection of ^{1}H NMR signals from the metal-coordinated histidines in Co(II) carbonic anhydrase [119]. D_2O solu-tions of several five-coordinated inhibitor derivatives of the enzyme show reasonably well resolved isotropically shifted signals of the histidine protons (Fig. 2). In H_2O three more signals are detected

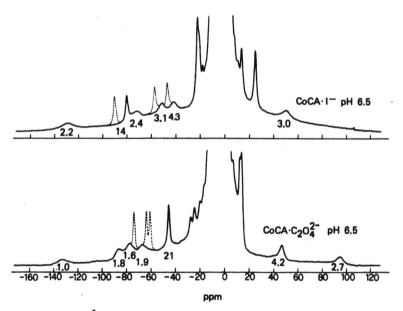

FIG. 2.   60-MHz ^{1}H NMR spectra of Co(II) bovine carbonic anhydrase B adducts with iodide and oxalate in D_2O solutions. Dotted lines refer to the three imino protons of the cobalt-bound histidines which are observed in H_2O solutions. The T_1 values (ms) for the signals are also shown [119].

arising from the three imino protons, whose exchange rate appears to
be slow on the NMR time scale. Although the four-coordinated noninhibited enzyme shows spectra of lower quality, the pH dependence
of the NH signals could be followed, allowing us to establish that
none of the NH protons are released in the pH range where the activity-
linked ionization takes place. A similar experiment has been carried
out on Co(II) carboxypeptidase [96]: the histidine signals display
T_1 values in agreement with the τ_s values calculated through water
proton relaxation, but T_2 values much shorter than expected. This
was attributed to line broadening due to non-completely fast interconversion among conformational isomers in solution; similar conclusions were also reached for the cadmium derivative (see Sec. 7).
^{1}H NMR spectra of cobalt-substituted metalloproteins seem to be a
promising tool, at least in systems where τ_s is short enough not to
cause too severe line broadening [120].

4.3. Electron Spin Resonance and Magnetic Circular Dichroism

The ESR spectra of several Co(II)-substituted metalloproteins have
been recorded at 4K (Fig. 3). All of them are characteristic of a
low-symmetry environment, which is not unexpected in metalloproteins.
There is no safe theoretical tool to relate the observed g values to
the cobalt stereochemistry, although attempts to do this have been
made [121,122]. An empirical criterion has been proposed to discriminate between "trigonal" spectra (pseudotetrahedral and trigonal
bipyramidal) from "tetragonal" (square pyramidal and pseudooctahedral)
on the grounds of the hyperfine splitting observed in the low-field
signal [123]. If this criterion is applied to the spectra in Fig. 3,
all the cobalt derivatives would be at least five coordinate, with
the exception of the cobalt carbonic anhydrase derivatives with
cyanate and sulfonamides. This conclusion would be in remarkable
agreement with the assignment made from the electronic spectra and
the ^{1}H NMR data. In the case of Co(II)$_2$-Cu(II)$_2$ superoxide dismutase,
the presence of Co(II) in the site of native zinc causes broadening

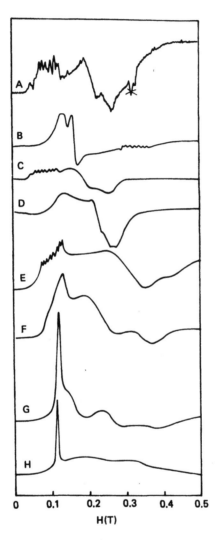

FIG. 3. ESR spectra of Co(II)-substituted metalloenzymes.
(A) Co(II)$_n$ transcarboxylase; the crossed signal is due to the
presence of Cu(II) [103]. (B) Co(II)$_2$ alkaline phosphatase
(catalytic site) [127]. (C) Co(II) alkaline phosphatase
(regulatory site) [127]. (D) Co(II) thermolysin [214].
(E-H) Co(II) bovine carbonic anhydrase B adducts with iodide,
thiocyanate, acetazolamide, and cyanate [123].

beyond detection of the EPR signal of Cu(II). For the Cu(I)-
containing reduced form, the EPR signal of the Co(II) has been
reported [124].

The MCD spectra of several of the foregoing cobalt-substituted
enzymes have also been recorded and discussed at some length [50,80,
83,94,125-127]. While the spectra are clearly indicative of highly
distorted geometries, the ellipticity associated with them has been
considered indicative of four-coordinated rather than five- or six-
coordinated chromophores. Again, the underlying theory is inadequate
to distinguish between four and five coordination in low-symmetry
chromophores, and the suggestions are empirically based on the comparison within a limited series of simple cobalt complexes. Synthetic efforts and theoretical studies aimed at a better understanding of the model systems [128] would in our opinion be highly desirable in order to have a unique picture from all of the spectroscopic
approaches. The occurrence of electronic absorptions in the d-d
spectrum of Co(II) ions has been exploited to estimate the distance
between zinc and calcium sites in thermolysin [129]. The native
zinc has been substituted with Co(II), and calcium with fluorescent
probes. The distance has been estimated by the quenching of fluorescence caused by Co(II) (for a note added in proof see Ref. 216).

5. COPPER(II) DERIVATIVES

The information obtained when Cu(II) is substituted for the native
metals in metalloproteins has been reviewed in Volume 12 of this
series [63]. Carbonic anhydrase, carboxypeptidase, thermolysin,
alcohol dehydrogenase, alkaline phosphatase, and superoxide dismutase have been covered in detail; we will therefore focus primarily on the most recent advances. A field-dependent study on
water proton relaxation in solutions of Cu(II) carbonic anhydrase
and some of its inhibitor derivatives has been performed in the
range 4-60 MHz [98]. The analysis confirmed the previous qualitative
finding [130] that inhibitors such as oxalate and some sulfonamides

reduce the metal-water proton interaction, whereas azide, like most
of the anions, has no such effect. However, the field dependence of
the relaxing properties of copper proteins is in general not as well
understood as those of Co(II) or Mn(II) [131]. This limitation prevents us from having a clear picture of Cu(II) carbonic anhydrase.
For example, some sulfonamides, depending on the particular isoenzyme,
give rise to four coordinated chromophores ($A_{//}$ = 60-70·10^{-4} cm^{-1})
without coordinated water, whereas others give rise to five-coordinated chromophores ($A_{//}$ = 120-130·10^{-4} cm^{-1}) with water in the coordination sphere [132]. The ligand HCO_3^- leaves some exchangeable
water protons interacting with the paramagnetic center but sensibly
less than the other monoanionic inhibitors [130,133]. The possibility exists that there are two nonequivalent water molecules coordinated to the metal in the noninhibited enzyme, and that inhibitors
and substrates may displace the first, the second, or both, affecting
the spectral properties and the proton relaxation to different extents. The problem of the exchange capability of a metal-coordinated
hydroxide, which is particularly relevant in carbonic anhydrase, has
been recently attacked by investigating $H_2^{17}O$ relaxation in the presence of Cu(II) carbonic anhydrase [134]. The data indicate that if
there is a coordinated hydroxide ion as the only oxygenated species
in the high-pH form of the enzyme, its exchange occurs even faster
than is required by the catalytic efficiency of the active zinc
enzyme. A tentative account of this behavior is given in Sec. 9.3.

        Liver alcohol dehydrogenase selectively substituted with Cu(II)
at the pseudotetrahedral catalytic sites has been studied through ^1H
NMR [99] in a wide range of magnetic fields either as pure copper
protein or as derivatives with NAD^+ and pyrazole, and with both
coenzyme and inhibitor; the results are compared with the extensive
ESR data on the same systems, reviewed elsewhere [63]. The main
findings are the presence of coordinated water, which is not removed
upon binding of substrate or inhibitor, but is absent in the ternary
adduct. Whereas NAD^+ is known not to bind the metal, pyrazole does,
so that the formation of a five-coordinate copper complex in the
enzyme-pyrazole adduct is proposed, consistent with the ESR data

[135]. Evidence is also presented for direct interaction of the
substrate methanol with the metal ion, although weaker than that
presumed to occur in the catalytically active native enzyme [136].

An additional interesting feature of the present systems is
the unusually short τ_s value (~10^{-11}) obtained for the Cu(II) ion,
which is ascribed to the delocalization of the unpaired spin onto
the orbitals of the sulfur ligands and to the large sulfur spin-
orbit interaction.

At variance with cobalt and manganese, apoalkaline phosphatase
selectively binds the first two copper ions at the catalytic sites
[137]; water proton relaxation measurements indicate that water seems
bound to the metal and not displaced in the monophosphoryl derivative
[102]. However, the latter finding still has to be reconciled with
a recent picture of the anticooperativity effect discussed in Sec. 7.

Copper is found to occupy partially the 12 metal sites of
transcarboxylase [20], the remaining sites being occupied by cobalt
and zinc. At variance with the results on the Co(II) derivative,
no evidence for fast-exchanging water molecules is obtained for the
inactive Cu(II) derivative [103].

It has recently been proposed that binding of anions such as
NCS^- [138], CN^-, NCO^-, N_3^- [139], and OH^- [140] to superoxide dismutase occurs through displacement of one of the copper-ligated
histidines [5]. Subsequent studies on $Cu(II)_2$-$Cu(II)_2$ superoxide
dismutase have shown that upon addition of thiocyanate, the metal-
metal interaction is abolished, suggesting that in this case the
displaced histidine is the bridging His 61 [141].

6. MANGANESE(II) DERIVATIVES

The Mn(II) ion is not a suitable probe for electronic spectroscopy,
since the low molar absorbance of Mn(II) complexes does not allow
the recording of electronic spectra at the relatively low protein
concentrations attainable. The ESR spectra of the Mn(II) derivatives of several zinc enzymes have been recorded and analysis of

the spin Hamiltonian parameters has been attempted [142]. The
tentative values of the zero-field splitting parameters proposed
are weakly related to the stereochemistry of the chromophore. A
general conclusion is that the rhombic distortions are sizable.

The wide use of Mn(II) as a substitute for Zn(II) arises
mainly from its high nuclear relaxing properties. In proton relaxation measurements on solutions of Mn(II) proteins, the correlation
time for the electron-nucleus interaction may be the rotation of
the molecule, the water proton exchange time, or the electronic
relaxation time, τ_s. When the latter is the dominant term, the
field dependence of the ^{1}H T_1^{-1} values is often complicated by
the fact that τ_s is also field dependent in the usual range of
investigation [64,143]. The underlying theory, and its possible
breakdown, are now well understood; a great deal of information
can be obtained provided one does not rely too much on the calculation of exact hydration numbers [143]. In carboxypeptidase A,
for example, it seems certain that water is present in the first
coordination sphere of manganese; this was already proposed from
single-frequency measurements, which also showed that the interaction
was reduced by the addition of inhibitors [144]. Later, analysis of
both ^{1}H and ^{19}F nuclear relaxation prompted the suggestion that water
and fluoride can be simultaneously bound to the metal ion [145]; in
the absence of fluoride, the second binding site could be occupied
by the weaker chloride ion, always present in high concentration to
solubilize the protein. Successive measurements in the magnetic
field range corresponding to Larmor frequencies of 10-100 MHz indicated the presence of only one water molecule in the coordination
sphere [146]. Finally, it has been elegantly demonstrated that the
parameters obtained from fitting the data to the standard relaxation
theory are critically dependent on the range of magnetic field
covered [147]. Measurements extended in the low-field region showed
that either one water molecule is unreasonably close to the metal,
or more than one water molecules is indeed bound [147]. Analogous
experiments performed on the cobalt derivative are qualitatively

consistent with the latter finding (see Sec. 4.2). Water has also
been found in the hydration sphere of Mn(II) carboxypeptidase B
[148].

Mn(II) carbonic anhydrase also shows $^1H\ T_1^{-1}$ enhancement typical
of coordinated water. The paramagnetic effect measured at high field
is substantially pH independent and is not altered by addition of the
inhibitor N_3^- [149]. On the other hand, the inhibitors p-toluenesul-
fonamide and oxalate drastically reduce the $^1H\ T_1^{-1}$ values. Although
limited to the high-field range, measurements at various magnetic
fields seem to confirm these findings [98]; analogous to what is
found for the Cu(II) [98] (and partially Co(II) [98]) derivatives
of carbonic anhydrase, monovalent anions still allow water to interact with the paramagnetic center.

Horse-liver alcohol dehydrogenase selectively depleted of zinc
does not seem to be able to bind manganese to the catalytic sites
[150]; instead, water proton relaxation measurements in a wide field
range have shown evidence of a relatively tight metal binding site
per dimer, the availability of which is unaffected by the presence
or absence of metals in the catalytic sites [150]. This novel site
displays a high affinity for manganese and a lower affinity for zinc
and cadmium. It is interesting to note that the process of binding
of zinc to the catalytic sites seems temporarily to displace manganese
from the former site, as if it were kinetically favored or even required at an intermediate binding stage. The binding of a metal in
this site could be connected with the tendency of alcohol dehydrogenase to polymerize and ultimately precipitate in the presence of
excess Zn(II) ions [151]. Additional weaker binding for manganese
has also been suggested [150].

The binding of manganese to apoalkaline phosphatase was followed
through water ^1H NMR by adding increasing amounts of metal ion [102].
Four manganese ions were taken up per protein molecule, with either
a similar binding constant for the two sites or similar proton relaxing capability. The analysis of the field dependence of the proton
relaxation is probably inadequate for obtaining quantitative

information. In fact, the field dependence of τ_s, although noticed, is not taken into account; furthermore, the authors appear to believe that all four sites contribute equally to the relaxivity, which is unlikely in the present system. Therefore, it can only be concluded that there is evidence of water interacting with the metal ion.

The field dependence of water proton relaxation has been recently investigated for Mn(II)-substituted yeast aldolase [113]. There is strong evidence of exchanging protons in the first coordination sphere of the metal, which is believed to be due to a single water molecule; substrates do not alter the water proton relaxation, with the exception of acetol phosphate, which apparently reduces it by 50%.

More limited information is available on Mn(II) thermolysin [152,153]. ^1H T_1^{-1} measurements at 100 MHz have shown that the binding of manganese is dependent on a deprotonation in the active site with a pK_a value of ~6. At higher pH values the data are consistent with the presence of metal-bound water; there is evidence for two further ionizations in the pH range 8-10, one of which is proposed to be due to the coordinated water molecule. The water molecule may or may not be replaced by inhibitors [152].

Mn(II) is probably the most widely used metal to study the interaction of substrates and inhibitors with the active site. Carboxylic inhibitors have been shown to bind through the carboxylic group to Mn(II) carboxypeptidase [154]; the exchange rate is generally slow on the NMR time scale, with the exception of the very weak inhibitor methoxyacetic acid. In several cases the exchange rate could be calculated. As anticipated, $^{19}F^-$ is also able to bind the metal, probably without displacing water from coordination, and is competitively removed by β-phenylpropionate [145].

Inhibitors of carbonic anhydrase also bind the manganese enzyme, as has been shown for acetate and sulfonamides. A second binding site was identified for the former [106,155], while T_2 measurements on several protons of N-acetylsulfanilamide allowed the authors to establish that binding occurs through the nitrogen atom, and to map the molecule in the active site [156].

^{31}P NMR has been used widely to study the binding of phosphate
[114,157-159] and phosphonates [158] to alkaline phosphatase, which
is known to occur only in the presence of metals. As far as the
manganese derivative is concerned, a small paramagnetic effect was
detected on T_1 and T_2 of 10^{-2} M phosphate solutions [157]. While
T_2 was exchange limited, T_1 showed an opposite temperature dependence
and was assumed suitable for distance calculations. Even though the
way used to calculate the correlation time was questionable, the distance obtained (18 Å) ruled out the possibility of direct binding.
It was noted, however, that a metal-bound phosphate with a very long
lifetime could have escaped detection. The problem was reexamined
later [158] using stoichiometric phosphate addition: as in the case
of Co(II) [114,115], the first mole of phosphate added could not
be detected by NMR, while the second gave rise to the unbroadened
phosphate resonance [158]. Thus the exchange from the phosphate
binding site, whatever it may be, was demonstrated to be very slow.
The discrepancy with the previous data [157] might be explained by
the presence of a second, much weaker binding site which becomes
partially populated only at higher concentrations. On the other
hand, p-aminobenzylphosphonate was found to be in fast exchange with
the enzyme, and a distance calculation could be attempted [158].
Using a reasonable range of τ_c values, the obtained values range
from 6 to 11 Å. The point is then made that the smaller and more
tightly interacting phosphate is likely to be nearer to the metal;
also, as already discussed, in the case of cobalt the line width
expected for a phosphate ion at the same distance as the phosphonate
could have been within the detection limit even in the stoichiometric adduct. Although no direct evidence could be obtained, the
possibility of binding of phosphate to the metal ion cannot be ruled
out.

The interaction of the nonhydrolyzable inhibitor D-N-trifluoro-
acetylphenylalanine with Mn(II) thermolysin has been studied through
^{19}F NMR [153]. The inhibitor binds with an affinity constant of the
order of 10^3, has a lifetime of about 10^{-3} s in the active site, and
displays a metal-fluorine distance of about 8 Å. The binding

observed is not the same as that found from inhibition experiments, which is much weaker, suggesting the existence of two binding modes for this inhibitor.

The binding of substrates to Mn(II) yeast aldolase has been investigated extensively through ^{1}H, ^{13}C, and ^{31}P NMR [113,160]. All the distances estimated between the acetolphosphate nuclei and the paramagnetic center rule out the possibility of direct coordination. As already discussed, measurements on the Co(II) derivative confirmed these results. High-resolution 360-MHz ^{1}H NMR of the C-2 proton resonances allowed establishment of the existence of at least three metal-coordinated histidines in yeast aldolase. The assignment was verified through manganese substitution, which wiped out the signals of the bound histidines [161]. Thiomethylation of one of the five cysteine residues abolished the enzymatic activity of the native enzyme but had no effect on the stability of the Mn(II)-apoenzyme complex and on the water proton relaxation rate, indicating that the active-site cysteine is not coordinated to the metal [161].

The interaction of the ligands succinate and cytidine 5'-triphosphate with Mn(II)-substituted aspartate transcarbamylase has been investigated through ^{1}H NMR. The ligand binding sites are proposed to be very far (15-20 Å) from the metal ions [162].

7. CADMIUM(II) DERIVATIVES

In the past few years considerable effort has been put into the preparation of cadmium-substituted zinc enzymes, with the aim of recording their NMR spectra. ^{111}Cd and ^{113}Cd are both I = 1/2 nuclei, with about the same sensitivity of 1% relative to protons. The chemical shift range for cadmium complexes is large, extending from 800 to -200 ppm relative to $Cd(OH_2)_6^{2+}$; therefore, cadmium is in principle a sensitive probe for the metal sites of enzymes as they change with pH and inhibitor or substrate binding. For the same reason, kinetic information on equilibria involving cadmium species with different chemical shifts may be obtained if the interconversion falls in the time scale 10^{-2}-10^{-4} S. In Fig. 4 the chemical shift values of ^{113}Cd are shown for the enzymes investigated so far.

From the limited data on cadmium complexes a trend for the donor atom dependence of chemical shifts can be established as follows [163-165]: O < N $\simeq$ X << S, in order of increasing frequency (or increasing deshielding or downfield shift). The downfield shift is also increased with decreasing coordination number. As a matter of fact, this trend seems to be followed qualitatively by the enzyme derivatives noted above; ^{113}Cd-substituted liver alcohol dehydrogenase, where sulfur atoms are present in the pseudotetrahedral coordination sphere of both sites, shows the most downfield shifted signals [166,167]; superoxide dismutase [65,168], carboxypeptidase [65], carbonic anhydrase [65,169,170], and disubstituted alkaline phosphatase [117,118] show ^{113}Cd signals in the intermediate region 330-130 ppm; cadmium in the structural site of alkaline phosphatase [117,118], which is believed to be octahedral with only one nitrogen and presumably five oxygen ligands (see Sec. 9.1), displays a signal at 55 ppm; and cadmium in the regulatory site (possibly the O_6 donor set) shows a resonance at 2 ppm, very close to the reference ^{113}Cd$(OH_2)_6^{2+}$.

On the basis of the foregoing considerations, the two signals observed at 751 and 483 ppm in the fully substituted cadmium$_4$ liver alcohol dehydrogenase have been assigned [166,167] to the structural and catalytic sites, respectively: the former is, in fact, believed to have an S_4 donor set, while the latter is probably S_2NO, with the oxygen belonging to a coordinated water molecule. Addition of a large excess of the inhibitor imidazole leaves the most downfield resonance unaltered, while the other signal is shifted 36 ppm downfield, thus confirming the assignment of the latter as the catalytic, solvent-exposed site [6]; the replacement of a water molecule with an imidazole nitrogen also yields 30- to 40-ppm downfield shifts in aqueous solutions of cadmium salts [171]. Interestingly, at lower imidazole concentrations no peak is observed for the catalytic cadmium, probably because of exchange broadening effects.

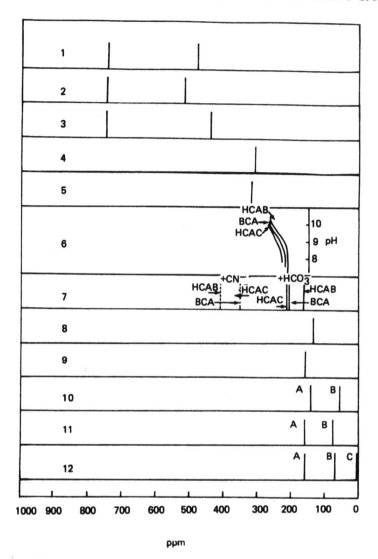

FIG. 4. ^{113}Cd NMR spectra of: Cd_4 liver alcohol dehydrogenase (1), and in the presence of imidazole (2), and NADH (3) [167]; Cd_2 superoxide dismutase (4), and Cd_2-$Cu(I)_2$ superoxide dismutase (5) [168]; cadmium carbonic anhydrase as a function of pH (6), human B (HCAB), human C (HCAC), and bovine (BCA) isoenzymes, and adducts with cyanide and bicarbonate (7) [170]; cadmium carboxypeptidase A, β-phenylpropionate adduct (8) [65]; Cd_2 alkaline phosphatase, catalytic sites (9), Cd_2 monophosphoryl derivative (10), Cd_2-Mg(II)$_2$ monophosphoryl derivative or Cd_4-Mg(II)$_2$ diphosphoryl derivative (11), and Cd_6 diphosphoryl derivative (12) [118]. The catalytic, structural, and regulatory sites are labeled A, B, and C, respectively.

The ^{113}Cd NMR spectra of superoxide dismutase substituted at the Zn(II) site have been reported from two different laboratories. In the first report [65] a resonance at 170 ppm was observed upon addition of 2 equivalents of ^{113}Cd to the apoenzyme. Successive reaction with Cu(II) wiped out the signal, which reappeared at 9 ppm upon reduction of the copper centers. Apparently, in the later study [168] the same protocol was followed, but the signal for the cadmium derivative in the absence of copper was observed at 311 ppm; the signal also disappeared upon addition of Cu(II) and was again detected at 320 ppm in the reduced enzyme. The discrepancy is too large to be ascribed to small changes in the experimental conditions; possibly cadmium was not properly substituted in the zinc site in one of the experiments. On the basis of the much smaller difference in chemical shift observed in the absence of copper and in the presence of Cu(I), which parallels the very small changes in the electronic spectrum of cobalt under the same circumstances, the latter report seems to be related to the actual zinc-substituted enzyme. Also, a chemical shift of 9 ppm seems unlikely for an N_3O chromophore. Further studies are needed, however, before drawing definite conclusions.

Similar discrepancies exist for carboxypeptidase A and, to a lesser extent, for carbonic anhydrase. In the former enzyme a chemical shift of 133 ppm is reported for the adduct with the inhibitor β-phenylpropionate at pH 6.5 [65], and values of 240 and 217 ppm with the inhibitor dibenzylsuccinate at pH 8, the only difference being the kind of inert salt used to solubilize the protein [172]. In this case the large difference between the two sets of measurements may depend on pH changes and/or different binding behavior of the ligands. No signal could be observed for the noninhibited enzyme at pH 6.5 in 1 M NaCl; slow fluctuations of the protein ligands were considered to be a possible cause [65].

^{113}Cd carbonic anhydrase has been studied in greater detail [65,169,170,173]. For all the isoenzymes investigated (i.e., bovine, human B, and human C), the chemical shift in CO_2-free solutions have been found to be pH dependent, with pK_a values close to those reported

for the esterase activity of the cadmium derivative. The low-pH
forms display chemical shifts in the range 205-230 ppm, while the
high-pH forms level off around 275 ppm in both cases. The cadmium
derivatives are also sensitive to the binding of anions, to different extents. Halides produce a small downfield shift [65]; cyanide
causes a much larger effect in the same direction [169,170]. The
reported values for cyanide are 354 ppm for both bovine and human C
enzymes and 411 ppm for the human B derivative [170]. ^{13}CN gives
rise to a doublet in all cases, arising from strong ^{13}C-^{113}Cd coupling (J $\simeq$ 1050 Hz). Measurements with ^{15}N-sulfonamides led to the
conclusion that these inhibitors bind through nitrogen (J $\simeq$ 200 Hz)
and presumably as anions [173]. The ^{113}Cd signal is shifted to
390 ppm. Bicarbonate has a high affinity for the cadmium derivative;
the chemical shifts for the adducts are upfield with respect to the
nonligated enzymes. The 50-ppm upfield shift in the human B enzyme
upon addition of bicarbonate could suggest an increase of coordination number in the adduct, since the simple substitution of the
water oxygen should not change the metal environment too much, as
seems to be the case for the other isoenzymes. The presence of
adventitious bicarbonate has been considered to be a possible cause
for measuring smaller shifts and even for losing the signal at pH
values lower than 9.

Otvos et al. have recently published a thorough investigation
of the anticooperativity phenomenon in alkaline phosphatase using
^{13}C and ^{113}Cd NMR [116-118]. After identifying the ^{13}C signals
arising from γ-^{13}C-labeled histidines, they proposed that four
of them are involved in metal coordination at the catalytic, or
A, site, one at the structural, or B, site, and none at the regulatory, or C, site. The assignment was aided by the observation
of resolved ^{113}Cd-^{13}C coupling (^{3}J = 12-19 Hz) in the cadmium
derivative [117].

The addition of two equivalents of cadmium to the apoenzyme
at pH 6.2 resulted in a sharp resonance at 169 ppm, which was assigned
as arising from the A site [118]. No appreciable population of B and
C sites was observed, indicating that in these conditions the affinity

of cadmium for the A site is at least one order of magnitude larger
than for the other sites. Successive additions of cadmium from two
to four equivalents resulted in the progressive disappearance of the
signal. This was attributed to exchange broadening, not arising from
exchange of free ^{113}Cd from B or C sites but rather from conformational fluctuations allowed by the binding of the second 2 equivalents
of cadmium.

As noted in a previous paper [65], addition of phosphate to the
cadmium$_2$ derivative at low pH produces a monophosphorylated enzyme
showing two distinct resonances at 141 and 56 ppm instead of the one
located at 169 ppm. A crucial point was then raised [118]: if these
resonances arise from two nonequivalent A sites, one bearing a covalently attached phosphate and the other perturbed by its presence,
as is usually suggested to explain anticooperativity, then the addition of 2 more equivalents of cadmium and 2 equivalents of magnesium
to fulfill the metal requirement should restore a single peak at
either 141 or 56 ppm, and eventually display a second peak somewhere
else due to the site B cadmium. This was not the case: addition of
magnesium caused a moderate downfield shift of both resonances,
which were then observed at 156 and 75 ppm. Addition of cadmium
essentially doubled their intensity; the formation of the symmetric
diphosphoryl enzyme under these conditions was independently checked
by ^{31}P NMR [174]. The conclusion is then straightforward: in the
cadmium$_2$ derivative, addition of phosphate causes the slow migration
of one cadmium from site A to site B in the phosphorylated subunit,
leaving the second subunit empty. Upon addition of the full complement of metal ions, the second subunit is populated in the same way,
giving rise to the symmetric diphosphorylated enzyme. Ancillary
experiments on the cadmium$_6$ derivative gave full support to this
picture [118]. As already anticipated, the ^{113}Cd chemical shifts
in inorganic complexes allowed the proposal of an N_4O donor set for
the A site, NO_5 for the B site, and O_6 for the C site. The presence
of at least three metal-coordinated nitrogens in the catalytic site
has been proposed independently [63,175]. An N_4O donor set for the
catalytic site is in agreement with the electronic spectra and NMR
data on the Co(II) derivative.

[111]Cadmium has been investigated in metal-substituted carbonic anhydrase, carboxypeptidase, and superoxide dismutase through the technique of perturbed angular correlation of γ rays [176-179]. Independent information has been obtained on these systems, often in agreement with the overall picture of the metalloproteins. Radioactive [109]Cd has also been used to monitor the binding sites in liver alcohol dehydrogenase through equilibrium dialysis [180]. The sulfur ligands in both sites give rise to a characteristic charge-transfer band.

8.   OTHER METAL DERIVATIVES

Co(II) in the active site of metalloenzymes can often be oxidized to Co(III) without damage to the protein backbone. The resulting chromophore has been defined as being substitution inert, emphasizing its ability to "freeze" the donor set in subsequent protein manipulations [181]. So far, reports have appeared on Co(III) carbonic anhydrase [182,183], carboxypeptidase [184-186], and alkaline phosphatase [127]. In all cases the resulting complexes are believed to be octahedral [181], due to the well-known chemical properties of Co(III). In Co(II)-substituted alkaline phosphatase a maximum of two Co(II) ions per molecule can be oxidized, even starting from the $Co(II)_6$ derivative [127]. Experiments performed on zinc, cobalt, and magnesium hybrid enzymes have established that (1) once the $Co(III)_2$ derivative is formed, the metal cannot be replaced by excess of other divalent ions; (2) only cobalt in the catalytic site can be oxidized; and (3) when a total of 2 equivalents of Co(II), distributed among the three types of sites, are treated with oxidizing agents, metal migration occurs and cobalt is fully recovered in the catalytic sites as Co(III) [127].

Oxovanadium(IV) has been substituted in the active site of carbonic anhydrase [187,188] and carboxypeptidase [189]. The electronic [188] and ESR [189] spectra of the first derivative have been interpreted on the basis of either square pyramidal or elongated

octahedral tetragonal symmetry; from ^{1}H NMR data water seems bound
to the metal [188], while the extra oxygen present in the coordination sphere apparently prevents the binding of inhibitors. On the
basis of the observed ESR parameters in oxovanadium(IV) carboxypeptidase, octahedral coordination is suggested. The carboxylate
ion of Glu 72 binds in the axial position opposite to the vanadyl
oxygen [189].

     A [199]Hg NMR spectrum has been reported for Hg(II) human car-
bonic anhydrase B [190]. The derivative, prepared by addition of
Hg(II) acetate to the apoenzyme, is capable of binding only one
chloride ion. The x-ray structure obtained on apoenzyme crystals
reacted with $HgCl_2$ [191] shows mercury in an octahedral environment,
the donors being the three histidine nitrogens, two chloride ions
in the cis position, and an oxygen from threonine 199 present in
the active site.

     The Ni(II) ion has also been used as a probe for the active
site. The coordination in $Ni(II)_6$ aspartate transcarbamylase [192],
where the metals probably are not solvent accessible, is tetrahedral;
the spectra are similar to those obtained for Ni(II)-substituted
liver alcohol dehydrogenase [193] and Ni(II) in the blue site of
azurin [194]. Low-lying charge-transfer bands are consistent with
the presence of sulfur atoms in the coordination sphere.

     The Ni(II) in carboxypeptidase is octahedral, indicating that
the coordination number is not essential for catalytic activity
since, for example, Co(II) and Ni(II) are both active but have different stereochemistries [69,70]. This metal derivative confirms
that the active site is quite flexible and capable of hosting up to
six donor atoms.

     Ni(II) has also been substituted in carbonic anhydrase [195].
Absorption bands at 15.5 and $25.5 \cdot 10^3$ cm^{-1} could be consistent with
six coordination; however, the relative high intensity of the transitions (ε = 30 and 70 M^{-1} cm^{-1}, respectively, at pH 9) does not rule
out the possibility of the occurrence of five coordination (for a
note added in proof see Ref. 217).

9. CONCLUDING REMARKS

The information obtained by investigating each metal-substituted
enzyme is quite valuable and contributes to a detailed picture of
the metalloprotein as a whole. Depending on the degree of knowledge
on a given metalloprotein, the metal substitution is aimed toward
different goals. We would like to illustrate this point with three
meaningful examples: (1) alkaline phosphatase, for which no high-
resolution x-ray data are available; (2) carboxypeptidase, for which
the x-ray refinement is at the level of 1.75 Å of resolution [24];
and (3) carbonic anhydrase, whose x-ray structure is known at 2.0 Å
of resolution and whose catalytic mechanism has been quite a controversial matter [39,40].

9.1. Alkaline Phosphatase

Metal substitution in this case has meant the beginning of an understanding of the structure of the metalloenzyme. Particularly meaningful are the results obtained with the cadmium-substituted derivatives, which have shown that the first two cadmium ions go in the
site A (i.e., the catalytic site), that phosphate induces migration
of one cadmium from site A to site B (the structural site), and that
excess cadmium can populate site C (the regulatory site) [118].
These data are consistent with the EPR of the $Mn(II)_2$ derivative,
which shows two overlapping signals in presence of phosphate [142].
Through ^{13}C NMR measurements on both native and cadmium ^{13}C-enriched
enzymes it has been possible to ascertain that four histidines are
present in the A site, one in the B site, and none in the C site
[116,117]. Although the distribution among the sites of the other
metal ions is not completely settled, the water ^1H NMR measurements
on paramagnetic metal ion derivatives have provided evidence that
at least one water molecule is involved in the coordination [103].
From the electronic spectra of the Co(II) derivative, tetra- or
pentacoordination is suggested for site A [83,93] and, in our

opinion, the latter coordination is more probable, whereas the
cobalt in sites B and C is six-coordinated [83]. Finally, the
chemical shifts of ^{113}Cd suggest the presence of several oxygen
donors in both the B and C sites.

The A and B sites are probably spatially close since the
binding of a single phosphate, both at high and low pH values,
affects the EPR data of the Cu(II)$_4$ derivative [52] and the NMR
shifts of the Cd$_2$ derivative [118]. This guess is consistent with
the low-resolution x-ray data, which have shown two metals to be
close [34]. In light of these conclusions, the cobalt and copper
derivatives should be reconsidered and further studies would be
highly desirable (for a note added in proof see Ref. 218).

9.2. Carboxypeptidase A

The structure of the protein is now known in great detail [24], and
much is known about the enzymatic mechanism [23]. Apparently, the
coordination polyhedron consists of two oxygens of a bidentate Glu
residue, two histidine nitrogens, and a water molecule. The dipeptide Gly-L-Tyr binds the metal with a carbonylic group substituting
the water molecule and the amino group binding at a sixth position
with a 50% probability [24].

Measurements on metal-substituted derivatives in solution, and
therefore in the presence of 1 M NaCl in order to solubilize the
protein, indicate a more complex picture and a more important role
of water. The Co(II) enzyme in the presence of the dipeptide Gly-
L-Tyr shows an electronic spectrum with intensity of the order of
that of the pure enzyme [50], thus ruling out the possibility of
large extent of octahedral species. When β-phenylpropionate is
bound to the Co(II) ion, the derivative is probably five-coordinated
and water still appears to be present in the coordination sphere
[96]. Even the Mn(II) derivative shows evidence of the simultaneous
presence of water and inhibitor in the coordination sphere [145].
Therefore, the metal substitution raises the problem of a different

role for water in the enzyme in solution, and indicates the possibility that two water molecules are present in the coordination
sphere.

9.3. Carbonic Anhydrase

This enzyme has been thoroughly studied through metal substitution,
and the very many data obtained have led to the development of several models, which do not, however, fully account for all the observations. We refer here to a model developed very recently [84],
which includes the characteristics of most previous models. It is
assumed that the chromophore of the active form of the enzyme is
pseudotetrahedral and consists of three histidine nitrogens and a
hydroxide group.

$$
\begin{array}{c}
N \\
\ \ \backslash \\
N - Co - O - H \\
\ \ / \\
N
\end{array}
$$

The enzyme allows the hydroxide group to exchange with the bulk
solution as part of a water molecule; this condition is required by
NMR detection of exchangeable protons for the Co(II) derivative at
high pH. At low pH values the predominant inactive species for the
bovine enzyme is

$$
\begin{array}{c}
N \\
\ \ \backslash \\
N - Co - O \begin{array}{c} H \\ H \end{array} \\
\ \ / \\
N
\end{array}
$$

possibly in equilibrium with a small amount of five-coordinate
species [87]. The $H_2O \rightleftharpoons OH^-$ equilibrium was first proposed by
Lindskog and Coleman [196]. The pK_a of such an equilibrium depends
on the particular isoenzyme: For the bovine B isoenzyme it may be
set around 6, and for the human B, around 7.5 [39,40]. In the case
of the latter Co(II) enzyme at low pH, the water ^{1}H relaxation rate

is sensibly lower than at high pH. Variable magnetic field investigations have allowed us to establish that this is due to a small τ_s
value of the acidic form [84]. This and the low intensity of the
electronic spectra have led us to propose that the low-pH form of
the human enzyme is predominantly five-coordinated with two water
molecules.

$$
\begin{array}{c}
N \\
\quad \diagdown \\
N \text{——} Co \overset{\diagup OH_2}{\diagdown OH_2} \\
\quad \diagup \\
N
\end{array}
$$

Thus the two major forms of carbonic anhydrase (high and low activity)
represented by the bovine and human B isoenzymes, respectively, differ
mainly in the average coordination number of the low-pH form, the
active high-pH species being spectroscopically very similar. The
puzzling difference in pK_a for the coordinated water (~1.5 pH units)
is consistent with the well-known decrease in acidity of the coordinated water by increasing the coordination number.

The availability of a fifth coordination position allows also
the proposal of a pathway for the anticipated fast exchange of the
hydroxo group present in the high-pH form as part of a water molecule
[84]. The exchange takes place through a five-coordinate intermediate
in which an incoming water molecule may transfer a proton to the coordinated hydroxide, allowing its release as a water molecule in the
bulk solution. This picture is consistent with the fast exchange of
^{17}O in the Cu(II) derivative even at high pH [134], and with the
kinetic data on the native enzyme of ^{18}O exchange during the dehydration reaction of HCO_3^- [197].

The acidic groups capable of affecting the electronic spectra
of the Co(II) enzymes are at least two [76]; therefore, it is proposed

that a histidine hanging into the cavity may modulate the spectra
with a pK_a of ~7. The NH groups of the coordinated histidines have
been shown to be present up to pH 9 [119].

Inhibitors bind the metal ion, giving rise either to pseudotetrahedral or five-coordinated species, or both [87]. The general
scheme may be represented as follows:

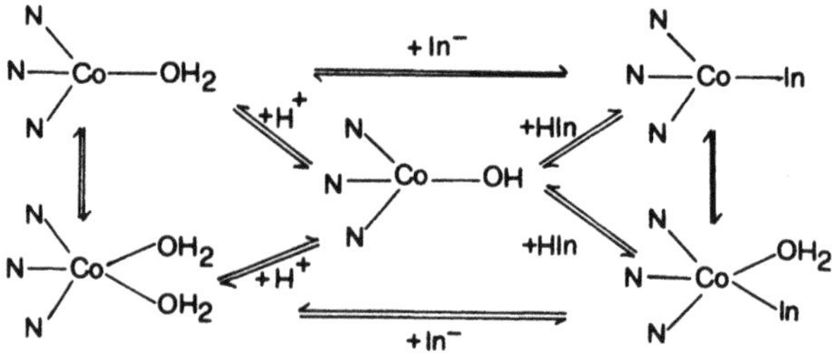

In the case of the copper derivative [130] all the anionic
inhibitors (N_3^-, CN^-, I^-, CH_3COO^-, etc.) give rise to chromophores
of the type

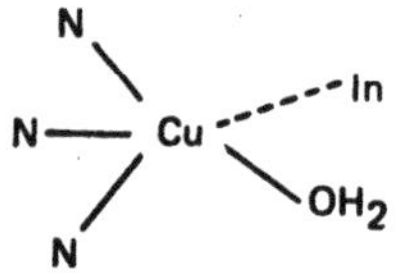

with the Cu-In bond somewhat longer. The bicarbonate ion probably
binds at the other site, giving rise to a chromophore of the type

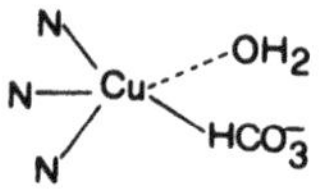

The other substrate CO_2 approximates the copper ion by the
side of H_2O and indeed is removed by monoanionic inhibitors, together
with bicarbonate [198].

From the kinetic point of view, HCN, sulfonamides, and possibly
hydrated aldehydes exchange as such [39,40] but bind as anions, in a
manner similar to the exchange mechanism for OH^- proposed above.
Therefore, all of them formally compete with OH^- just like the other
anionic inhibitors. The CO_2 species would react with the high-pH
species, giving rise to a bicarbonate adduct.

$$
\begin{array}{c}
N\!\!\diagdown \\
N\!-\!Co\!-\!O\diagup^{H} \\
N\diagup
\end{array}
\xrightleftharpoons{+CO_2\ +H_2O}
\begin{array}{c}
N\!\!\diagdown \quad \diagup HCO_3^- \\
N\!-\!Co \\
N\diagup \quad \diagdown OH_2
\end{array}
\xrightleftharpoons{-HCO_3^-\text{-}H^+}
\begin{array}{c}
N\!\!\diagdown \\
N\!-\!Co\!-\!O\diagup^{H} \\
N\diagup
\end{array}
$$

Apparently, such a mechanism is consistent with the pH independence of k_{cat} of CO_2 hydration at pH values [199] above the pK_a value of the coordinated water. The proton transfer has been found to be buffer assisted [39].

The release of HCO_3^- and H^+ (or their uptake as substrates in the reverse reaction) may be either stepwise [39] or simultaneous [200]; in the former case the enzyme is temporarily transformed in the low-pH form; in the latter a concerted mechanism formally similar to the binding of OH^- and CN^- would allow no change of charge in the active site and no need for the low-pH form to be involved as an intermediate [200]. Further experimental data are required to discriminate between the two pathways.

REFERENCES

1. C.-I. Brändén, J. Jörnvall, H. Eklund, and B. Furugren, *Enzymes*, 3rd ed., Vol. 11 (P. D. Boyer, ed.), Academic Press, New York, 1975, p. 103.

2. C.-I. Brändén and H. Eklund, *Mol. Interact. Proteins, Ciba Found. Symp.*, *63* (1978).

3. J. V. Bannister and H. A. O. Hill, eds., *Chemical and Biochemical Aspects of Superoxide and Superoxide Dismutase*, Elsevier/North-Holland, New York, 1980.

4. W. H. Bannister and J. V. Bannister, eds., *Biological and Clinical Aspects of Superoxide and Superoxide Dismutase*, Elsevier/North-Holland, New York, 1980.

5. J. S. Richardson, K. A. Thomas, B. H. Rubin, and D. C. Richardson, *Proc. Natl. Acad. Sci. USA*, *72*, 1349 (1975).

6. C.-I. Brändén, *Biochem. Soc. Trans.*, *5*, 612 (1977).

7. A. S. Mildvan, *Enzymes*, 3rd ed., Vol. 2 (P. D. Boyer, ed.), Academic Press, New York, 1970, p. 445.

8. T. Boiwe and C.-I. Brändén, *Eur. J. Biochem.*, *77*, 173 (1977).

9. I. Giannini, V. Baroncelli, G. Boccalon, and P. Renzi, *J. Mol. Catal.*, *2*, 39 (1977).

10. M. C. De Traglia, J. Schmidt, M. F. Dunn, and J. T. McFarland, *J. Biol. Chem.*, *252*, 3493 (1977).

11. P. Andersson, J. Kvassman, A. Lindström, B. Oldén, and G. Pettersson, *Eur. J. Biochem.*, *113*, 425 (1981).

12. P. B. Gaber, R. D. Brown, S. H. Koenig, and J. A. Fee, *Biochim. Biophys. Acta*, *271*, 1 (1972).

13. M. W. Pantoliano, P. J. McDonnell, and J. S. Valentine, *J. Am. Chem. Soc.*, *101*, 6454 (1979).

14. J. S. Valentine, M. W. Pantoliano, P. J. McDonnell, A. R. Burger, and S. J. Lippard, *Proc. Natl. Acad. Sci. USA*, *76*, 4245 (1979).

15. H. L. Monaco, J. C. Crawford, W. N. Lipscomb, *Proc. Natl. Acad. Sci. USA*, *75*, 5276 (1978).

16. J. P. Rosenbusch and K. Weber, *J. Biol. Chem.*, *246*, 1644 (1971).

17. M. E. Nelbach, V. P. Pigiet, J. C. Gerhart, and H. K. Schachman, *Biochemistry*, *11*, 315 (1972).

18. H. G. Wood, *Enzymes*, 3rd ed., Vol. 6 (P. D. Boyer, ed.), Academic Press, New York, 1972, p. 83.

19. N. M. Green, R. C. Valentine, N. G. Wrigley, F. Ahmad, B. Jacobson, and H. G. Wood, *J. Biol. Chem.*, *247*, 6284 (1972).

20. D. B. Northrop and H. G. Wood, *J. Biol. Chem.*, *244*, 5801 (1969).

21. B. L. Vallee, *Trends Biochem. Sci.*, *1*, 88 (1976).

22. C.-W. Wu, F. Y.-H. Wu, and D. C. Speckhard, *Biochemistry*, *16*, 5449 (1977).

23. M. L. Ludwig and W. N. Lipscomb, in *Inorganic Biochemistry*, Vol. 1 (G. L. Eichhorn, ed.), Elsevier, Amsterdam, 1973, p. 438ff.

24. D. C. Rees, M. Lewis, R. B. Honzatko, W. N. Lipscomb, and K. D. Hardman, *Proc. Natl. Acad. Sci. USA*, *78*, 3408 (1981).

25. J. A. Hartsuck and W. N. Lipscomb, *Enzymes*, 3rd ed., Vol. 3 (P. D. Boyer, ed.), Academic Press, New York, 1971, p. 1.

26. D. C. Rees, R. B. Honzatko, and W. N. Lipscomb, *Proc. Natl. Acad. Sci. USA*, *77*, 3288 (1980).

27. S. A. Latt, B. Holmquist, and B. L. Vallee, *Biochem. Biophys. Res. Commun.*, *37*, 333 (1969).

28. P. M. Colman, J. N. Jansonius, and B. W. Matthews, *J. Mol. Biol.*, *70*, 701 (1972).

29. W. R. Kester and B. W. Matthews, *J. Biol. Chem.*, *252*, 7704 (1977).

30. T. W. Reid and I. B. Wilson, *Enzymes*, 3rd ed., Vol. 4 (P. D. Boyer, ed.), Academic Press, New York, 1971, p. 373.

31. J. F. Chlebowski and J. E. Coleman, in *Metal Ions in Biological Systems,* Vol. 6 (H. Sigel, ed.), Marcel Dekker, New York, 1976, p. 1ff.

32. J. E. Coleman and J. F. Chlebowski, in *Advances in Inorganic Biochemistry,* Vol. 1 (G. L. Eichhorn and L. G. Marzilli, eds.), Elsevier/North-Holland, New York, 1979, p. 1ff.

33. J. R. Knox and H. W. Wyckoff, *J. Mol. Biol., 74,* 533 (1973).

34. J. M. Sowadski, B. A. Foster, and H. W. Wyckoff, *J. Mol. Biol., 150,* 245 (1981).

35. J. M. T. Hamilton-Miller and J. T. Smith, *Beta-Lactamases,* Academic Press, London, 1979.

36. R. B. Davies and E. P. Abraham, *Biochem. J., 143,* 129 (1974).

37. J. E. Coleman, in *Inorganic Biochemistry,* Vol. 1 (G. L. Eichhorn, ed.), Elsevier, Amsterdam, 1973, p. 488ff.

38. Y. Pocker and S. Sarkanen, *Adv. Enzymol., 47,* 149 (1978).

39. S. Lindskog, in *Advances in Inorganic Biochemistry,* Vol. 4, Elsevier, New York, in press, and references therein.

40. I. Bertini, C. Luchinat, and A. Scozzafava, *Struct. Bonding (Berl.), 48,* 45 (1982), and references therein.

41. D. Keilin and T. Mann, *Biochem. J., 34,* 1163 (1940).

42. K. K. Kannan, in *Biophysics and Physiology of Carbon Dioxide* (C. Bauer, G. Gros, and H. Bartels, eds.), Springer-Verlag, Berlin, 1980, p. 184ff., and references therein.

43. B. L. Horecker, O. Tsolas, and C. Y. Lai, *Enzymes,* 3rd ed., Vol. 7 (P. D. Boyer, ed.), Academic Press, New York, 1972, p. 213.

44. E. A. Noltmann, *Enzymes,* 3rd ed., Vol. 6 (P. D. Boyer, ed.), Academic Press, New York, 1972, p. 271.

45. R. W. Gracy and E. A. Noltmann, *J. Biol. Chem., 243,* 5410 (1968).

46. M. C. Scrutton and M. R. Young, *Enzymes,* 3rd ed., Vol. 6 (P. D. Boyer, ed.), Academic Press, New York, 1972, p. 1.

47. M. C. Scrutton, M. R. Young, and M. F. Utter, *J. Biol. Chem., 245,* 6220 (1970).

48. S. Lindskog and B. G. Malmström, *J. Biol. Chem., 237,* 1129 (1962).

49. J. B. Hunt, M. J. Rhee, and C. B. Storm, *Anal. Biochem., 55,* 617 (1977).

50. S. A. Latt and B. L. Vallee, *Biochemistry, 10,* 4263 (1971).

51. S. A. Latt, B. Holmquist, and B. L. Vallee, *Biochem. Biophys. Res. Commun., 37,* 333 (1969).

52. C. Lazdunski, D. Chappelet, C. Petitclerc, F. Letterier, P. Douzou, and M. Lazdunski, *Eur. J. Biochem.*, *17*, 239 (1970).

53. J. A. Fee, *Biochim. Biophys. Acta*, *295*, 87 (1973).

54. A. S. Mildvan, R. D. Kobes, and W. J. Rutter, *Biochemistry, 10*, 1191 (1971).

55. K. M. Beem, D. C. Richardson, and K. V. Rajagopalan, *Biochemistry, 16*, 1930 (1977).

56. L. Calabrese, G. Rotilio, and B. Mondovì, *Biochim. Biophys. Acta*, *263*, 827 (1972).

57. D. E. Drum and B. L. Vallee, *Biochem. Biophys. Res. Commun.*, *41*, 33 (1970).

58. W. Maret, I. Andersson, H. Dietrich, H. Schneider-Bernlöhr, R. Einarsson, and M. Zeppezauer, *Eur. J. Biochem.*, *98*, 501 (1979).

59. M. E. Nelbach, V. P. Pigiet, J. C. Gerhart, and H. K. Schachman, *Biochemistry, 11*, 315 (1972).

60. D. C. Speckhard, F. Y.-H. Wu, and C.-W. Wu, *Biochemistry, 16*, 5228 (1977).

61. R. L. Carlin, *Trans. Met. Chem.*, *1*, 1 (1966).

62. S. Lindskog, *Struct. Bonding (Berl.)*, *8*, 153 (1970).

63. I. Bertini and A. Scozzafava, in *Metal Ions in Biological Systems*, Vol. 12 (H. Sigel, ed.), Marcel Dekker, New York, 1981, p. 31ff.

64. A. S. Mildvan and R. K. Gupta, *Methods Enzymol.*, *F49*, 322 (1978), and references therein.

65. I. M. Armitage, A. J. M. Schoot Uiterkamp, J. F. Chlebowski, and J. E. Coleman, *J. Magn. Reson.*, *29*, 375 (1978).

66. P. H. Haffner and J. E. Coleman, *J. Biol. Chem.*, *250*, 996 (1975).

67. S. Lindskog and A. Ehrenberg, *J. Mol. Biol.*, *24*, 133 (1967).

68. R. Aasa, M. Hanson, and S. Lindskog, *Biochim. Biophys. Acta*, *453*, 211 (1976).

69. R. C. Rosenberg, C. A. Root, R.-H. Wang, M. Cerdonio, and H. B. Gray, *Proc. Natl. Acad. Sci. USA*, *70*, 161 (1973).

70. R. C. Rosenberg, C. A. Root, and H. B. Gray, *J. Am. Chem. Soc.*, *97*, 21 (1975).

71. T. H. Moss and J. A. Fee, *Biochem. Biophys. Res. Commun.*, *66*, 799 (1975).

72. H. Dietrich, W. Maret, L. Wallén, and M. Zeppezauer, *Eur. J. Biochem.*, *100*, 267 (1979).

73. C. K. Jørgensen, *Absorption Spectra and Chemical Bonding in Complexes*, Pergamon Press, New York, 1952, p. 125ff.

74. G. Rotilio, L. Calabrese, B. Mondovì, and W. E. Blumberg, *J. Biol. Chem.*, *249*, 3157 (1974).

75. G. S. Baldwin, A. Galdes, H. A. O. Hill, S. G. Waley, and E. P. Abraham, *J. Inorg. Biochem.*, *13*, 189 (1980).

76. I. Bertini, C. Luchinat, and A. Scozzafava, *Inorg. Chim. Acta*, *46*, 85 (1980).

77. G. S. Baldwin, A. Galdes, H. A. O. Hill, B. E. Smith, S. G. Waley, and E. P. Abraham, *Biochem. J.*, *175*, 441 (1978), and references therein.

78. A. Galdes, H. A. O. Hill, G. S. Baldwin, W. G. Waley, and E. P. Abraham, *Biochem. J.*, *187*, 789 (1980).

79. R. T. Simpson, R. D. Kobes, R. W. Rutter, and B. L. Vallee, *Biochemistry*, *10*, 2466 (1971).

80. B. Holmquist and B. L. Vallee, *J. Biol. Chem.*, *249*, 4601 (1974).

81. M. L. Applebury and J. E. Coleman, *J. Biol. Chem.*, *244*, 709 (1969).

82. R. T. Simpson and B. L. Vallee, *Biochemistry*, *7*, 4343 (1968).

83. R. A. Anderson, F. S. Kennedy, and B. L. Vallee, *Biochemistry*, *15*, 3710 (1976).

84. I. Bertini, R. D. Brown, S. H. Koenig, and C. Luchinat, *Biophys. J.*, in press.

85. I. Bertini, G. Canti, C. Luchinat, and F. Mani, *Inorg. Chem.*, *20*, 1670 (1981).

86. I. Bertini, C. Luchinat, and A. Scozzafava, *J. Am. Chem. Soc.*, *99*, 581 (1977).

87. I. Bertini, G. Canti, C. Luchinat, and A. Scozzafava, *J. Am. Chem. Soc.*, *100*, 4873 (1978).

88. I. Bertini, G. Canti, C. Luchinat, and P. Romanelli, *Inorg. Chim. Acta*, *46*, 211 (1980).

89. P.-C. Bergstén, I. Vaara, S. Lövgren, A. Liljas, K. K. Kannan, and U. Bengtsson, in *Proceedings of the Alfred Benzon Symposium IV* (M. Rorth and P. Astrup, eds.), Munksgaard, Copenhagen, 1972, p. 363.

90. I. Bertini, C. Luchinat, and A. Scozzafava, *Bioinorg. Chem.*, *9*, 93 (1978).

91. J. Hirose and Y. Kidani, *J. Inorg. Biochem.*, *14*, 313 (1981).

92. I. Bertini, C. Luchinat, and A. Scozzafava, *Inorg. Chim. Acta*, *22*, L23 (1977).

93. J. S. Taylor, C. Y. Lau, M. L. Applebury, and J. E. Coleman, *J. Biol. Chem.*, *248*, 6216 (1973).

94. L. Calabrese, D. Cocco, L. Morpurgo, B. Mondovì, and G. Rotilio, *Eur. J. Biochem.*, *64*, 465 (1976).

95. I. Bertini, in *ESR and NMR of Paramagnetic Species in Biological and Related Systems* (I. Bertini and R. S. Drago, eds.), D. Reidel, Dordrecht, Holland, 1980, p. 221ff.

96. I. Bertini, G. Canti, and C. Luchinat, *J. Am. Chem. Soc.*, *104*, 4943 (1982).

97. J. W. Wells, S. I. Kandel, and S. H. Koenig, *Biochemistry*, *10*, 1989 (1979).

98. I. Bertini, G. Canti, and C. Luchinat, *Inorg. Chim. Acta*, *56*, 99 (1981).

99. I. Andersson, W. Maret, M. Zeppezauer, R. D. Brown, and S. H. Koenig, *Biochemistry*, *20*, 3424 (1981).

100. A. Bencini, D. Gatteschi, and C. Zanchini, *J. Am. Chem. Soc.*, *102*, 5234 (1980).

101. A. Rigo, M. Terenzi, C. Franconi, B. Mondovì, L. Calabrese, and G. Rotilio, *FEBS Lett.*, *39*, 154 (1974).

102. R. S. Zukin and D. P. Hollis, *J. Biol. Chem.*, *250*, 835 (1975).

103. C.-H. Fung, A. S. Mildvan, and J. S. Leigh, Jr., *Biochemistry*, *13*, 1160 (1974).

104. I. Bertini, C. Luchinat, and A. Scozzafava, *Biochim. Biophys. Acta*, *452*, 239 (1976).

105. I. Bertini, E. Borghi, G. Canti, and C. Luchinat, *J. Inorg. Biochem.*, *11*, 49 (1979).

106. I. Bertini, C. Luchinat, and A. Scozzafava, *J. Chem. Soc.*, *Dalton*, 1962 (1977).

107. G. Alberti, I. Bertini, C. Luchinat, and A. Scozzafava, *Biochim. Biophys. Acta*, *668*, 16 (1981).

108. P. W. Taylor, J. Feeney, and A. S. V. Burgen, *Biochemistry*, *10*, 3866 (1971).

109. I. Bertini, C. Luchinat, and A. Scozzafava, *FEBS Lett.*, *93*, 251 (1978).

110. P. L. Yeagle, C. H. Lochmüller, and R. W. Henkens, *Proc. Natl. Acad. Sci. USA*, *72*, 454 (1975).

111. P. J. Stein, S. H. Merrill, and R. W. Henkens, *J. Am. Chem. Soc.*, *99*, 3194 (1977).

112. D. L. Sloan, J. M. Young, and A. S. Mildvan, *Biochemistry*, *14*, 1998 (1975).

113. G. M. Smith, A. S. Mildvan, and E. T. Harper, *Biochemistry*, *19*, 1248 (1980).

114. J. L. Boch and B. Sheard, *Biochem. Biophys. Res. Commun.*, *66*, 2430 (1975).

115. J. F. Chlebowski, I. M. Armitage, P. P. Tusa, and J. E. Coleman, *J. Biol. Chem.*, *251*, 1207 (1976).

116. J. D. Otvos and D. T. Browne, *Biochemistry*, *19*, 4011 (1980).

117. J. D. Otvos and I. M. Armitage, *Biochemistry*, *19*, 4021 (1980).

118. J. D. Otvos and I. M. Armitage, *Biochemistry*, *19*, 4031 (1980).

119. I. Bertini, G. Canti, C. Luchinat, and F. Mani, *J. Am. Chem. Soc.*, *103*, 7784 (1981).

120. H. A. O. Hill, B. E. Smith, C. B. Storm, and R. P. Ambler, *Biochem. Biophys. Res. Commun.*, *70*, 783 (1976).

121. S. A. Cockle, S. Lindskog, and E. Grell, *Biochem. J.*, *143*, 703 (1974).

122. A. Desideri, L. Morpurgo, J. B. Raynor, and G. Rotilio, *Biophys. Chem.*, *8*, 267 (1978).

123. A. Bencini, I. Bertini, G. Canti, D. Gatteschi, and C. Luchinat, *J. Inorg. Biochem.*, *14*, 81 (1981).

124. G. Rotilio, L. Calabrese, B. Mondovì, and W. E. Blumberg, *J. Biol. Chem.*, *249*, 3157 (1974).

125. T. A. Kaden, B. Holmquist, and B. L. Vallee, *Biochem. Biophys. Res. Commun.*, *46*, 1654 (1972).

126. B. Holmquist, T. A. Kaden, and B. L. Vallee, *Biochemistry*, *14*, 1454 (1975).

127. R. A. Anderson and B. L. Vallee, *Biochemistry*, *16*, 4388 (1977).

128. W. DeW. Horrocks, Jr., J. N. Ishley, B. Holmquist, and J. S. Thompson, *J. Inorg. Biochem.*, *12*, 131 (1980).

129. W. DeW. Horrocks, Jr., B. Holmquist, and B. L. Vallee, *Proc. Natl. Acad. Sci. USA*, *72*, 4764 (1975).

130. I. Bertini, G. Canti, C. Luchinat, and A. Scozzafava, *J. Chem. Soc., Dalton*, 1269 (1978).

131. S. H. Koenig and R. D. Brown, *Ann. N.Y. Acad. Sci.*, *222*, 752 (1973).

132. I. Bertini, C. Luchinat, R. Monnanni, and A. Scozzafava, *J. Inorg. Biochem.*, *16*, 155 (1982).

133. I. Bertini, E. Borghi, G. Canti, and C. Luchinat, *J. Inorg. Biochem.*, in press.

134. I. Bertini, G. Canti, and C. Luchinat, *Inorg. Chim. Acta*, *56*, 1 (1981).

135. W. Maret, H. Dietrich, H.-H. Ruf, and M. Zeppezauer, *J. Inorg. Biochem.*, *12*, 241 (1980).

136. R. L. Brooks and J. S. Shore, *Biochemistry*, *10*, 3855 (1971).

137. H. Csopak and K. E. Falk, *Biochim. Biophys. Acta*, *359*, 22 (1974).

138. I. Bertini, C. Luchinat, and A. Scozzafava, *J. Am. Chem. Soc.*, *102*, 7349 (1980).

139. I. Bertini, E. Borghi, C. Luchinat, and A. Scozzafava, *J. Am. Chem. Soc.*, *103*, 7779 (1981).

140. I. Bertini, C. Luchinat, and L. Messori, *Biochem. Biophys. Res. Commun.*, *577*, 101 (1981).

141. H. G. Strothkamp and S. J. Lippard, *Biochemistry*, *20*, 7488 (1981).

142. P. H. Haffner, F. Goodsaid-Zalduondo, and J. E. Coleman, *J. Biol. Chem.*, *249*, 6693 (1974).

143. S. H. Koenig and R. D. Brown, in *ESR and NMR of Paramagnetic Species in Biological and Related Systems* (I. Bertini and R. S. Drago, eds.), D. Reidel, Dordrecht, Holland, 1980, p. 89ff.

144. R. G. Shulman, G. Navon, B. J. Wyluda, D. C. Douglass, and T. Yamane, *Proc. Natl. Acad. Sci. USA*, *56*, 39 (1966).

145. G. Navon, R. G. Shulman, B. J. Wyluda, and T. Yamane, *J. Mol. Biol.*, *51*, 15 (1970).

146. G. Navon, *Chem. Phys. Lett.*, *7*, 390 (1970).

147. S. H. Koenig, R. D. Brown, and J. Studebaker, *Cold Spring Harbor Symp. Quant. Biol.*, *36*, 551 (1971).

148. N. Zisapel, G. Navon, and M. Sokolovsky, *Eur. J. Biochem.*, *52*, 487 (1975).

149. I. Bertini, C. Luchinat, and A. Scozzafava, *FEBS Lett.*, *87*, 92 (1978).

150. I. Andersson, W. Maret, M. Zeppezauer, R. D. Brown, and S. H. Koenig, *Biochemistry*, *20*, 3433 (1981).

151. W. Maret, I. Andersson, H. Dietrich, H. Schneider-Bernlöhr, R. Einarsson, and M. Zeppezauer, *Hoppe Seyler's Z. Physiol. Chem.*, *359*, 1116 (1978).

152. W. L. Bigbee and F. Dahlquist, *Biochemistry*, *13*, 3542 (1974).

153. W. L. Bigbee and F. Dahlquist, *Biochemistry*, *16*, 3798 (1977).

154. G. Navon, R. G. Shulman, B. J. Wyluda, and T. Yamane, *Proc. Natl. Acad. Sci. USA*, *60*, 86 (1968).

155. A Lanir and G. Navon, *Biochim. Biophys. Acta*, *341*, 75 (1974).

156. A. Lanir and G. Navon, *Biochemistry*, *11*, 3536 (1972).

157. R. S. Zukin, D. P. Hollis, and G. A. Gray, *Biochem. Biophys. Res. Commun.*, *53*, 238 (1973).

158. J. F. Chlebowski, I. M. Armitage, P. P. Tusa, and J. E. Coleman, *J. Biol. Chem.*, *251*, 1207 (1976).

159. W. E. Hull, S. E. Halford, H. Gutfreund, and B. D. Sykes, *Biochemistry*, *15*, 1547 (1976).

160. A. S. Mildvan and M. Cohn, *Adv. Enzymol.*, *33*, 1 (1970).

161. G. M. Smith and A. S. Mildvan, *Biochemistry*, *20*, 4340 (1981).

162. S. Fan, L. W. Harrison, and G. G. Hammes, *Biochemistry*, *14*, 2219 (1975).

163. R. J. Kostelnik and A. A. Bothner-By, *J. Magn. Reson.*, *14*, 141 (1974).

164. A. D. Cardin, P. D. Ellis, J. D. Adam, and J. W. Howard, *J. Am. Chem. Soc.*, *97*, 1672 (1975).

165. R. A. Haberkorn, L. Que, W. O. Gillum, R. H. Holm, C. S. Liu, and R. C. Lord, *Inorg. Chem.*, *15*, 2408 (1976).

166. B. R. Bobsein and R. J. Myers, *J. Am. Chem. Soc.*, *102*, 2454 (1980).

167. B. R. Bobsein and R. J. Myers, *J. Biol. Chem.*, *256*, 5313 (1981).

168. D. B. Bailey, P. D. Ellis, and J. A. Fee, *Biochemistry*, *19*, 591 (1980).

169. J. L. Sudmeier and S. J. Bell, *J. Am. Chem. Soc.*, *99*, 4499 (1977).

170. N. B.-H. Johnsson, L. A. E. Tibell, J. L. Evelhoch, S. J. Bell, and J. L. Sudmeier, *Proc. Natl. Acad. Sci. USA*, *77*, 3269 (1980).

171. J. B. Jensen, *Acta Chem. Scand.*, *Ser. A*, *29*, 250 (1975).

172. D. B. Bailey and P. D. Ellis, unpublished experiments, as quoted in Ref. 168.

173. J. L. Evelhoch, D. F. Bocian, and J. L. Sudmeier, *Biochemistry*, *20*, 4951 (1981).

174. J. D. Otvos, I. M. Armitage, J. F. Chlebowski, and J. E. Coleman, *J. Biol. Chem.*, *254*, 4707 (1979).

175. S. McCracken and E. A. Meighen, *J. Biol. Chem.*, *256*, 3495 (1981).

176. R. Bauer, P. Limkilde, and J. T. Johansen, *Biochemistry*, *15*, 334 (1976).

177. R. Bauer, P. Limkilde, and J. T. Johansen, *Carlsberg Res. Commun.*, *42*, 325 (1977).

178. R. Bauer, I. Demeter, V. Haseman, and J. T. Johansen, *Biochem. Biophys. Res. Commun.*, *94*, 1296 (1980).

179. R. Bauer, C. Christensen, J. T. Johansen, and B. L. Vallee, *Biochem. Biophys. Res. Commun.*, *90*, 679 (1979).

180. A. J. Sytkowski and B. L. Vallee, *Biochemistry*, *18*, 4095 (1979).

181. J. I. Legg, *Coord. Chem. Rev.*, *25*, 103 (1978).

182. H. Shinar and G. Navon, *Biochim. Biophys. Acta*, *334*, 471 (1974).

183. H. Shinar and G. Navon, *Eur. J. Biochem.*, *93*, 313 (1979).

184. E. P. Kang, C. B. Storm, and F. W. Carson, *J. Am. Chem. Soc.*, *97*, 6723 (1975).

185. M. S. Urdea and J. I. Legg, *Biochemistry*, *18*, 4984 (1979).

186. M. S. Urdea and J. I. Legg, *J. Biol. Chem.*, *254*, 11868 (1979).

187. J. J. Fitzgerald and N. D. Chasteen, *Biochemistry*, *13*, 4338 (1974).

188. I. Bertini, G. Canti, C. Luchinat, and A. Scozzafava, *Inorg. Chim. Acta*, *36*, 9 (1979).

189. R. J. DeKoch, D. J. West, J. C. Cannon, and N. D. Chasteen, *Biochemistry*, *13*, 4347 (1974).

190. J. L. Sudmeier and T. G. Perkins, *J. Am. Chem. Soc.*, *99*, 7732 (1977).

191. K. K. Kannan, in *Proceedings on Biomolecular Structure, Conformation, Function, and Evolution* (R. Srinivasan, ed.), Pergamon Press, Oxford, 1978.

192. R. S. Johnson and H. K. Schachman, *Proc. Natl. Acad. Sci. USA*, *77*, 1995 (1980).

193. H. Dietrich, W. Maret, H. Kozłowski, and M. Zeppezauer, *J. Inorg. Biochem.*, *14*, 297 (1981).

194. D. L. Tennent and D. R. McMillin, *J. Am. Chem. Soc.*, *101*, 2307 (1979).

195. I. Bertini, E. Borghi, and C. Luchinat, *Bioinorg. Chem.*, *9*, 495 (1978).

196. S. Lindskog and J. E. Coleman, *Proc. Natl. Acad. Sci. USA*, *70*, 2505 (1973).

197. D. N. Silverman, C. K. Tu, S. Lindskog, and G. C. Wynns, *J. Am. Chem. Soc.*, *101*, 6734 (1979).

198. I. Bertini, E. Borghi, and C. Luchinat, *J. Am. Chem. Soc.*, *101*, 7069 (1979).

199. Y. Pocker, T. L. Deits, and N. Tanaka, in *Advances in Solution Chemistry* (I. Bertini, L. Lunazzi, and A. Dei, eds.), Plenum Press, New York, 1981, p. 253ff.

200. S. H. Koenig, R. D. Brown III, and G. S. Jacob, in *Biophysics and Physiology of Carbon Dioxide* (C. Bauer, G. Gros, and H. Bartels, eds.), Springer-Verlag, Berlin, 1980, p. 238ff.

201. G. Rotilio, L. Calabrese, F. Bossa, D. Barra, A. Finazzi-Agrò, and B. Mondovì, *Biochemistry, 11*, 2182 (1972).

202. K. M. Beem, W. E. Rich, and K. V. Rajagopalan, *J. Biol. Chem., 249*, 7298 (1974).

203. J. H. Griffin, J. P. Rosenbusch, E. R. Blout, and K. K. Weber, *J. Biol. Chem., 248*, 5057 (1973).

204. J. E. Coleman and B. L. Vallee, *J. Biol. Chem., 235*, 390 (1960).

205. J. E. Coleman and B. L. Vallee, *J. Biol. Chem., 236*, 2244 (1961).

206. J. D. McConn, D. Tsuru, and K. T. Yasunobu, *J. Biol. Chem., 239*, 3706 (1964).

207. J. D. McConn, D. Tsuru, and K. T. Yasunobu, *Arch. Biochem. Biophys., 120*, 479 (1967).

208. M. I. Harris and J. E. Coleman, *J. Biol. Chem., 243*, 5063 (1968).

209. M. Gottesman, R. T. Simpson, and B. L. Vallee, *Biochemistry, 8*, 3776 (1969).

210. R. B. Davies and E. P. Abraham, *Biochem. J., 143*, 129 (1974).

211. J. E. Coleman, *Nature (Lond.), 214*, 193 (1967).

212. R. D. Kobes, R. T. Simpson, B. L. Vallee, and W. J. Rutter, *Biochemistry, 8*, 585 (1969).

213. F. H. Bruns and E. Noltman, *Nature (Lond.), 181*, 1467 (1958).

214. F. S. Kennedy, H. A. O. Hill, T. A. Kaden, and B. L. Vallee, *Biochem. Biophys. Res. Commun., 48*, 1533 (1972).

215. A new technique has been developed by Kuo and Makinen (L. C. Kuo and M. W. Makinen, *J. Biol. Chem., 257*, 24 (1982)) to detect water coordinated in cobalt(II) enzymes. If ^{17}O enriched H_2O is used and water is bound to the metal, then the electronic relaxation times of cobalt(II) decrease. This can be measured through the microwave power necessary to saturate the EPR signal.

216. Another criterion has been suggested to monitor the coordination geometry, i.e., the magnitude of zero field splitting (ZFS) of the quadruplet ground level. In particular, small ZFS values have been taken as indicative of fourcoordination and large ZFS values indicative of five or sixcoordination (M. B. Yim, L. C. Kuo, and M. W. Makinen, *J. Magn. Reson., 46*, 247 (1982)). This agrees with the general chemical common sense, and may be often true; however, the relative energies of the two lowest lying

Kramers' doublets are very sensitive to small geometrical variations and to the donor properties of the ligands (L. Banci, A. Bencini, C. Benelli, D. Gatteschi, and C. Zanchini, *Struct. Bonding, 52*, 37 (1982)) so that in every coordination either the ±3/2 or ±1/2 Kramers' doublets can be the ground states. Therefore in our opinion there is no theoretical justification for the above criterion. Furthermore the determination of the ZFS is based on the assumption that the dominant relaxation process is the Orbach process involving the first excited Kramers' doublet; the effect of other low lying excited states through possible concomitant mechanisms is not taken into consideration.

217. A recent report of electronic and CD spectra points out the possibility for Ni(II) carbonic anhydrase of being fivecoordinate (I. Bertini, E. Borghi, C. Luchinat, and R. Monnanni, *Inorg. Chim. Acta, 67*, 99 (1982)).

218. A study on $Cu(II)_2$ alkaline phosphatase with phosphate has pointed out that the latter binds in a 1:1 molar ratio. Since the two copper ions are still equivalent within the sensitivity of the X-band EPR spectrum, the migration model does not seem to hold for this derivative (I. Bertini, C. Luchinat, A. Maldotti, A. Scozzafava, and O. Traverso, *Inorg. Chim. Acta,* in press)).

Chapter 4

THE ROLE OF ZINC IN DNA AND RNA POLYMERASES

Felicia Ying-Hsiueh Wu and Cheng-Wen Wu
Department of Pharmacological Sciences
State University of New York at Stony Brook
Stony Brook, New York

1. INTRODUCTION

DNA and RNA polymerases are nucleotidyl transferases that catalyze
the replication and transcription of the cellular genome. The reactions of these enzymes require the presence of extrinsic divalent

157

metal ions, Mg(II) or Mn(II). In 1971, the DNA [1] and RNA [2]
polymerase from *Escherichia coli* were the first two nucleotidyl
transferases shown to be zinc metalloenzymes, which contain 1 [3]
and 2 [2] g-atoms of tightly bound zinc per mol enzyme, respectively.
Subsequently, it has been demonstrated that a variety of DNA and RNA
polymerases from both prokaryotic and eukaryotic sources possess a
constant and stoichiometric amount of intrinsic Zn(II) ions. Although
the importance of zinc in the transfer of genetic information has been
recognized, the detailed mechanisms involving these zinc ions in DNA
and RNA polymerases have not yet been established.

    This chapter reviews the up-to-date knowledge about the intrin-
sic zinc ions in polymerases isolated from both prokaryotic and eukaryotic cells. The role of zinc ions in the mechanism of DNA polymerases
has recently been reviewed by Mildvan and Loeb [4,5]. Therefore, only
a short description pertaining to this topic is presented here. Emphasis is placed on the recent developments of the metal studies of *E.
coli* RNA polymerase, as this is the enzyme in which the role of zinc
ions has been most extensively studied. The biological implications
of the roles of these intrinsic metal ions in gene replication and
transcription are also discussed.

2. UNIVERSAL PRESENCE OF ZINC IN DNA AND RNA POLYMERASES

The zinc contents of DNA and RNA polymerases isolated from various
prokaryotes and eukaryotes are listed in Tables 1 and 2. All DNA
and RNA polymerases examined thus far are inhibited by the metal
chelator, 1,10-phenanthroline (OP) [4-6]. As seen in Tables 1 and
2, 50% inhibition of polymerase activity is achieved at a concentration of 10^{-5}-10^{-3} M. An essential role for zinc in all polymerases
has been suggested [1-3,7-10] on the basis of their inhibition by
1,10-phenanthroline but not by its nonchelating analogue, 1,7-
phenanthroline. However, Sigman and co-workers [11,12] have reported
that the inhibition of DNA and RNA polymerases by OP is due to the

TABLE 1

Zinc in DNA Polymerases

Source	Zinc content (g-atoms/mol)	Molecular weight	Number of subunits	OP[a] for 50% inhibition[b] (mM)	References
Bacteria					
E. coli DNA pol. I	1.8	110,000	1	0.04	1
E. coli DNA pol. I	1.0 ± 0.15	110,000	1	+	3
Bacteriophage					
T4 phage	1.0	112,000	1	+	3
Eukaryotes					
Sea urchin	4.2	150,000		0.4	1
RNA tumor viruses					
Avian myeloblastosis	1.3	$1.65-1.85 \cdot 10^5$	2	0.4	7,8
Avian myeloblastosis	1.8-2.0	$1.6-1.8 \cdot 10^5$	2	0.1	9
Murine leukemia	1.4			0.01	10
Woolly monkey	1.0			0.01	10

[a]Ortho- or 1,10-phenanthroline.

[b]+, inhibited by OP, but the concentration for 50% inhibition is unavailable.

TABLE 2

Zinc in RNA Polymerases

Source[a]	Zinc content (g-atoms/mol)	Molecular weight	Number of subunits	OP[b] for 50% inhibition[c] (mM)	Reference
Bacteria					
E. coli	2	493,000	5	0.8~0.9	2
B. subtilis	2	451,000	5	1.2	16
Bacteriophage					
T7 phage	2~4	107,000	1	+	14
T3 phage	1	105,000	1	+	59
Eukaryotes					
E. gracilis					
RPase I	2.2	634,000	10	0.01	17
RPase II	2.2	700,000	5	0.3	60
Yeast					
RPase I	2.4	650,000	11~12	0.3	18
RPase II	1	460,000	9	0.6	19
RPase III	2	380,000	11~12	0.1	62
Wheat germ					
RPase II	7	550,000	10	+	20

[a]RPase, RNA polymerase.

[b]Ortho- or 1,10-phenanthroline.

[c]+, inhibited by OP, but the concentration for 50% inhibition is unavailable.

formation of an inhibitory phenanthroline/cuprous ion complex in the assay mixture. Therefore, the criterion for the presence of essential zinc in polymerases based on OP inhibition should be exercised with caution. Some polymerases reported to be inhibited by OP are not included in these tables because their actual zinc contents have not been determined. Thus we consider here only those DNA and RNA polymerases that can be classified as zinc metalloenzymes according to the criteria of Vallee [13].

One direct way to establish the essential role of zinc in
polymerase catalysis is to demonstrate enzyme inactivation by the
removal of zinc and reactivation upon zinc replacement. Of the
various polymerases listed in Tables 1 and 2, the essential requirement of zinc was shown only in *E. coli* DNA polymerase I [3], avian
myeloblastosis virus DNA polymerase [7,9], T7 RNA polymerase [14],
and *E. coli* RNA polymerase (see Secs. 3, 4.1.1, and 5.1). Most
other polymerases were irreversibly inactivated upon removal of the
intrinsic zinc ions [15-20]. Besides using the intrinsic metal as
a probe to study its environment, one can also obtain useful information about the metal-substituted enzymes from various physical studies.
The diamagnetic zinc is a silent metal in a biophysical sense, in that
it possesses neither optical nor magnetic properties. Therefore, in
the course of studying zinc metalloenzymes, zinc is usually replaced
by other paramagnetic metals, such as cobalt and manganese, which
have properties sensitive to agents that can affect enzyme activity.
Metal-substitution methods, either in vitro or in vivo, have been
successfully employed for a number of zinc metalloenzymes [6,21].
Metal substitution has thus far been applied only to the *E. coli*
RNA polymerase, which will be discussed in some detail.

3. DNA POLYMERASE

The finding by Chang and Bollum [22] that terminal deoxynucleotidyl
transferase was inhibited by 1,10-phenanthroline has led to the
examination of the zinc content in some DNA polymerases. So far,
six DNA polymerases from various sources, including reverse tran-
scriptases (RNA-dependent DNA polymerases) from RNA tumor virus
[7-10], have been found to contain stoichiometric quantities .of
tightly bound zinc (Table 1).

The necessity of zinc for catalysis in DNA polymerase has
been demonstrated in only two enzymes: DNA polymerase I from *E. coli*
[1,3] and reverse transcriptase from avian myeloblastosis virus
(AMV) [7-10]. No mechanistic studies have been pursued for the
other four DNA polymerases. From the studies of *E. coli* DNA

polymerase [3], there are two indirect lines of evidence suggesting
that the intrinsic zinc ion interacts with the DNA template-primer
complex. First, the nuclear quadrupolar relaxation studies of the
role of bound zinc using ^{79}Br$^-$ as a halide probe [3] showed that the
bound zinc ion has a large effect on the relaxation rate of Br$^-$.
This effect is reduced by 70% upon the addition of one molecule of
polydeoxynucleotide primer per enzyme molecule, but is not altered
by the addition of the substrate, dTTP. This result suggests that
the DNA primer but not the substrate competes out 70% of the Br$^-$ from
the enzyme-bound zinc. Second, the kinetic analysis of the inhibition
of DNA polymerase by 1,10-phenanthroline showed that the metal chelator competes with DNA but not substrate. The same observation was
obtained for DNA polymerase from sea urchin nuclei [3]. However,
since the inhibition of AMV DNA polymerase by 1,10-phenanthroline is
noncompetitive [7], it may be attributable either to the loss of
enzyme activity by removal of zinc or to a structural role of the
zinc ion in catalysis. In view of the findings by Sigman and co-
workers [11,12], these inhibition studies with metal chelator deserve
reinvestigation.

A catalytic role of bound zinc in DNA polymerase was proposed
by Abboud et al. [23] (Fig. 1). The role of the zinc ion in this
mechanism is analogous to that proposed for terminal nucleotidyl
transferase [22] and for carbonic anhydrase [24] in that zinc coordi-
nates with the 3'-OH primer terminus, thereby facilitating its deprotonation and preparing it for a nucleophilic attack on the α-phosphorus
atom of the incoming nucleotide. The deprotonation of the 3'-OH group
could be assisted by a nearby general base, as supported by a study of
the suicidal inactivation of DNA polymerase I by 2',3'-epoxy ATP [23].

AMV DNA polymerase is inhibited by the metal chelator 1,10-
phenanthroline, but not by its nonchelating isomers. It has also
been noted that DNA polymerases from Rauscher murine leukemia virus
and avian myeloblastosis virus are inactivated by 1,10-phenanthroline
at a much faster rate than the bacterial (*E. coli*) and animal (sea
urchin and human lymphocyte) DNA polymerases [7]. It may, therefore,

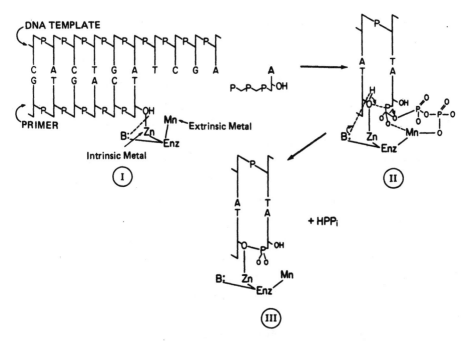

FIG. 1. Proposed catalytic role of intrinsic zinc ion of DNA
polymerase I from *E. coli*. (Adapted from Ref. 23.)

be possible to inactivate DNA polymerase from animal tumor viruses
selectively by brief exposure to the appropriate metal chelators.

4. RNA POLYMERASE

4.1. Bacterial RNA Polymerase

4.1.1. E. coli *RNA Polymerase*

E. coli RNA polymerase is an oligomeric enzyme consisting of five
subunits ($\alpha_2\beta\beta'\sigma$) [25], possessing two substrate-binding sites, the
initiation and elongation site [26,27]. The initiation of RNA
chains occurs primarily with purine nucleotides [28]. The binding
of a purine nucleotide to the initiation site does not require
exogenous metal, such as Mg(II) or Mn(II) [26], suggesting that

the polymerase may contain a bound metal. Atomic absorption spectro-
metric analysis revealed that purified RNA polymerase contains 2
g-atoms zinc per mol enzyme [2]. These two zinc ions are tightly
bound to the enzyme and cannot be removed by dialysis against buffer
containing chelating agents such as EDTA or OP [2,15]. Moreover, the
ratio of zinc content to enzyme activity approaches a constant value
upon the progressive purification of the enzyme.

Some indirect evidence from earlier studies [2] indicates that
intrinsic zinc ions may be involved in the initiation of RNA synthesis.
1,10-Phenanthroline appears to block the initiation and competes with
the binding of purine nucleotide at the initiation site. Although
these observations suggest that 1,10-phenanthroline may inhibit polymerase by interacting with the essential zinc ion at the initiation
site, this possibility requires further investigation because the
inhibition may also be caused by the formation of a phenanthroline/
cuprous ion complex [11,12].

As stated previously, a direct proof of the functional role of
zinc ion in a metalloenzyme is the removal and readdition of intrinsic
metal with the concurrent loss and restoration of enzymatic activity.
However, this approach could not be easily applied to the E. coli RNA
polymerase [2,15], since the bound zinc ion could not be removed by
dialysis against 10 mM EDTA at pH 8 for as long as 2 weeks, and prolonged dialysis of polymerase against 1,10-phenanthroline at pH 6
yielded inactive apoenzyme which could not be revived by the addition
of exogenous Zn(II) ions [15]. Because of these difficulties, metal-
substitution methods, both in vivo [15] and in vitro [29], were used
to study the functional role of the intrinsic metal in polymerase.

(a) *Substitution of both Zinc Ions in Vivo with Cobalt Ions.*
The substitution of Zn(II) by Co(II) was first achieved by an in vivo
method [15], in which E. coli cells were grown in zinc-depleted
(residual zinc ion concentration of 10^{-7}-10^{-8} M) and cobalt-enriched
(5 · 10^{-6} M) media. Although there was an initial lag period of
2-3 h, the growth resumed at a normal rate comparable to that of
E. coli grown in standard minimal medium (5 · 10^{-5} M Zn).

RNA polymerase purified from these cells [30,31] contained
approximately two bound Co(II) ions, as determined by atomic absorption spectrometry. The metal content and specific activity of RNA
polymerases from *E. coli* grown in standard and cobalt-enriched media
are shown in Table 3. The total content of intrinsic metals (Zn +
Co) was constant (2.1 ± 0.2 g-atoms per mol enzyme) in all preparations, indicating that more than 90% of the intrinsic Zn(II) has
been replaced with Co(II) by biosynthetic incorporation [15]. Re-
sembling the zinc ions, these two cobalt ions were tightly bound to
the enzyme and could not be removed by dialysis against EDTA or OP.

The naturally occurring Zn-Zn RNA polymerase and biosyntheti-
cally substituted Co-Co enzyme were physically similar with respect
to their subunit compositions, aggregation properties, pH, and temperature stability profiles. The two differences observed between
these two enzymes were their metal contents and absorption spectra.
Whereas the zinc enzyme manifests no absorption spectra in the visible region, the cobalt enzyme exhibits two major absorption bands at
584 and 703 nm, with molar absorptivity of 200 and 335 M^{-1}/cm, respectively (Fig. 2) [15]. From these spectral properties, an irregular
tetrahedral coordination of Co(II) in the polymerase may be deduced
[21] based on the wavelengths and molar absorptivities of known
Co(II)-substituted enzymes [32-35]. Although the actual coordination geometries require further verification, it seems likely that
two cobalt ions are situated at heterogeneous sites generating two
absorption bands.

The absorption spectra of Co-Co RNA polymerase can be perturbed
by addition of either substrate (PuTP) or template (DNA) (Fig. 2).
Addition of substrate (ATP or GTP) selectively alters the 703-nm
peak, whereas the addition of a template analogue, $d(pT)_{10}$, perturbs
both peaks. These differential changes brought about by substrate
and template imply that the binding of the substrate may involve one
of the two intrinsic metals, whereas the binding of the template may
involve both. However, these spectral changes may be due to the
direct involvement of the intrinsic metal in substrate or template

TABLE 3

Metal Contents and Specific Activity of RNA Polymerase from *E. coli* Grown in Standard and Cobalt-Enriched, Zinc-Depleted Media

Media	Strain	Zinc content (g-atoms/mol)	Cobalt content (g-atoms/mol)	Zn + Co (g-atoms/mol)	Specific activity (units[a]/mg)
Standard	MRE600	2.0 ± 0.1	0	2.0	1150
	MRE600	2.1 ± 0.2	0	2.1	930
	K12	2.2 ± 0.3	0	2.2	880
	K12	1.9 ± 0.2	0	1.9	1020
	Average	2.1 ± 0.2	0	2.1 ± 0.2	995 ± 118
Cobalt-enriched	MRE600	0.1 ± 0.1	2.2 ± 0.3	2.3	1095
	MRE600	0.2 ± 0.1	1.8 ± 0.2	2.0	1105
Zinc-depleted	K12	0.2 ± 0.1	1.8 ± 0.3	2.0	990
	K12	0.2 ± 0.1	2.1 ± 0.1	2.3	1180
	Average	0.2 ± 0.1	1.9 ± 0.2	2.1 ± 0.2	1043 ± 166

[a]One unit is defined as the nmol of a radioactive-labeled nucleoside monophosphate incorporated in 20 min at 37°C using calf-thymus DNA as template.

Source: Reprinted by permission of Ref. 15.

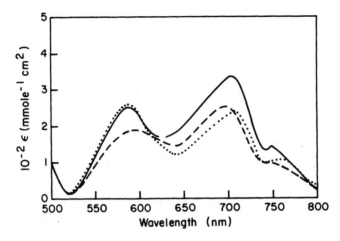

FIG. 2.   Difference spectra between Co-Co and Zn-Zn RNA polymerase:
enzyme alone (——); enzyme plus ATP or GTP (•••); enzyme plus d(pT)$_{10}$
(---).   (Reprinted by permission from Ref. 15.)

binding or some indirect effect such as a conformational change of
the enzyme induced by substrate or template binding at sites away
from the metal.   This uncertainty has recently been resolved by the
selective substitution of one of the two zinc ions with cobalt in
vitro [29,36] (see Secs. 4.1.1.(c) and 4.1.1(d)).

The Co-Co RNA polymerase exhibits the same catalytic activity
on various DNA templates as the native Zn-Zn enzyme (Table 3), which
is not surprising as the cobalt enzyme was isolated from *E. coli*
growing normally except for the initial lag period.   In fact, Zn-Zn
and Co-Co RNA polymerases are similar in many of their biochemical
properties: pH-activity profile, extrinsic metal requirements for
activity, inhibition patterns by metal chelators (EDTA, 1,10-phen-
anthroline, etc.), substrate affinities, and fidelity of transcription on synthetic template [15].   However, the apparent K_m values
for T7 DNA are twice as high for native Zn-Zn as for Co-Co polymerase.   Although the difference is not great, it does tend to imply
that the intrinsic metal may be involved in template binding.

Three different approaches have been used to study the effect
of Co(II) substitution into RNA polymerase on RNA chain initiation

[15]. First, the incorporation of γ-^{32}P-ATP or GTP into the 5'-
terminus of the RNA chain [28] was measured using T7 DNA as a
template. There are three major promoters (A_1, A_2, A_3) on T7 DNA,
two of which (A_1 and A_3) start RNA chains with ATP, while the other
(A_2) starts with GTP [37]. The finding that the Co-Co polymerase
initiates less efficiently with GTP than the Zn-Zn enzyme at the
A_2 promoter is indicative of a differential promoter selection due
to the Co(II) substitution. Second, the relative efficiency of the
formation of initiation complexes at a specific promoter on T7 DNA
was measured using various combinations of dinucleotide and nucleoside triphosphate [38]: CpC + ATP preferentially selected for promoter
A_1, CpG + CTP selected for promoter A_2, and ApC + ATP for promoter A_3.
The ratio of the measured efficiency for initiation at (A_1 + A_3) versus
A_2 is 3~4 for cobalt enzyme as compared to 1~2 for zinc enzyme, in
agreement with the results obtained in the first study. Finally, the
E. coli lactose operon system was used to compare the relative abilities of zinc and cobalt enzymes in the specific initiation of lac
mRNA, which is stimulated by the presence of the cAMP and cAMP receptor (CRP) [39]. The in vitro transcription by Co-Co polymerase of a
restriction fragment of λplac5 DNA (lac-796), which contains intact
lac operon [40,41], is less sensitive to cAMP and CRP than is transcription by native zinc enzyme. The results from these three studies
indicate that the intrinsic metal ions in RNA polymerase may be involved in promoter recognition and specific RNA chain initiation.

(b) Subunit Locations of the Intrinsic Metal Ions. The subunit location of the intrinsic metal ions has been studied by an
indirect approach [42] since direct demonstration by x-ray crystallography cannot be achieved at the present time due to the structural
complexity of RNA polymerase and the lack of suitable crystals. The
indirect approach involves the isolation of individual subunits of
RNA polymerase and determination of their zinc contents by atomic
absorption spectrometry. The regulatory subunit σ was first separated from the core polymerase ($\alpha_2\beta\beta'$) by chromatography of the
holoenzyme on a phosphocellulose column [25]. The core enzyme was

further separated in the presence of 7 M urea into either individual
subunits by phosphocellulose column chromatography [43] or the β'
subunit and $\alpha_2\beta$ complex by Affi-Gel Blue column chromatography [42].
No subunit or subunit complex binds zinc in the presence of urea.
After the removal of urea by the dialysis of the individual subunits
against reconstitution buffer, which contained 10^{-5} M zinc ions, some
subunits were capable of being reconstituted into active holoenzyme
and of tightly binding zinc. The zinc contents of the renatured
individual subunits are given in Table 4. In the native enzyme,
one zinc ion is located in the β' subunit while the other may be
located in either the β or β' subunit, or at the contact domain
between these two subunits. A more clear-cut result which was later
obtained from cobalt polymerase by in vitro substitution [29] (Sec.
4.1.1.(c)) indicates that the β and β' subunits each contain one
zinc ion. This is in agreement with the report from Kimball and his
co-workers [44].

A question may be raised regarding the issue that the isolated
subunits may assume conformations dissimilar from the native enzyme,

TABLE 4

Tightly Bound Zinc Ions in RNA
Polymerase and Its Subunits

Preparations	Zinc content[a] (g-atoms/mol protein)
Holoenzyme	2.1 ± 0.2
Core enzyme	2.0 ± 0.1
σ	<0.1
α	<0.1
β	0.6 ± 0.3
β'	$1.4 \pm 0.5 \ (1.5 \pm 0.3)$
$\alpha_2\beta$	$0.7 \pm 0.3 \ (0.6 \pm 0.2)$

[a]The numbers given in parentheses are zinc contents of
subunits or subunit complexes isolated by Affi-Gel Blue
chromatography. Zinc contents of subunits separated by
phosphocellulose chromatography are given without parentheses.

Source: Reprinted by permission of Ref. 42.

with different metal-binding properties. To confirm the validity
of the subunit location of the intrinsic metal in native enzyme
deduced from the binding of Zn(II) ions to individual subunits, the
intrinsic metal ions are frozen in their respective binding sites
in the native enzyme by the oxidation of Co-Co RNA polymerase with
peroxide [42]. The oxidation of Co(II) in the cobalt-substituted
enzymes transforms Co(II) from the exchange-labile to the exchange-
inert Co(III) state, thereby immobilizing the cobalt ions in their
binding sites (assuming that the location of metal on the enzyme is
not changed upon oxidation). That Co(II) in RNA polymerase is oxidized to Co(III) is indicated by the absorption spectrum of the
Co(III)-RNA polymerase which is distinct from that of Co(II)-enzyme
(Fig. 3). After the denaturation of Co(III)-enzyme and the separation of the subunits by Affi-Gel Blue chromatography, it was found
that one Co(III) ion still bound to the β' subunit in the presence
of 7 M urea. This result clearly indicated that one of the intrinsic
metal ions was located in the β' subunit in the native enzyme [42].

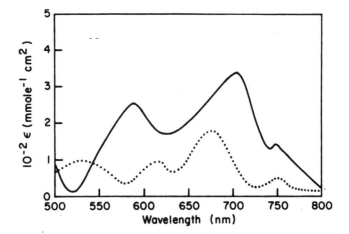

FIG. 3. Visible absorption spectra of Co-Co RNA polymerase in the
presence (•••) and absence (——) of hydrogen peroxide. (Reprinted
by permission from Ref. 42.)

   *(c) Selective Substitution in Vitro of One Zinc Ion with
Various Divalent Metal Ions.* In vitro metal substitution has recently been achieved [29] by a sequential denaturation-reconstitution
method in which purified core polymerase was partially denatured with
7 M urea to remove the intrinsic zinc ion. The denatured core enzyme
was then reconstituted by dialysis against reconstitution buffer containing 10^{-5} M Co(II), Mn(II), Ni(II), or Cu(II) ions. Analysis of
the metal contents of the reconstituted core enzymes by atomic absorption spectrometry indicated that only one of the two intrinsic zinc
ions had been replaced by the corresponding divalent metal ion. This
stoichiometry was confirmed by the radioactivity measurement of the
core polymerase reconstituted in the presence of ^{57}Co(II). Thus, by
in vitro substitution, various metal hybrid (Co-Zn, Mn-Zn, Ni-Zn, and
Cu-Zn) core polymerases were obtained. Metal hybrid holoenzymes
could be produced by the addition of isolated σ subunit to the metal
hybrid core enzymes.

   The question of which of the two zinc ions was substituted was
answered by oxidation of ^{57}Co(II)-Zn(II) core polymerase with H_2O_2
to secure the cobalt ion at its appropriate binding site in the
enzyme. The oxidized enzyme was denatured with 7 M urea and chromatographed on Affi-Gel Blue column to separate the β' subunit from
the $α_2β$ complex. The finding that the ^{57}Co counts coincided with
the $α_2β$ complex indicated that the zinc in the β subunit was replaced
by a cobalt ion.

   The reconstituted Co-Zn, Mn-Zn, Ni-Zn, and Cu-Zn holoenzymes
were found to possess 100, 100, 60, and 17% of the specific activity,
respectively, of the reconstituted Zn-Zn holoenzyme [29] using T7
DNA as template. The metal hybrid enzymes also decreased in the
same order (Zn > Co > Mn > Ni > Cu) in their abilities to catalyze
the synthesis of pppApU on the same template in the presence of only
ATP and UTP (abortive initiation [45]). These studies suggested
that the variation in the enzymatic activities of metal-substituted
polymerases resides at the level of RNA chain initiation. Kinetic
analysis indicated that the reduced activity of abortive initiation

by metal substitution was due primarily to the difference in the V_{max} of the enzyme rather than the K_m of the substrate [29]. In addition, all metal hybrid enzymes exhibited no significant difference in their abilities to incorporate noncomplementary or erroneous (deoxyribo)nucleotides into RNA product.

Replacement of zinc in the β subunit with cobalt or nickel resulted in the formation of intense absorption bands in the range 400-500 nm [29] due to charge-transfer transitions [46] (Fig. 4). The Co-Zn enzyme exhibits two major peaks at 400 nm (ε = 3000) and 475 nm (ε = 2700), and the Ni-Zn enzyme has a major peak at 462 nm (ε = 8000). A similar intense charge-transfer band has been reported for nickel-aspartate transcarbamylase [47]. The visible absorption spectra of Co-Zn or Ni-Zn RNA polymerase can be perturbed by the addition of purine nucleoside triphosphates in the absence of template and Mg(II) ions (Fig. 5). This is consistent with the contention that the substituted metal is located at the initiation site in the β subunit of enzyme [29].

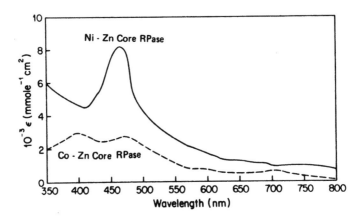

FIG. 4. Visible and near-ultraviolet absorption spectra of Co-Zn (---) and Ni-Zn(——) core RPases. (Reprinted by permission from Ref. 29.)

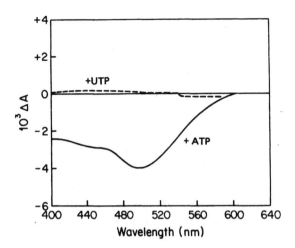

FIG. 5.   Effect of ATP and UTP on the visible spectrum of the Ni-Zn core RPase.   (Reprinted by permission from Ref. 29.)

   (d)   *Geometry of the Intrinsic Metal and the Nucleotide Bound at the Initiation Site.*   The perturbation of the visible absorption spectrum of Ni-Zn RNA polymerase by the addition of nucleoside tri-phosphates [29] suggests that the intrinsic metal ion in the β sub-unit may be involved in substrate binding at the initiation site. One may question whether this spectral change is due to a direct interaction of the metal with substrate or an indirect effect such as the conformational change of the enzyme induced by substrate binding at a site distant from the metal.   This question can be answered by the distance measurement between the metal and substrate bound at the initiation site of the enzyme.   Since Co(II) is para-magnetic and Co-RNA polymerase possesses visible absorption spectra [15,29], the distance between the intrinsic metal ion and the initi-ating substrate can be determined by NMR spectroscopy [48] or the fluorescence energy transfer technique [49].   The techniques have recently been applied to elucidate the spatial arrangement of the initiation site on *E. coli* RNA polymerase [36,50].

   NMR studies were performed with Co-Zn core RNA polymerase in the absence of template [36].   The paramagnetic effects of the Co-Zn polymerase on the relaxation rates of fast-exchanging water protons indicate that the cobalt ion was accessible to solvent. There are two fast-exchanging water protons in the inner coordination sphere of Co(II) ion. The addition of saturating concentration of ATP reduced this number to one suggesting that ATP replaces one H_2O directly coordinating to the intrinsic Co(II) ion. Initiator ApA (but not UTP) is also capable of replacing one water molecule coordinated to the cobalt ion (Fig. 6) [36]. The effects of ATP and ApA do not require the presence of DNA or Mg(II) ions, and their respective K_d values (0.15 and 0.075 mM) for RNA polymerase derived from NMR studies are in agreement with those obtained by equilibrium dialysis and fluorescence quenching techniques [36,50]. These results confirm the earlier conclusion [29] that the cobalt ion is at the initiation site on the β subunit.

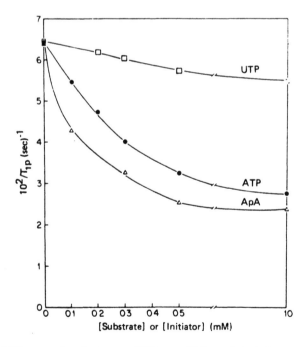

FIG. 6. Effects of substrate (UTP or ATP) and initiator (ApA) binding on the relaxivity of water protons at 80 MHz in the presence of Co-Zn core RPase. (Reprinted by permission from Ref. 36.)

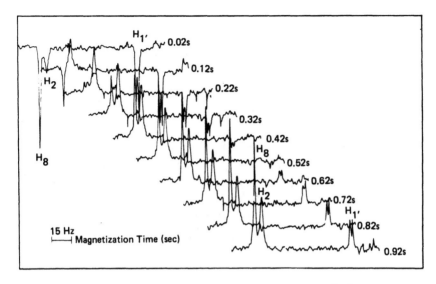

FIG. 7. Longitudinal magnetic relaxation spectra of the proton resonance of ATP in a complex with Co-Zn core RNA polymerase. (Reprinted by permission from Ref. 36.)

        A more definitive proof that substrate ATP is actually located at the inner coordination sphere comes from the distance measurements between cobalt and ATP [36]. The paramagnetic effect of Co-Zn core polymerase on the $1/T_{1p}$ of ATP (Fig. 7) can be used to estimate the distances from the intrinsic Co(II) ion to various proton and phosphorus nuclei of ATP on the enzyme from ^1H and ^{31}P NMR measurements. A model illustrating the spatial relationship between the intrinsic cobalt and the bound ATP is given in Fig. 8 [36]. The distances from cobalt to H_2 and H_8 are 3.6 and 4.1 Å, respectively. These short distances indicate the direct coordination of Co(II) to the base moiety of ATP, possibly through the unpaired electrons of N_7 and/or other nearby nitrogen atoms. The relative position of cobalt with respect to the base moiety suggests that the stacking of cobalt with the aromatic ring may form a π complex. Such a stacking interaction may constitute the structural basis for the recognition of initiating nucleotide by the enzyme. Furthermore, the triphosphate group of ATP is positioned more than 10 Å away from the metal center, suggesting

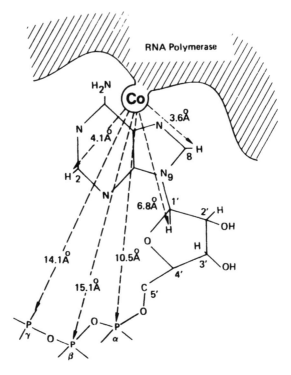

FIG. 8. Distances from the intrinsic Co(II) to various proton and phosphorus nuclei of ATP on Co-Zn RPase. (Reprinted by permission from Ref. 36.)

that it is not essential for recognition. This is consistent with the previous findings that AMP, ADP, UpA, or ApA can also bind to the initiation site to initiate RNA chains [51,52]. The distances from Co(II) to the sugar proton $H_{1'}$ and the three phosphorus nuclei are too long to allow direct coordinations of the metal ion to these atoms. These larger distances are appropriate for a second sphere complex in which a coordinated water molecule or other ligand may intervene between the metal and the substrate. Unfortunately, the distance from Co(II) to the 3'-OH cannot be determined since the strong HDO signal overshadows the $H_{2'}$ and $H_{3'}$ peaks of the sugar ring. As can be seen from the model, the distance from cobalt to the 3'-OH is more than 7 Å. Therefore, it appears unlikely that

the intrinsic metal on the β subunit directly participates in
catalysis by coordinating with the 3'-OH as postulated by Mildvan
and Loeb [5] (see below). An alternative hypothesis involves the
coordination of the intrinsic metal with 2'-OH [6], which may serve
as a basis for discrimination between 2'- and 3'-deoxyribonucleotide
by RNA polymerase. The validity of this hypothesis cannot be substantiated by these NMR studies [36] due to the unavailability of
data for the distance from cobalt to $H_{2'}$.

(e) *Structural and Mechanistic Studies of RNA Polymerase
using Intrinsic Metal as a Probe.* The intrinsic metal in hybrid
Co-Zn RNA polymerase can be used in NMR studies to investigate the
conformation of a nucleotide substrate bound at the initiation site
on the enzyme, in the presence of DNA template in a manner similar
to the study described above where template was absent. In addition,
the intrinsic metal can also be used as a probe to reveal other
structural information about the enzyme which may be relevant to
its function. For example, recently the spatial relationship between
the cobalt ion in the β subunit and a fluorescent probe covalently
attached to a specific site on the σ subunit has been studied by
energy transfer techniques [50]. The isolated σ subunit covalently
labeled with N-(1-pyrene)maleimide (PM-σ) at a specific cysteine
residue has a fluorescence spectrum which overlaps the absorption
spectrum of Co-Zn core RNA polymerase. When Co-Zn core enzyme was
added to an equimolar amount of isolated PM-σ, 50% and 39% quenching
of the PM-σ fluorescence was observed, respectively, in the absence
and presence of a short d(A-T)-copolymer template of 60 base pairs.
In contrast, the corresponding fluorescence quenching caused by
Zn-Zn core polymerase was 17% and 14%, in the absence and presence
of template. The difference in the extent of fluorescence quenching
was due to the excited-state energy transfer from the PM fluorophore
to the cobalt ion in the enzyme. From the measured energy transfer
efficiencies, the distance between the intrinsic metal ion in the β
subunit and the specific cysteine residue in the σ subunit was calculated to be 22 Å in the absence of DNA template. In the presence

of DNA template, this distance is increased to 33 Å, which in turn decreases to 29 Å on the addition of substrate ATP [50]. These distance changes are indicative of a template- or substrate-induced conformational change of RNA polymerase.

4.1.2. Bacillus subtilis *RNA Polymerase*

B. subtilis RNA polymerase resembles the *E. coli* enzyme with respect to molecular weight, subunit composition [53,54], inhibition by 1,10-phenanthroline, as well as containing 2 g-atoms zinc per mol enzyme (Table 2) [16]. One zinc ion is bound more tightly than the other since prolonged dialysis against EDTA removed 50% of the zinc from the enzyme. This is similar to what has been observed for the *E. coli* polymerase in that during in vitro metal substitution [29] one zinc ion is tightly bound and cannot be replaced by other divalent metal.

When the *B. subtilis* polymerase was denatured with 7 M urea and chromatographed on a Blue Dextran-Sepharose column, the "β" subunit (designated analogous to the nomenclature for *E. coli* RNA polymerase because it appears as the second largest polypeptide on SDS-polyacrylamide gel electrophoresis) was retained on the column [16]. It was found that this "β" subunit contains two zinc ions. However, the reconstitution studies performed later [55] showed that rifampicin resistance, which is functionally associated with the β subunit of *E. coli* RNA polymerase [56], is determined rather by the largest polypeptide of *B. subtilis* RNA polymerase. It can be suggested, then, that the current nomenclature for the subunits of *B. subtilis* RNA polymerase is misleading and that the designations should be based on function, rather than size, with the largest subunit termed β, and the subunit retained on the blue Dextran-Sapharose column should be called β'. Therefore, in contrast to *E. coli* RNA polymerase, the two intrinsic zinc ions in *B. subtilis* polymerase are located in the β' subunit.

4.2. Phage RNA polymerase

4.2.1. *T7 RNA Polymerase*

The RNA polymerase of bacteriophage T7 is a single polypeptide chain with a molecular weight of 107,000 [57]. Homogeneous T7 RNA polymerase contains 2-4 g-atoms zinc per mol enzyme. Inactive enzymes, which can be separated from the active molecules by repeated chromatography, contain a smaller amount of zinc ion (0.4-1 zinc ion per enzyme) [14]. Moreover, the purified enzyme is relatively unstable, with specific activity varying between different preparations. The zinc content of the purified polymerase increases concommitantly with the increasing specific activity, and the polymerase with low activity can be activated by exogenous zinc ions. This rough correlation between zinc content and specific activity implies a functional role for the intrinsic zinc in T7 RNA polymerase. For T7 RNA polymerase, the inhibition by 1,10-phenanthroline, which correlates well with the loss of zinc from the enzyme, is not caused by the inhibitory phenanthroline/cuprous ion complex [11,12], since the enzyme is also inhibited in a time-dependent manner by other metal chelators (e.g., EDTA, CN^-, SH^-, and N_3^-) [14].

4.2.2. *T3 RNA Polymerase*

Bacteriophage T3 RNA polymerase is also a single polypeptide chain [58] with a molecular weight similar to that of the T7 enzyme. This enzyme has recently been characterized as a zinc metalloenzyme containing 1 g-atom zinc per mol enzyme [59]. Unlike the T7 enzyme, the intrinsic zinc ion of T3 RNA polymerase is tightly bound and cannot be removed by dialysis against 5 mM 1,10-phenanthroline at 4°C for a week. However, the enzyme can be inhibited by 1 mM 1,10-phenanthroline, due to the formation of phenanthroline/cuprous ion complex [11,12]. Little inhibition by OP is observed if the reaction is carried out in the presence of some cuprous ion specific chelating agents (e.g., 2,2',2''-terpyridine or 2,9-dimethyl-1,10-phenanthroline).

4.3. Eukaryotic RNA Polymerase

Few advances have been made in the studies of the intrinsic metal
ions in eukaryotic RNA polymerase. One major difficulty is to obtain
sufficient quantities of eukaryotic polymerases in highly purified
form. Thus far, eukaryotic RNA polymerases from three sources have
been established to be zinc metalloenzymes: *Euglena gracilis* [17,60],
yeast [18,19,61,62], and wheat germ [20], and it has also been suggested that RNA polymerases from rat liver and sea urchin [63] contain zinc based on 1,10-phenanthroline inhibition, but without the
direct determination of the metal content.

4.3.1. E. gracilis *RNA Polymerase*

Vallee and his co-workers [17,60] observed derangements in RNA
metabolism due to zinc deficiency, and subsequently, two RNA polymerases from *E. gracilis* were purified and shown to be zinc metalloenzymes [17,60].

E. *gracilis* RNA polymerase I (α-amanitin-insensitive) [17] and
II (α-amanitin sensitive) [60] are multisubunit enzymes (10 and 5
subunits, respectively) which contain 2.2 g-atoms zinc per mol enzyme
[17,60] as determined by microwave-induced emission spectrometry [64,
65]. The subunit locations of the intrinsic zinc ions in *E. gracilis*
RNA polymerase have not as yet been determined. Inhibition of polymerase II by 1,10-phenanthroline is instantaneous and fully reversible by dilution [60]. In addition to OP, the *E. gracilis* enzymes
are also inhibited by other chelating agents, such as 8-hydroxy-
quinoline-5-sulfonic acid and 8-hydroxyquinoline, but nonchelating
analogues of OP, 1,7- and 4,7-phenanthroline, have no effect. Thus
it appears that the inhibition of polymerase is due to chelation of
the intrinsic zinc ion in the enzyme. The demonstration that RNA
polymerase I and II from *E. gracilis* are zinc metalloenzymes has
extended the role of zinc in gene expression in addition to its
role in cell metabolism in eukaryotic cells.

4.3.2. Yeast RNA Polymerase

Yeast nuclear RNA polymerases I, II, and III were also found to be
zinc metalloenzymes possessing 2.4 [18], 0.98 [19], and 2 [62]
g-atoms zinc per mol enzyme, and all three of the enzymes can be
inhibited by 1,10-phenanthroline. Addition of exogenous Zn(II)
ions can partially reverse the inhibition of polymerase II by 1,10-
phenanthroline. Treatment of polymerase II with this inhibitor
leads to a loss of 75% of its zinc content, with a parallel decrease
in enzyme activity [66]. In contrast, the intrinsic zinc ions in
polymerase III cannot be removed by chelating agents or exchanged
with exogenous metal ions [62].

4.3.3. Wheat Germ RNA Polymerase

Wheat germ is a rich source for the amanitin-sensitive RNA polymerase
II [67-69], which has a molecular weight of 550,000 and consists of
10 polypeptide chains [68,69]. This polymerase contains 7.0 ± 0.7
g-atoms of tightly bound zinc per mol enzyme as determined by atomic
absorption spectrometry [20]. Chromatography of purified polymerase
on a number of ion-exchange columns or prolonged dialysis against
low-zinc buffers does not affect the zinc content. Surprisingly,
as opposed to all the previous studies discussed in this chapter,
little immediate inhibition of enzyme activity by 1,10-phenanthroline
is observed.

As is evident from Table 2, wheat germ RNA polymerase II contains more intrinsic zinc ions than all other RNA polymerases reported
so far. Although the function of these bound zinc ions is unknown,
one may speculate that they may be responsible for the unusual stability of this polymerase.

5. POSSIBLE ROLE OF INTRINSIC ZINC

The precise roles of the intrinsic metal ions in all DNA and RNA
polymerases have not been established. Nevertheless, the presence

of multiple zinc ions in many of the polymerases implies the possibility of multiple functions. As may be evinced from the material
presented here, the intrinsic metal ions play at least three possible
roles in the function of the polymerases: (1) a catalytic role in the
binding of substrate, primer, or template; (2) a regulatory role in
the specificity of gene replication and transcription; and (3) a
structural role in the maintenance of the proper conformation of
the polymerase.

5.1. Catalytic Role

The catalytic function of the intrinsic zinc ion has been suggested
for many DNA and RNA polymerases based simply on the inhibition by
1,10-phenanthroline, studies that may require a reexamination, as
discussed earlier. It should be noted, however, that more detailed
studies have been carried out with $E. coli$ DNA polymerase I, T7 RNA
polymerase, and $E. coli$ RNA polymerase, with greater characterization
of the role of the intrinsic metal ions.

A possible catalytic role for the intrinsic zinc ion in the
mechanism of DNA polymerase I has been proposed in which the zinc
ion acts as a Lewis acid to coordinate with the oxygen atom of the
3'-OH group of the DNA primer, thereby facilitating its deprotonation,
a necessary step for nucleophilic attack on the α-phosphorus of the
incoming nucleotide (Fig. 1). However, this hypothesis should not
be considered conclusive because it is based on indirect competition
studies which suggest that the intrinsic zinc ion in DNA polymerase
interacts with the DNA primer.

In the case for T7 RNA polymerase, there is a rough correlation
between the zinc content and the specific activity of various enzyme
preparations. Although some preparations of the enzyme with lower
activities can be activated by addition of exogenous zinc ions,
other possible effects of the exogenous zinc ion should also be
taken into consideration. For example, the zinc ion may affect the

activity by interacting with the sulfhydryl groups in T7 RNA polymerase. Furthermore, zinc ions have been known to assist the reversible unwinding and rewinding of native double-stranded DNA [70],
which may alter the polymerase activity.

The strongest evidence that the intrinsic zinc ions in *E. coli*
RNA polymerase are required for enzymatic activity comes from the
finding that the addition of zinc is necessary to reconstitute active
enzyme from urea-denatured inactive apoenzyme [71]. In addition, the
participation of the intrinsic metal ions in substrate or template
binding are evident from the characteristic spectral changes of Co-Co
[15], Co-Zn, and Ni-Zn [29] enzymes induced by nucleotides or DNA
template (Figs. 2 and 4). NMR studies of Co-Zn RNA polymerase and
substrate ATP [36] have confirmed that the cobalt ion is located at
the initiation site in the β subunit of the polymerase and that ATP
is in direct coordination with the intrinsic metal ion. Furthermore,
the distance between the cobalt and various atoms of ATP bound at
the initiation site indicates that the intrinsic metal ion in the
β subunit plays a recognition role in discriminating the initiating
nucleotide and may orient the nucleotide in a stereospecific position
suitable for catalysis.

For eukaryotic RNA polymerases, this catalytic function of the
intrinsic metal ions is again implicated mainly from inhibition
studies with 1,10-phenanthroline [66]. The binding of DNA to yeast
RNA polymerase II or the formation of the enzyme-DNA-nucleotide complex was not reduced significantly by 1,10-phenanthroline [67].
However, this chelator inhibits DNA-dependent pyrophosphate exchange
to the same extent as the overall RNA synthesis reaction. On the
assumption that the inhibition is due to the chelation of the inhibitor with the intrinsic metal, these results imply that enzyme bound
zinc is involved in the transfer of the entering nucleotide to the
3'-terminus of the growing RNA chains.

A mechanism similar to that described above has previously
been proposed for terminal deoxynucleotidyl transferase [22] and
DNA polymerase I [3]. In this mechanism, the intrinsic zinc ion of

DNA polymerase may directly participate in catalysis by its interaction with the 3'-OH of the primer [23] (Fig. 1). Because RNA polymerase does not require a primer for initiation, a similar hypothesis
for RNA polymerase would require that the intrinsic zinc ions be
directly coordinated to the 3'-OH group of the initiating nucleotide
or the 3'-terminal nucleotide of the growing RNA chain. Such a
hypothesis may be ruled out for E. coli RNA polymerase, since the
distance between the intrinsic metal and 3'-OH of ATP bound at the
initiation site is more than 7 Å away (Fig. 8). Other hypotheses
for a catalytic role of the intrinsic zinc ion involve coordination
of zinc to 2'-OH of sugar moiety or the oxygen atoms of phosphate
group of the initiating nucleotide [6]. All of these are possibilities that require further investigation before they can be verified.

5.2. Regulatory Role

Comparative studies of E. coli Co-Co and Zn-Zn RNA polymerase from
the observations that these two enzymes have different efficiencies
in utilizing the A_2 promoter versus the A_1 + A_3 promoters of T7 DNA
[15] suggest a regulatory role for the intrinsic metal ion. In addition, the in vitro transcription of the lac operon by Co-polymerase
is less sensitive to cAMP and CRP than is the transcription by zinc
enzyme [15]. These results suggest that the intrinsic metal may
play a role in promoter recognition and specific initiation. The
physical basis for these observations may involve the stabilization
of the DNA-enzyme complex or a conformational change of the DNA
mediated through the intrinsic metal.

Two mutant E. coli RNA polymerases with temperature-sensitive
mutations in the β' subunit have been reported to alter their properties in promoter binding, with one decreasing the ability of DNA
binding [72] and the other unable to melt out a promoter site [73].
The β' subunit has also been implicated in the kinetic facilitation
of promoter search [74]. Further studies of intrinsic zinc ions in
these mutant RNA polymerases may shed some light on the role of
intrinsic metal in template binding and promoter selection.

5.3. Structural Role

The intrinsic zinc ions in DNA and RNA polymerases can affect the
enzymatic activity indirectly by playing a structural role in maintaining proper enzyme conformation. The removal and restoration of
the intrinsic metal ions has proven much more difficult for a multisubunit polymerase than a single-chain enzyme, suggesting that some
of the metal ions in the multisubunit enzymes may affect the quaternary structure of the polymerase.

The structural role of metal ions may be more important in
eukaryotic RNA polymerases, which contain a larger number of subunits than do prokaryotic enzymes, where for example, wheat germ
RNA polymerase II has 7 zinc ions per mol enzyme. It is probable
that most of these zinc ions will have a structural rather than
catalytic role.

6. CONCLUSIONS AND PERSPECTIVE

The universal presence of zinc as an integral part of RNA polymerase
has been well established. Although the precise functions of these
metal ions are still unknown, the three possible roles that the
intrinsic metal ions may play have been derived from studies with
polymerases in which one or more of the intrinsic zinc ions is
replaced by other divalent metals. These types of studies have been
successfully performed on other metalloenzymes, such as horse-liver
alcohol dehydrogenase [75-79] and E. coli alkaline phosphatase [80-
82]. For E. coli RNA polymerase, NMR studies of the selective substitution of the zinc ion in the β subunit with paramagnetic cobalt
has led to the revelation that this intrinsic metal is directly
coordinated to the nucleotide substrate bound at the initiation
site of the enzyme. Furthermore, the paramagnetic metal ions can
also be used to probe the conformation of substrate and the spatial
relationship between the metal binding site and other active sites
on the enzyme by NMR, EPR, and fluorescence energy-transfer

techniques. This information is crucial for the delineation of a detailed catalytic mechanism of DNA and RNA polymerases.

Zinc is essential for normal growth of living species; it is present in all living tissues and appears necessary for the synthesis of nucleic acids [83-85] and proteins [86-88]. It also plays a role in maintenance of the conformation of nucleic acid [70,89-92] and stabilization of the secondary and tertiary structure of RNA [93]. Moreover, zinc deficiency results in growth arrest and major abnormalities of composition and function of the cell [94]. In zinc-deficient cells, both biochemical and morphological processes of the cell cycle are markedly disturbed [95,96]. Undoubtedly, mechanistic studies of the intrinsic metal ions in DNA and RNA polymerase should lead to a better understanding of the regulation and mechanism of gene expression as well as the role of zinc in the cell growth, division, and differentiation.

ACKNOWLEDGMENTS

This study was supported in part by research grants from the National Institutes of Health (GM 28057-02) and the National Science Foundation (PCM 8119326) to F. Y.-H. Wu, and N.I.H. (GM 28069-02) and the American Cancer Society (NP 309G) to C.-W. Wu.

ABBREVIATIONS

ADP	adenosine 5'-diphosphate
Affi-Gel Blue	Cibacron blue dye covalently cross-linked to agarose
AMP	adenosine 5'-monophosphate
ApA	adenylyl-(3'-5')-adenosine
ApC	adenylyl-(3'-5')-cytidine
ATP	adenosine 5'-triphosphate
cAMP	cyclic adenosine 3'-5'-monophosphate

Co-Zn, Mn-Zn, Ni-Zn, or Cu-Zn RNA polymerase	RNA polymerase contains one intrinsic Zn ion and one intrinsic Co, Mn, Ni, or Cu ion, respectively
CpC	cytidylyl-(3'-5')-cytidine
CpG	cytidylyl-(3'-5')-guanosine
CRP	cAMP receptor
$d(pT)_6$ and $d(pT)_{10}$	oligo-(deoxythymidylic acid) with 6 and 10 deoxythymidylic acid residues
dTTP	deoxythymidine 5'-triphosphate
EDTA	ethylenediaminetetraacetic acid
EPR	electron paramagnetic resonance
GTP	guanosine 5'-triphosphate
lac-796	the 796 base pair, Hind-II and III restriction fragment of λplac5 DNA
NMR	nuclear magnetic resonance
NTP	nucleoside triphosphate
OP	ortho- or 1,10-phenanthroline
poly[d(A-T)]	alternating copolymer of deoxyadenylate·deoxythymidylate
PuTP	purine nucleoside triphosphate
$(rI)_n$	polyriboinosinic acid
RPase	RNA polymerase
SDS	sodium dodecyl sulfate
Tris	tris(hydroxymethyl)aminomethane
UpA	uridylyl-(3'-5')-adenosine
UTP	uridine 5'-triphosphate
Zn-Zn or Co-Co RNA polymerase	RNA polymerase contains two intrinsic Zn or Co ions, respectively

REFERENCES

1. J. P. Slater, A. S. Mildvan, and L. A. Loeb, *Biochem. Biophys. Res. Commun.*, *44*, 27 (1971).

2. M. C. Scrutton, C.-W. Wu, and D. A. Goldthwait, *Proc. Natl. Acad. Sci. USA*, *68*, 2497 (1971).

3. C. F. Springgate, A. S. Mildvan, R. Abramson, J. L. Engle, and L. A. Loeb, *J. Biol. Chem.*, *249*, 5987 (1973).

4. A. S. Mildvan and L. A. Loeb, in *CRC Critical Reviews in Biochemistry* (G. D. Fasman, ed.), CRC Press, Boca Raton, Fla., 1979, p. 219.

5. A. S. Mildvan and L. A. Loeb, in *Advances in Inorganic Biochemistry*, Vol. 3: *Metal Ions in Genetic Information Transfer* (G. L. Eichhorn and L. G. Marzilli, eds.), Elsevier, New York, 1981, p. 103.

6. F. Y.-H. Wu and C.-W. Wu, in *Advances in Inorganic Biochemistry*, Vol. 3: *Metal Ions in Genetic Information Transfer* (G. L. Eichhorn and L. G. Marzilli, eds.), Elsevier, New York, 1981, p. 143.

7. B. J. Poiesz, G. Seal, and L. A. Loeb, *Proc. Natl. Acad. Sci. USA*, *71*, 4892 (1974).

8. B. J. Poiesz, J. Battula, and L. A. Loeb, *Biochem. Biophys. Res. Commun.*, *56*, 959 (1974).

9. D. S. Auld, H. Kawaguchi, D. M. Livingston, and B. L. Vallee, *Proc. Natl. Acad. Sci. USA*, *71*, 2091 (1974).

10. D. S. Auld, H. Kawaguchi, D. M. Livingston, and B. L. Vallee, *Biochem. Biophys. Res. Commun.*, *62*, 296 (1975).

11. V. D'Aurora, A. M. Stern, and D. S. Sigman, *Biochem. Biophys. Res. Commun.*, *78*, 170 (1977).

12. V. D'Aurora, A. M. Stern, and D. S. Sigman, *Biochem. Biophys. Res. Commun.*, *80*, 1025 (1978).

13. B. L. Vallee, in *Enzymology in the Practice of Laboratory Medicine* (P. Blume and E. F. Freier, eds.), Academic Press, New York, 1974, p. 95.

14. J. E. Coleman, *Biochem. Biophys. Res. Commun.*, *60*, 641 (1974).

15. D. C. Speckhard, F. Y.-H. Wu, and C.-W. Wu, *Biochemistry*, *16*, 5228 (1977).

16. S. M. Halling, F. J. Sanchez-Anzaldo, R. Fukuda, R. H. Doi, and C. F. Meares, *Biochemistry*, *16*, 2880 (1977).

17. K. H. Falchuk, L. Ulpino, B. Mazus, and B. L. Vallee, *Biochem. Biophys. Res. Commun.*, *74*, 1206 (1977).

18. D. S. Auld, I. Atsuya, C. Campino, and P. Valenzuela, *Biochem. Biophys. Res. Commun.*, *69*, 548 (1976).

19. H. Lattke and U. Weser, *FEBS Lett.*, *65*, 288 (1976).

20. P. Petranyi, J. J. Jendrisak, and R. R. Burgess, *Biochem. Biophys. Res. Commun.*, *74*, 1031 (1977).

21. S. Lindskog, *Struct. Bonding (Berl.)*, *8*, 153 (1970).

22. L. M. S. Chang and F. J. Bollum, *Proc. Natl. Acad. Sci. USA*, *65*, 1041 (1970).

23. M. M. Abboud, W. J. Sim, L. A. Loeb, and A. S. Mildvan, *J. Biol. Chem.*, *253*, 3415 (1978).

24. R. R. Davis, in *The Enzymes*, 2nd ed., Vol. 5 (P. D. Boyer, ed.), Academic Press, New York, 1961, p. 545.

25. R. R. Burgess, *J. Biol. Chem.*, *244*, 6168 (1969).

26. C.-W. Wu and D. A. Goldthwait, *Biochemistry*, *8*, 4450 (1969).

27. C.-W. Wu and D. A. Goldthwait, *Biochemistry*, *8*, 4458 (1969).

28. U. Maitra and J. Hurwitz, *Proc. Natl. Acad. Sci. USA*, *54*, 815 (1965).

29. D. Chatterji and F. Y.-H. Wu, *Biochemistry*, *21*, 4651 (1982).

30. F. Y.-H. Wu and C.-W. Wu, *Biochemistry*, *12*, 4343 (1973).

31. R. R. Burgess and J. J. Jendrisak, *Biochemistry*, *14*, 4634 (1975).

32. J. E. Coleman and B. L. Vallee, *J. Biol. Chem.*, *235*, 390 (1960).

33. M. L. Applebury and J. E. Coleman, *J. Biol. Chem.*, *244*, 308 (1969).

34. A. Curdel and M. Iwatsubo, *FEBS Lett.*, *1*, 133 (1968).

35. J. D. Shore and D. Santiago, *J. Biol. Chem.*, *250*, 2008 (1975).

36. D. Chatterji and F. Y.-H. Wu, *Biochemistry*, *21*, 4657 (1982).

37. D. Pribnow, *Proc. Natl. Acad. Sci. USA*, *72*, 784 (1975).

38. J.-P. Dausse, A. Sentenac, and P. Fromageot, *Eur. J. Biochem.*, *57*, 578 (1975).

39. I. Pastan and R. Perlman, *Science*, *169*, 339 (1970).

40. A. Landy, E. Olchowski, and W. Ross, *Mol. Gen. Genet.*, *133*, 273 (1974).

41. W. Gilbert, J. Gralla, J. Majors, and A. Maxam, in *Protein-Ligand Interactions* (H. Sund and G. Blaues, eds.), Walter de Gruyter, Berlin, 1975, p. 353.

42. C.-W. Wu, F. Y.-H. Wu, and D. C. Speckhard, *Biochemistry*, *16*, 5449 (1977).

43. L. R. Yarbrough and J. Hurwitz, *J. Biol. Chem.*, *249*, 5400 (1974).

44. J. A. Miller, G. F. Serio, R. A. Howard, J. L. Bear, J. E. Evans, and A. P. Kimball, *Biochim. Biophys. Acta*, *579*, 291 (1979).

45. D. E. Johnston and W. R. McClure, in *RNA Polymerase* (R. Losick and M. Chamberlain, eds.), Cold Spring Harbor Laboratory, Cold Spring Harbor, N.Y., 1976, p. 413.

46. P. Day and C. K. Jorgensen, *J. Chem. Soc.*, *6226* (1964).

47. R. Johnson and H. K. Schachman, *Proc. Natl. Acad. Sci. USA*, *77*, 1995 (1980).

48. A. S. Mildvan and M. Cohn, *Adv. Enzymol.*, *33*, 1 (1970).

49. T. Förster, *Ann. Phys.*, *2*, 55 (1947).

50. D. Chatterji and F. Y.-H. Wu, *Biochemistry*, in preparation (1982).

51. K. M. Downey and A. G. So, *Biochemistry*, *9*, 2520 (1970).

52. K. M. Downey, B. S. Furmak, and A. G. So, *Biochemistry*, *10*, 4970 (1971).

53. J. Arila, J. M. Hermoso, E. Vinuela, and M. Salas, *Eur. J. Biochem.*, *21*, 526 (1971).

54. R. G. Shorenstein and R. Losick, *J. Biol. Chem.*, *248*, 6163 (1977).

55. S. M. Halling, K. C. Burtis, and R. H. Doi, *J. Biol. Chem.*, *252*, 9024 (1977).

56. W. Zillig, P. Palm, and A. Heil, in *RNA Polymerase* (R. Losick and M. Chamberlin, eds.), Cold Spring Harbor Laboratory, Cold Spring Harbor, N.Y., 1976, p. 101.

57. M. Chamberlin and J. Ring, *J. Biol. Chem.*, *248*, 2235 (1973).

58. P. R. Chakraborty, P. Sarkar, H. Huang, and U. Maitra, *J. Biol. Chem.*, *248*, 6637 (1973).

59. F. Y.-H. Wu and K. Gruber, manuscript in preparation.

60. K. H. Falchuk, B. Mazus, L. Ulpino, and B. L. Vallee, *Biochemistry*, *15*, 4468 (1976).

61. T. M. Wandzilak and R. W. Benson, *Biochem. Biophys. Res. Commun.*, *76*, 247 (1977).

62. T. M. Wandzilak and R. W. Benson, *Biochemistry*, *17*, 426 (1978).

63. P. Valenzuela, R. W. Morris, A. Faras, W. Levinson, and W. J. Rutter, *Biochem. Biophys. Res. Commun.*, *53*, 1036 (1973).

64. H. Kawaguchi and B. L. Vallee, *Anal. Chem.*, *47*, 1029 (1975).

65. H. Kawaguchi and D. S. Auld, *Clin. Chem.*, *21*, 591 (1975).

66. H. Lattke and U. Weser, *FEBS Lett.*, *83*, 297 (1977).

67. J. J. Jendrisak and W. M. Becker, *Biochim. Biophys. Acta*, *319*, 48 (1973).

68. J. J. Jendrisak and R. R. Burgess, *Biochemistry*, *14*, 4639 (1975).

69. J. J. Jendrisak, P. W. Petranyi, and R. R. Burgess, in *RNA Polymerase* (R. Losick and M. Chamberlin, eds.), Cold Spring Harbor Laboratory, Cold Spring Harbor, N.Y., 1976, p. 779.

70. Y. A. Shin and G. L. Eichhorn, *Biochemistry*, *7*, 1026 (1968).

71. D. Solaiman and F. Y.-H. Wu, unpublished observations.

72. R. Panny, A. Heil, B. Mazus, P. Palm, W. Zillig, S. Z. Mindlin, T. S. Ilyina, and R. S. Khesin, *FEBS Lett.*, *48*, 241 (1974).

73. G. Gross, D. A. Fields, and E. K. F. Bautz, *Mol. Gen. Genet.*, *147*, 337 (1976).

74. Z. Hillel and C.-W. Wu, *Biochemistry*, *17*, 2954 (1978).

75. D. E. Drum and B. L. Vallee, *Biochem. Biophys. Res. Commun.*, *41*, 33 (1970).

76. B. L. Vallee and W. E. C. Wacker, in *The Proteins: Composition, Structure, and Function*, 2nd ed., Vol. 5 (Hans Neurath, ed.), Academic Press, New York, 1970.

77. J. M. Young and J. H. Wang, *J. Biol. Chem.*, *246*, 2815 (1971).

78. M. Takahashi and R. A. Harvey, *Biochemistry*, *12*, 4743 (1973).

79. D. L. Sloan, J. M. Young, and A. S. Mildvan, *Biochemistry*, *14*, 1998 (1975).

80. R. T. Simpson and B. L. Vallee, *Biochemistry*, *7*, 4343 (1968).

81. D. J. Plocke and B. L. Vallee, *Biochemistry*, *1*, 1039 (1962).

82. R. A. Anderson and B. L. Vallee, *Proc. Natl. Acad. Sci. USA*, *72*, 394 (1975).

83. M. Fujioka and I. Lieberman, *J. Biol. Chem.*, *239*, 1164 (1964).

84. R. B. Williams, C. F. Mills, J. Quarterman, and A. C. Dalgarno, *Biochem. J.*, *95*, 290 (1965).

85. H. H. Sandstead and R. A. Rinaldi, *J. Cell. Physiol.*, *73*, 81 (1969).

86. R. C. Theuer and W. G. Hoekstra, *J. Nutr.*, *89*, 448 (1966).

87. J. M. Hsu, W. L. Anthony, and R. J. Buchanan, *J. Nutr.*, *99*, 425 (1969).

88. J. M. Hsu and W. L. Anthony, *J. Nutr.*, *100*, 1189 (1970).

89. W. E. C. Wacker and B. L. Vallee, *J. Biol. Chem.*, *234*, 3257 (1959).

90. H. Altmann, F. Fetter, and K. Kaindl, *Z. Naturforsch.*, *234*, 395 (1968).

91. M. Tal, *Biochim. Biophys. Acta*, *195*, 76 (1969).

92. G. L. Eichhorn, N. A. Berger, J. J. Butzow, P. Clark, J. Heim, J. Pitha, C. Richardson, J. M. Rifkind, Y. Shin, and E. Tarien, *Adv. Exp. Med. Biol.*, *40*, 43 (1973).

93. K. Fuwa, W. E. C. Wacker, R. Druyan, A. F. Bartholomay, and B. L. Vallee, *Proc. Natl. Acad. Sci. USA*, *46*, 1298 (1960).

94. B. L. Vallee, in *Biological Aspects of Inorganic Chemistry* (A. W. Addison, W. R. Cullen, D. Dolphin, and B. R. James, eds.), Wiley, New York, 1976, p. 37.

95. W. E. C. Wacker, *Biochemistry*, *1*, 859 (1962).

96. K. H. Falchuk, D. Fawcett, and B. L. Vallee, *J. Cell Sci.*, *17*, 57 (1975).

Chapter 5

THE ROLE OF ZINC IN SNAKE TOXINS

Anthony T. Tu
Department of Biochemistry
Colorado State University
Fort Collins, Colorado

1. INTRODUCTION

Venom is a complex mixture of proteins with different biological
activities. Both composition and activities of venoms vary among
the species. Usually, the closer the phylogenetic relationship of
the snakes, the more similar are the venom properties and composition. Of the nearly 2000 different types of snakes that exist,
about 300 are known to be venomous. These are classified according
to morphological characteristics and comprise five families:

Crotalidae (crotalids, pit vipers), Viperidae (vipers, viperids),
Elapidae (elapids), Hydrophiidae (hydrophids, sea snakes), and
Colubridae (colubrids, rear fang snakes). Venoms of Elapidae and
Hydrophiidae have been extensively studied, but few studies have
been made on the venoms of Colubridae. The fangs of Colubridae are
located in the rear of the mouth rather than in the front; thus it
is difficult to extract venom from the snakes of this family. Consequently, we know little about the venoms of Colubridae. Most
venoms contain more than one toxic principle which tends to act
synergistically in an actual poisoning. Some snake venoms, especially those of Elapidae and Hydrophiidae (sea snakes), contain
potent neurotoxins which are major lethal toxins. Cardiotoxins are
also powerful lethal toxins found in the venoms of Elapidae. Snake
venoms also possess a variety of enzymes. Crotalidae venoms contain
a greater array of biologically active components than the venoms of
other snakes and have a more complex spectrum of pharmacological and
enzymatic activities [1].

About 90% of the dry weight of snake venom is protein. The
rest is composed of inorganic compounds, free amino acids, nucleo-
sides, carbohydrates, biogenic amines, lipids, and small peptides.
Because snake venom is a mixture of different proteins, it exhibits
a variety of pharmacological activities. Some protein components
are highly lethal neurotoxins (pre- and postsynaptic types), cardiotoxins, cytotoxins, hemorrhagic toxins, and myotoxins. Some components are less toxic and include hemolytic factors, proteins
affecting the blood coagulation system (anti- and procoagulants),
bradykinin releasing enzyme, histamine releasing enzyme, and cobra
venom factor (anticomplementary factor). Enzymes reported in snake
venoms are phospholipase A_2, phosphodiesterases, 5'-nucleotidase,
nonspecific phosphomonoesterase (phosphatases), acetylcholinesterase,
proteolytic enzymes (specific and nonspecific types), arginine
esterase, hyaluronidase, NAD nucleosidase (NAD glycohydrolase), and
L-amino acid oxidase. Snake venom also contains some enzyme inhibitors (bradykinin potentiating factor). There are many additional

biologically active proteins, many of which have not been identified
or isolated as yet.

2. METALS IN SNAKE VENOMS

2.1. Crude Venoms

Analyses of metals indicate that crude venoms contain a large amount
of metals. They are especially rich in the sodium ion [2-9]. On
extensive dialysis against distilled water, most of the metals are
removed; however, even after this treatment, some metal ions still
remain in venoms (Table 1). The major constituents of snake venoms
are proteins which are charged macromolecules. Thus it is natural
that snake venoms contain various cations and anions to neutralize
the charges. In other words, metal ions are present to serve as
counterions. Probably monovalent metal ions serve this purpose.
Actually, sodium is present in any venom in far greater quantity
than other cations. Potassium ions can be detected but in significantly less quantity than sodium ions. Investigations to date indicate no particular contribution from monovalent metal ions to the
biological activities of snake venoms.

2.2. Activation and Inhibition of Enzyme Activities

Another purpose of metal ions, especially divalent cations in snake
venom, is their requirement as cofactors for many different enzymatic
and biological activities. Phospholipase A_2 action requires Ca(II),
and, to a lesser extent, can also be activated with Mg(II) but inhibited with Cd(II) and Zn(II) (Table 2) [10]. 5'-Nucleotidase and
L-amino acid oxidase activities can be activated with Mg(II). When
Hg(II) is added to a mixture of phospholipase A_2 inhibitor and phospholipase A_2, an apparent enzyme activation occurs. Phospholipase
A_2 itself is not affected by the presence of Hg(II), but Hg(II)

TABLE 1

Metal Contents of Snake Venoms Before and After Dialysis
Analyzed by Atomic Absorption (μg Metal/g Venom)

Venom (origin)	Hr^a	Ca	Zn	Mg	Na	K	Cu	Mn	Fe	Other metals[b]
Crotalidae										
A. acutus (Formosa)	0	3,000	1,200	450	36,977	1,070	175	0	0	0
	48	2,668	522	409	12,780	965	42	0	0	
C. atrox (U.S.A.)	0	4,196	1,394	701	57,300	410	0	0	0	0
	48	3,780	1,093	344	24,600	320				
C. adamanteus (U.S.A.)	0	1,610	773	107	42,300	750	0	0	0	0
	48	1,604	452	97	8,400	750				
C. basciliscus (Mexico)	0	1,989	1,400	376	16,800	670	0	0	0	0
	48	1,990	990	310	10,200	638				
C. durissus (Central America)	0	3,003	1,203	1,470	36,700	13,500	0	0	0	0
	48	2,968	700	775	12,800	3,970				
C. durissus terrificus (South America)	0	2,390	1,856	342	45,700	1,660	0	0	0	0
	48	2,280	1,380	204	1,780	1,440				
C. durissus totonacus (Mexico)	0	1,633	840	117	28,800	590	0	0	0	0
	48	1,590	680	100	1,500	550				
C. horridus horridus (U.S.A.)	0	4,930	980	973	53,000	420	0	0	0	0
	48	3,629	800	406	21,900	400				

	Time[a]									
C. horridus atricaudatus (U.S.A.)	0	150	680	129	49,900	350	0	0	0	0
	48	97	657	91	10,010	240	0	0	0	0
C. viridis viridis (U.S.A.)	0	4,560	1,847	240	26,400	710	0	0	0	0
	48	2,730	1,050	209	1,200	600	0	0	0	0
S. milarius barbouri (U.S.A.)	0	4,000	2,010	446	39,500	2,540	200	0	0	0
	48	2,750	1,525	297	1,550	2,159	90	0	0	0
Viperdae										
B. arietans (South Africa)	0	2,306	1,000	700	41,500	500	0	500	0	0
	48	1,200	846	274	500	439	0	52	0	0
B. gabonica (South Africa)	0	2,900	690	636	36,400	220	0	0	0	0
	48	1,080	680	277	750	220	0	0	0	0
V. russelli siamensis (Thailand)	0	1,987	1,800	976	34,100	760	0	0	0	0
	48	1,306	809	306	654	310	0	0	0	0
Elapidae										
N. naja (India)	0	1,000	1,600	840	60,200	150	0	200	0	0
	48	105	360	650	24,800	100	0	521	0	0
N. naja atra (Formosa)	0	1,000	380	650	43,600	300	0	13	0	0
	48	138	170	317	25,250	109	0	3	0	0
B. fasciatus (Thailand)	0	1,620	196	810	26,500	391	0	0	0	0
	48	137	139	500	24,700	110	0	0	0	0

[a]Length of time that crude venoms were dialyzed against distilled water before analysis.

[b]Mo, Bi, Se, Pt, Pd, Ag, and Au.

Source: Ref. 6.

TABLE 2

Effect of Divalent Metal Ions on
Snake Venom Enzyme Activities

Activity	Activator	Inhibitor	References
Phospholipase A_2	Ca(II)		1
			48
			49, 50
			10
	Mg(II)	Cd(II)	10
		Zn(II)	10
5'-Nucleotidase	Mg(II)		51
			52
	Mn(II)	Ni(II)	52
		Zn(II)	52
L-Amino acid oxidase	Mg(II)	Ca(II)	53
Exonuclease	Zn(II)		16

combines with phospholipase A_2 inhibitor and the inhibitor's activity
is lost [11].

Venom proteases are known to contain metal ions but additional
cations of various types did not increase the proteolytic activities
of purified or semipurified enzymes [12-15]. Exonuclease (phospho-
diesterase)'s activity can be enhanced by the addition of Zn(II) and
the affinity of substrate for the enzyme increases with Mg(II) and
Ca(II). The explanation is that the enzyme is a metalloprotein and
contains Zn(II) which is important for catalytic activity while
Ca(II) and Mg(II) are required for proper substrate binding [16,17].

Nerve growth factor (NGF) is found in the mouse submaxillary
gland and is commonly found in snake venoms. NGFs from the two
sources usually have homologous amino acid regions in their sequence
[1]. NGF is a special protease having specificity for lysyl and
arginyl bonds. NGF is a metalloenzyme containing 1 mol Zn(II) per
mol. The function of zinc is to prevent the activation of NGF.
Thus, upon removing Zn(II) with ethylenediaminetetraacetic acid
(EDTA), the zymogen undergoes autocatalytic activation [18]. The

presence of zinc was reported for the NGF from the mouse source.
However, since NGF from both sources is very similar, it is reasonable to assume that snake venom NGF is also a zinc enzyme.

3. EFFECT OF CHELATING AGENTS

Some snakes, especially the families Crotalidae and Viperidae, induce
severe local hemorrhage on envenomation. Since these venoms are
strongly hemorrhagic and also contain proteolytic enzymes, as early
as 1930 [19] it was thought that venom proteases were responsible for
inducing hemorrhage. However, there is little direct experimental
evidence to support this supposition. The first evidence, provided
by Flowers [20] and Goucher and Flowers [21], observed that a chelating agent, EDTA, reduced both the hemorrhagic and protease activity
of *Agkistrodon piscivorus piscivorus* (eastern cottonmouth) venom.
This experiment provided the first clue that metal ions may be essential for both hemorrhagic and proteolytic enzyme activities. Metal
analysis of the EDTA-treated venom was performed by Friederich and
Tu [6], who reported that practically all the calcium and magnesium
ions disappeared while some zinc remained, suggesting that zinc has
the highest affinity for snake venom proteins. The EDTA-treated
venom lost both hemorrhagic and proteolytic activities. Hemorrhagic
activity was slightly restored on addition of magnesium or zinc to
the EDTA-treated venom but not by the calcium ion. Another chelating
agent, DTPA (diethylenetriamine pentaacetic acid), has a stronger
antihemorrhagic activity than EDTA [22]. With all these findings,
the circumstantial evidence has been built that some divalent metal
ions are probably responsible for both hemorrhagic and proteolytic
activities. But real evidence must await the isolation of venom
proteases, the exact determination of metal content in a purified
component and the study of the role of metals in hemorrhagic and
proteolytic actions. Proteinase activity of many snake venoms is
EDTA dependent, suggesting the metalloprotein nature of the enzyme
[23-26], although hemorrhagic activity and metal content were not

determined. Proteinase activity from the venom of *Crotalus atrox*,
Lachesis muta, and *Bitis arietans* was inhibited by addition of EDTA
[27]. Similarly, a proteolytic enzyme and hemorrhagic activities
of several proteinases isolated from the venom of *Agkistrodon acutus*
were inhibited with EDTA [28-32].

4. ZINC IN PURIFIED HEMORRHAGIC TOXINS AND PROTEOLYTIC ENZYMES

4.1. Presence of Zinc

The first report of zinc in a purified proteolytic enzyme was made
on leucostoma peptidase A from the venom of *Agkistrodon piscivorus
leucostoma* [33]. The enzyme contains a unimolar amount of zinc.
The following year, 1 mol zinc per mol of proteinase was reported
for the enzyme isolated from *Trimeresurus flavoviridis* [15]. Hemorrhagic activity was not tested by these investigators. In 1978
Bjarnason and Tu [34] isolated five hemorrhagic toxins to purity
with associated proteolytic activity from the venom of *Crotalus
atrox*. All five hemorrhagic toxins contain a unimolar ratio of
zinc per mol of toxin (Table 3). In 1977 a proteolytic enzyme,
AC_1-proteinase, with hemorrhagic activity was isolated from the
venom of *Agkistrodon acutus* [28-31]. Zinc content was not analyzed
at the time, but recent analyses in our laboratory indicate that
the hemorrhagic proteinase does contain 1 mol zinc per mol protein
[35]. A proteinase from the venom of *T. flavoviridis* was found to
have 1 mol of zinc per mol protein, but hemorrhagic activity was not
assayed. Recent investigations in our laboratory show this enzyme
to be a strong hemorrhagin [36]. Fibrinogenase, a proteolytic
enzyme which renders fibrinogen unclottable, is found to have 1 mol
zinc per mol protein. Moreover, the apoenzyme's activity can be
restored 87% by incubation with $ZnCl_2$ [37]. It appears that all
venom hemorrhagic toxins studied contain unimolar amounts of zinc
and also exhibit proteolytic enzyme activities. Usually, a venom
contains several hemorrhagic toxins and proteolytic enzymes.

TABLE 3

Zinc Content and Properties of Purified Hemorrhagic Toxins and
Proteolytic Enzymes Isolated from Snake Venoms

Venoms (origin)	Name	Mol zinc/mol protein	Molecular weight	Biological activity tested	References
Agkistrodon acutus (Formosa)	AC_1 proteinase	1.15	24,000	Hemorrhagic, proteolytic	28
A. contortrix mokasen (U.S.A.)	Fibrinogenase	1.0	22,900	Proteolytic	37
A. piscivorus leucostoma (U.S.A.)	Leucostoma peptidase A	0.98	22,500	Proteolytic	33
Crotalus atrox (U.S.A.)	Hemorrhagic toxin: *a*	0.99	68,000	Hemorrhagic, proteolytic	34
	b	0.82	24,000	Hemorrhagic, proteolytic, myotoxic	34
	c	0.86	24,000	Hemorrhagic, proteolytic	34
	d	0.86	24,000	Hemorrhagic, proteolytic	34
	e	1.03	25,700	Hemorrhagic, proteolytic	34
	f	1.15	64,000	Hemorrhagic, proteolytic	Unpublished data
Trimeresurus flavoviridis (Japan)	Proteinase	1.0	22,000	Proteolytic, hemorrhagic	15

Whether all venom proteolytic enzymes also have hemorrhagic activity is not firmly established. Many venom proteolytic enzymes have been isolated but few have been tested for hemorrhagic activity. The presence of zinc in proteolytic enzymes other than snake venoms is not uncommon. The best examples are zinc in carboxypeptidase A and B [38,39], thermolysin [40], neutral proteases from *Bacillus subtilis*, *Streptomyces naraensis*, *Aspergillus sojae* [41-43], leucine aminopeptidase [44], and aminopeptidase from porcine kidney [45].

4.2. Structure-Function Relationship

The role of zinc for hemorrhagic and proteolytic activity was extensively studied by Bjarnason and Tu [34]. When zinc was removed from hemorrhagic toxin e (from *C. atrox* venom) with 1,10-phenanthroline, both the proteolytic and hemorrhagic activity were equally inhibited. From circular dichroism (CD) spectroscopy in the 190- to 250-nm region (Fig. 1), it was calculated that the native toxin consists of 23% α-helix, 6% β-structure, and 71% random coil. When over 90% of the zinc was removed the α-helix content dropped from 23% to 7%. The Raman spectroscopic investigation supports this result. The native toxin has a doublet at 1655-1665 cm^{-1}, indicating the presence of both α-helix and unordered structure (Fig. 2). Apohemorrhagic toxin e has no 1655-cm^{-1} line in the amide I-band region. Instead, an amide I band at 1671 cm^{-1} indicates the presence of more unordered structure. As contrasted to the change in the peptide backbone region (Fig. 1), the 260- to 330-nm region of CD shows a very dramatic change (Fig. 3). The spectral region of 260-330 nm is due to the aromatic side chain and disulfide bonds. The fact that the CD spectra show such a drastic change upon the removal of zinc suggests that zinc attaches to the tyrosine side chain. With the addition of zinc ion to apotoxin, the CD spectrum becomes more like the native toxin (Figs. 1 and 3) and hemorrhagic and proteolytic activities are restored. Ultraviolet absorption spectra in the 275- to 290-nm

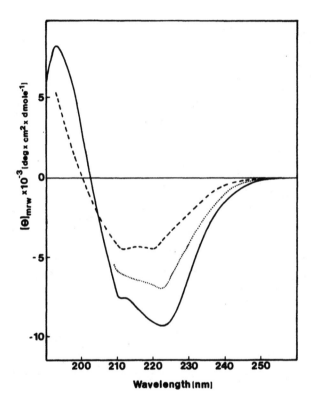

FIG. 1. Circular dichroic spectra of native hemorrhagic toxin *e*
(——) isolated from *Crotalus atrox* venom, hemorrhagic apotoxin *e*
(---), and zinc-regenerated hemorrhagic apotoxin *e* (•••) in the
peptide region. Mean residue ellipticities (θ_{mrw}) based on mean
residue weight of 115. (Reprinted with permission from Ref. 34.
© 1978 American Chemical Society.)

204

ANTHONY T. TU

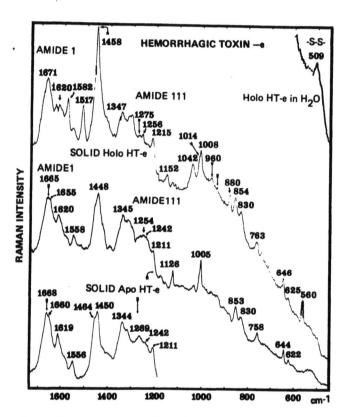

FIG. 2. Laser Raman spectra of native hemorrhagic toxin e (two bottom curves) and hemorrhagic apotoxin e (top curve) in the solid state, and the disulfide region of aqueous native hemorrhagic toxin e (upper right-hand corner). (Reproduced with permission from Ref. 34. © 1978 American Chemical Society.)

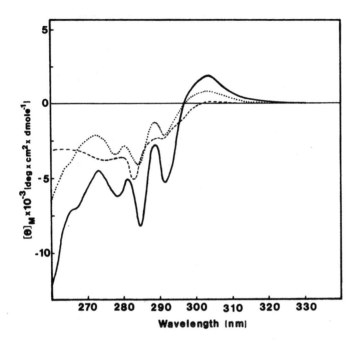

FIG. 3. Circular dichroic spectra of native hemorrhagic toxin e
(——), hemorrhagic apotoxin e (---), and zinc-regenerated hemorrhagic
toxin e (•••) in the aromatic region. Molecular ellipticities (θ_M)
are based on a molecular weight of 25,000. (Reproduced with permission from Ref. 34. © 1978 American Chemical Society.)

region also shows the change from native toxin to apotoxin. The
addition of Zn(II) restores the shape of the absorption spectrum to
that of the native toxin (Fig. 4). Since the ultraviolet (UV)
absorption spectra of proteins originates from the aromatic amino
acid residues, most notably tyrosine, this evidence is highly suggestive that the zinc atom chelates with the phenolate ion of
tyrosine.

When hemorrhagic toxin (AC_1-proteinase) from *Agkistrodon acutus*
was examined by Raman spectroscopy, there is a shift of the amide I
from 1665 cm^{-1} to 1670 cm^{-1} by removing zinc(II) from the native
toxin (Fig. 5). This also suggests that the removal of zinc increases
the content of unordered structure. However, the most dramatic change

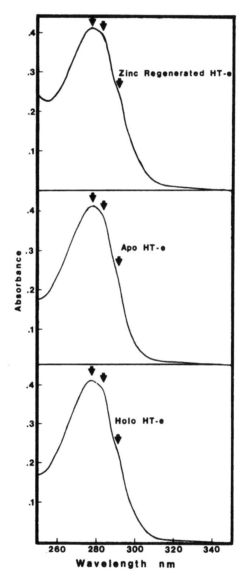

FIG. 4. Ultraviolet spectra of native hemorrhagic toxin e (bottom), hemorrhagic apotoxin e (middle), and zinc-regenerated hemorrhagic toxin e (top) in the aromatic region. Concentration was 0.37 mg/ml in all cases. Note the shoulders at 291 and 284 nm, which were greatly diminished upon zinc removal and regenerated when the apotoxin was incubated with zinc. (Reprinted with permission from Ref. 34. © 1978 American Chemical Society.)

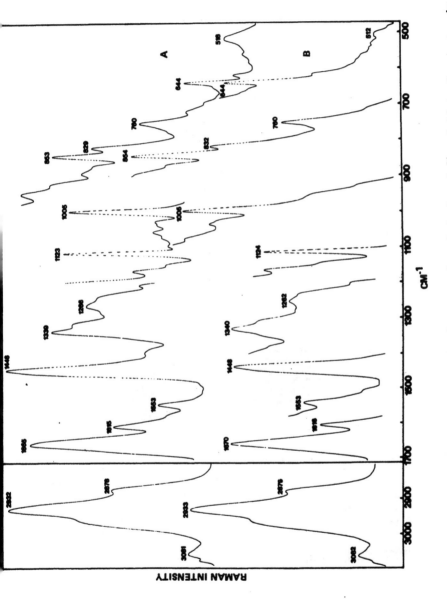

Fig. 5. Raman spectra of hemorrhagic toxin (AC$_1$-proteinase) isolated from *Agkistrodon acutus* in the range 500-1750 cm^{-1} and 2800-3100 cm^{-1}: (A) native hemorrhagic toxin; (B) apohemorrhagic toxin.

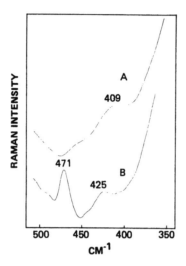

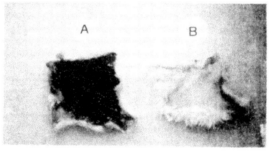

FIG. 6. Raman spectra of native (A) and apo (B) hemorrhagic toxins in the range 340-500 cm^{-1}. The bottom photograph indicates the inactivation of hemorrhagic activity by removal of zinc from the hemorrhagic toxin of *A. acutus*. (A) Injection of 55 μg of purified hemorrhagic toxin. Note the extensive hemorrhagic area. (B) Injection of 150 μg of purified apohemorrhagic toxin (hemorrhagic toxin without zinc). Notice that the skin is clean, with no trace of a hemorrhagic area. Even a threefold increase in concentration produced no observable hemorrhage.

was observed in the low-frequency region, which is most likely the
vibrational band for the zinc ligand (Fig. 6). It is not possible
to definitely determine the nature of the 409-cm^{-1} band; however,
it is probably the zinc-ligand bond with the possibility of tyrosine
as the ligand. The Zn-O stretching vibration was identified at
368 cm^{-1} for horse-liver alcohol dehydrogenase, a zinc-containing
enzyme [46]. Also, $Zn(H_2O)_6^{2+}$ is known to have a Zn-O band at
390 ± 5 cm^{-1} [47]. Histidine residues frequently serve as ligands
in many metalloenzymes and the possibility of a Zn-N ligation cannot
be ruled out. The change in the Raman spectra may reflect a change
in the microenvironment of the metal active site following the
removal of zinc. The C-H stretching vibration bands (2800-3100 cm^{-1})
show little change after removal of zinc. This suggests that there
is no drastic change in the microenvironment of the hydrophobic side
chains.

ACKNOWLEDGMENT

This investigation was supported by National Institutes of Health
Grant GM15591.

REFERENCES

1. A. T. Tu, *Venoms: Chemistry and Molecular Biology*, Wiley, New
   York, 1977.

2. E. Ueda, T. Sasaki, and M. T. Peng, *Mem. Fac. Med. Natl. Taiwan
   Univ.*, *1*, 194 (1951).

3. S. Gitter, S. Amiel, G. Gilat, T. Sonnino, and Y. Welwart,
   Nature (Lond.), *197*, 383 (1963).

4. A. Devi, in *Venomous Animals and Their Venoms*, Vol. 1 (W. Bucherl,
   E. E. Buckley, and V. Deulofeu, eds.), Academic Press, New York,
   1968, pp. 119-165.

5. B. Moav, S. Gitter, Y. Welwart, and S. Amiel, *Meth. Anal. Proc.
   Symp.* (Salzburg), *1*, 205 (1964).

6. C. Friederich and A. T. Tu, *Biochem. Pharmacol.*, *20*, 1549 (1971).

7. Y. Hirakawa, *Acta Med. Univ. Kagoshima*, *26*, 611 (1974).

8. O. G. Rodriquez, H. R. Scannone, and N. D. Parra, *Toxicon*, *12*, 297 (1974).

9. M. F. S. El-Hawary and F. Hassan, *Egypt. J. Physiol. Sci.*, *1*, 9 (1974).

10. A. T. Tu, R. B. Passey, and P. M. Toom, *Arch. Biochem. Biophys.*, *140*, 96 (1970).

11. J. C. Vidal and A. O. M. Stoppani, *Arch. Biochem. Biophys.*, *145*, 543 (1971).

12. A. Ohsaka, *Jpn. J. Med. Sci. Biol.*, *13*, 33 (1960).

13. G. Pfleiderer and G. Sumyk, *Biochim. Biophys. Acta*, *51*, 482 (1961).

14. T. Takahashi and A. Ohsaka, *Biochim. Biophys. Acta*, *198*, 293 (1970).

15. S. Hagihara, *Acta Med. Univ. Kagoshima*, *16*, 127 (1974).

16. R. A. Vassileva and L. Dolapchiev, *Dokl. Bolg. Akad. Nauk*, *32*, 1109 (1979).

17. L. B. Dolapchiev, R. A. Vassileva, and K. S. Koumanov, *Biochim. Biophys. Acta*, *622*, 331 (1980).

18. M. Young and M. J. Koroly, *Biochemistry*, *19*, 5316 (1980).

19. B. A. Houssay, *C.R. Seances Soc. Biol. Fil.*, *105*, 308 (1930).

20. H. H. Flowers, *Toxicon*, *1*, 131 (1963).

21. C. R. Goucher and H. H. Flowers, *Toxicon*, *2*, 139 (1964).

22. C. L. Ownby, A. T. Tu, and R. A. Kainer, *J. Clin. Pharmacol.*, *15*, 419 (1975).

23. I. Chinen, Y. Maeda, K. Yasura, and H. Yomo, *Ryukyu Univ. Agric. Coll. Bull.*, *25*, 203 (1978).

24. B. J. Morris and C. H. Lawrence, *Biochim. Biophys. Acta*, *612*, 137 (1980).

25. L. F. Kress and J. Catanese, *Biochim. Biophys. Acta*, *615*, 178 (1980).

26. M. Homma, F. Kubota, T. Nikai, and H. Sugihara, *Kitakanto Med.*, *30*, 485 (1980).

27. L. F. Kress and E. A. Paroski, *Biochem. Biophys. Res. Commun.*, *83*, 649 (1978).

28. H. Sugihara, T. Nikai, and T. Tanaka, *Yakugakuzasshi*, *97*, 507 (1977).

29. H. Sugihara, T. Nikai, T. Kawaguchi, and T. Tanaka, *Yakugakuzasshi*, *98*, 1523 (1978).

30. H. Sugihara, T. Nikai, H. Umeda, and T. Tanaka, *Yakugakuzasshi,*
    99, 1161 (1979).

31. H. Sugihara, T. Nikai, G. Kinoshita, and T. Tanaka, *Yakugakuzas-*
    shi, 100, 855 (1980).

32. X. Xu and Z. Y. Xu, *Chung-Kuo K'O Hsueh Chi Shu Ta Hsueh Hsueh*
    Pao, 10, 105 (1980).

33. A. M. Spiekerman, K. K. Fredericks, F. W. Wagner, and J. M.
    Prescott, *Biochim. Biophys. Acta, 293,* 464 (1973).

34. J. B. Bjarnason and A. T. Tu, *Biochemistry, 17,* 3395 (1978).

35. T. Nikai, H. Ishizaki, A. T. Tu, and H. Sugihara, *Comp. Biochem.*
    Physiol., 72C, 103 (1982).

36. A. T. Tu and S. Hagihara, unpublished data.

37. J. B. Moran and C. R. Geren, *Biochim. Biophys. Acta, 659,* 161
    (1981).

38. B. L. Vallee and H. Neurath, *J. Biol. Chem., 217,* 253 (1955).

39. E. Wintersberger, D. J. Cox, and H. Neurath, *Biochemistry, 1,*
    1069 (1962).

40. J. D. McConn, D. Tsuru, and K. T. Yasunobu, *J. Biol. Chem.,*
    239, 3706 (1964).

41. D. Tsuru, J. D. McConn, and K. T. Yasunobu, *Biochem. Biophys.*
    Res. Commun., 15, 367 (1964).

42. A. Hiramatsu and T. Ouchi, *Agric. Biol. Chem., 42,* 1309 (1978).

43. H. Sekine, *Agric. Biol. Chem., 36,* 2343 (1972).

44. U. Kettman and H. Hanson, *FEBS Lett., 10,* 17 (1970).

45. H. Wacker, P. Lehky, E. H. Fisher, and E. A. Stein, *Helv. Chim.*
    Acta, 54, 473 (1971).

46. P. W. Jagodzinski and W. L. Peticolas, *J. Am. Chem. Soc., 103,*
    234 (1981).

47. H. Kanno and J. Hiraishi, *J. Raman Spectrosc., 9,* 85 (1980).

48. M. A. Wells, *Biochemistry, 11,* 1030 (1972).

49. C. C. Viljoen, J. C. Schabort, and D. P. Botes, *Biochim. Bio-*
    phys. Acta, 360, 156 (1974).

50. C. C. Viljoen, D. P. Botes, and J. C. Schabort, *Toxicon, 13,*
    343 (1975).

51. I. I. Nikol'skaya, O. S. Kislina, and T. I. Tikhonenko, *Dokl.*
    Akad. Nauk SSSR, 157, 475 (1964).

52. Y. Chen and T. B. Lo, *J. Chin. Chem. Soc., 15,* 84 (1968).

53. W. K. Paik and S. Kim, *Biochim. Biophys. Acta, 139,* 49 (1967).

Chapter 6

SPECTROSCOPIC PROPERTIES OF METALLOTHIONEIN

Milan Vašák and Jeremias H. R. Kägi
Institute of Biochemistry
University of Zürich
Zürich, Switzerland

1. INTRODUCTION

Metallothioneins are unusual, low-molecular-weight proteins of
extremely high sulfur and metal content. They occur in varying
abundance in parenchymatous tissues of vertebrates and invertebrates
and have been identified also in some microorganisms [1]. They were
discovered initially in the search for the tissue constituent responsible for the natural accumulation of cadmium in equine [2] and human
[3] kidney and they are still the only defined macromolecular biological compound known to contain this metal. However, even so,
cadmium is most often only a minor metallic component in these
proteins. In vertebrates, the major and occasionally the sole
metallic component is zinc. Metallothioneins are, in fact, the zinc
metalloproteins with the highest known zinc content. This led to
the suggestion that the primary function of the metallothioneins
must be sought in relation to zinc [4,5]. The favored current view
is that its major functions are to serve as a zinc-supplying protein
in the biosynthesis of the numerous zinc enzymes and to maintain
zinc homeostasis within the organism [6,7].

Early studies showed that the unusual metal-binding properties
of metallothioneins are linked to the abundance of cysteine residues
in the polypeptide chain [8]. A stoichiometry of about three cysteine residues per bivalent metal ion is found in all well-characterized forms [9]. The spectroscopic features indicate that all

cysteine residues participate in metal binding through the formation
of metal-thiolate complexes [8,10]. Recently, it was shown that
several of these complexes are joined together to form oligonuclear
clusters [11,12]. Metallothioneins have not as yet been crystallized.
Thus all information concerning the structure of these proteins and
the mode of metal binding is derived from sequence analysis and from
spectroscopic studies and is thus of an indirect nature. The principal results of the study of their primary structure, their biosynthesis, and their potential biological roles have been the subject
of recent reviews [13,14], as have the studies on copper-containing
forms of metallothionein [15]. The aim of this chapter is to give
an up-to-date account of the spectroscopic features of mammalian
metallothioneins containing bivalent metals.

2. COVALENT STRUCTURE

All native mammalian metallothioneins are single polypeptide chain
proteins with a molecular weight ranging from 6500 to 7000 depending
on the metal composition [16]. This value is consistent with a chain
length of 61 amino acid residues and a chain weight of approximately
6100 as deduced from sequence data [1,17-20]. A most unique composi-
tional feature is the high content of cysteine (approximately 33
residue %), which exceeds even that of the high-sulfur proteins of
wool. In mammalian forms, there is also a relatively large proportion of serine (approximately 14 residue %) and of basic amino acids
(approximately 13 residue % lysine plus arginine). Also typical of
all mammalian metallothioneins is the complete lack of aromatic
amino acids and of histidine [1]. At present, the primary structure
of eight mammalian metallothioneins is known [13]. A representative
sequence, given in Fig. 1, reveals a fairly uniform distribution of
the cysteinyl residues along the polypeptide chain which is completely preserved in evolution [1]. The most striking feature is the
predominance of -Cys-X-Cys- sequences (where X stands for an amino
acid residue other than Cys), which occur seven times within the

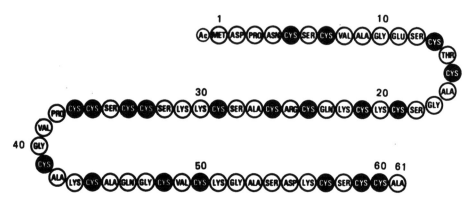

FIG. 1.   Amino acid sequence of equine metallothionein-1A.   (Adapted by permission from Ref. 38.)

polypeptide chain [17].   There are no disulfide bonds in the native protein.   All mammals examined thus far contain at least two major variants of metallothionein (isometallothioneins), which differ in a number of amino acid residues and which are thought to arise from gene duplication [21].   In addition, in some species there are minor variants with single amino acid substitutions probably attributable to allelic heterogeneity [22].

Native metallothioneins are nonglobular proteins.   This is clearly reflected in the large discrepancy of the molecular weight estimate by gel filtration (i.e., ∿10,000 for all native mammalian forms) and of the molecular weight deduced from shape-independent measurements (i.e., 6500-7000).   The molecular mass and the measured Stokes radius (16.1 Å) are consistent with the measures of a prolate ellipsoid that has an axial ratio of 6 [16].

All native mammalian metallothioneins characterized thus far are negatively charged at neutral pH [4].   Measurements by free boundary electrophoresis of unresolved native equine renal metallo-thioneins gave a value of -1.9 charges per molecule [23].   Compara-tive starch gel electrophoresis at pH 8.6 indicates that there are also two negative charges in the metallothionein-1 variant of mouse, rat, rabbit, and humans.   The variants metallothionein-2 of the same species carry three negative charges [4,21]. The isoelectric points have been reported for several metallothioneins to be close to 4 [24,25].

3. METAL COMPOSITION, REMOVAL, AND RECONSTITUTION: PREPARATION OF TRANSITION METAL DERIVATIVES

Metallothionein is, besides ferritin, the metalloprotein with the highest metal content known. Unlike most other metalloproteins, the metallothioneins are, however, exceptional by their widely varying metal ion composition. Thus preparations of metallothionein judged to be pure by chromatographic criteria are often heterogeneous in metal-ion composition exhibiting nonintegral and varying ratios of zinc, cadmium, mercury, and copper. However, the total content of bivalent metals is reasonably constant, reaching 6-7 mol per mol protein in the most purified preparations, thereby implying the existence of seven binding sites for group 2B metal ions per molecule (Table 1) [1]. In part, the metal-ion composition is determined by the extent of the exposure of the organism to the different metals.

TABLE 1

Typical Metal Composition of Human and Equine
Metallothioneins Isolated from Liver and Kidney

Metal	Human[a]		Equine[a]	
	Liver [4]	Kidney[b] [3]	Liver [16]	Kidney [16]
Zinc	6.40	3.05	5.82	3.02
Cadmium	0.14	3.52	0.34	3.09
Copper	0.13	0.34	0.18	0.14
Mercury	c	0.23	c	c
Total	6.67	7.14	6.34	6.25

[a]Moles of metal per 6100 g of apometallothionein.
[b]Recalculated on the basis of 20 Cys per molecule.
[c]Not determined.

Thus exposure to cadmium or mercury yields metallothioneins rich in these ions [1]. However, large variations in metal composition are also seen between preparations isolated from different tissues [22]. As summarized in Table 1, there is more zinc and less cadmium in metallothionein from human and equine liver than in the corresponding preparations from kidney cortex. Analogous differences exist with respect to the zinc and copper content between hepatic and renal metallothionein of rodents [26,27]. Since equine liver and kidney contain the same isometallothioneins [22], such differences in metal distribution probably have no structural basis but must be determined by local physiological circumstances and by the mechanisms controlling the flow of metals through the organism. It must be primarily for such a reason that copper is normally only a minor component of adult human and equine metallothionein, although it binds to metallothionein much more firmly than either zinc or cadmium [28]. However, copper becomes the major metallic component of metallothionein in patients where its normal metabolism is impaired (e.g., in Wilson's disease [29] and in inherited copper toxicosis in dogs [30]).

The mode of metal binding has been clarified by chemical and spectroscopic studies [8,10]. All metal ions are bound through thiolate groups with affinities increasing in the order $Zn(II) < Cd(II) < Cu(I)$, $Ag(I)$, $Hg(II)$. Thus zinc which is bound much less firmly than cadmium [8] is readily displaced by an excess of the latter and both of them by copper, silver, and mercury. Zinc and cadmium are also readily removed by exposure to low pH, yielding metal-free apometallothionein (thionein). The apoprotein can be obtained either by gel filtration or by dialysis at pH 2 [3]. However, this treatment does not remove copper. To displace this metal, exposure to 0.5 M HCl is required [28]. The apoprotein is stable against air oxidation at moderately low pH (1.5-3.5). It can be quantified by measuring absorbance at 220 nm (ϵ_{220} = 47,300) [31].

Metallothioneins containing stoichiometric quantities (7 mol metal per mol protein) of $Zn(II)$, $Cd(II)$, $Hg(II)$, $Co(II)$, and $Ni(II)$

have been prepared from the apoprotein and the appropriate metal
salts.   To reconstitute metallothionein from apometallothionein with
Zn(II) or Cd(II), a slight molar excess of the metal salt is added to
a solution of the apoprotein in 0.1 M HCl and the pH adjusted stepwise to neutrality with Trizma base [=2-amino-2(hydroxymethyl)-1,3-
propanediol].   To avoid air oxidation of the apoprotein, reconstitution is carried out preferably in an argon atmosphere. Excess metal
is removed either by addition of a small amount of moist Chelex 100
resin (sodium form) to the neutralized solution or by gel filtration.
Hg(II)-metallothionein is prepared by adding to the apoprotein at pH
2 exactly 7 mol $HgCl_2$ per mol [32]. Metallothionein containing the
transition metal Co(II) is prepared in analogy to the Zn(II)- and
Cd(II)-containing derivatives by adding at pH 1 a twofold molar excess
of $CoSO_4$. However, in this case very stringent precautions have to be
taken in order to avoid its oxidation. This is accomplished by carry-
ing out all preparations in an argon-purged glove box and by using
carefully degassed solutions [33]. Ni(II)-metallothionein is prepared
in an analogous fashion but without the addition of Chelex 100 resin
to remove excess of metal [33]. The Ni(II) derivative is not as sensitive to air oxidation as Co(II)-metallothionein, but it decomposes
on standing with partial release of metal.

4.   SPECTROSCOPIC FEATURES OF NATIVE AND RECONSTITUTED METALLOTHIONEIN

The unusual metal and amino acid composition of metallothionein is
reflected in its equally unusual spectroscopic features.   In fact,
it is from the study of these properties that most information on
their molecular organization has been obtained thus far.   Thus by
the application of a variety of spectroscopic techniques--electronic
absorption spectroscopy, circular dichroism (CD), magnetic circular
dichroism (MCD), electron spin resonance (ESR) spectroscopy, x-ray
photoelectron spectroscopy (ESCA), perturbed angular correlation of
γ-ray (PAC) spectroscopy, extended x-ray absorption fine-structure

(EXAFS) spectroscopy, and [113]Cd NMR spectroscopy--to both native and reconstituted metallothioneins, a consistent model of metal binding in these proteins has been elaborated. In addition, ^{1}H NMR studies and infrared (IR) spectroscopy have given important insight into their secondary and tertiary folding.

4.1. Electronic Absorption Spectra

The absorption spectra of metallothioneins are dominated by the transitions associated with metal binding. The spectra of metallo-thioneins containing Zn(II), Cd(II), or Hg(II) are shown in Fig. 2. In the metal-free form, these specific features are abolished, yield-ing the plain absorption spectrum of apometallothionein, characterized by a single absorption band at 5.25 μm^{-1} (190 nm) and complete trans-parency above 4.0 μm^{-1} (250 nm), reflecting the absence of aromatic amino acid residues in the polypeptide chain. The 5.25-μm^{-1} band includes in addition to the expected strong π-π* and the weak n-π*

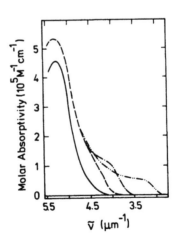

FIG. 2. UV absorption spectra of equine metallothionein (MT); apometallothionein (——) in 0.1 N HCl; Zn(II)-MT (---); Cd(II)-MT (-·-); Hg(II)-MT (-··-), in 0.001 M sodium phosphate buffer, pH 7.5. (Reprinted with permission from Ref. 34. © 1981 American Chemical Society.)

amide transitions the lowest thiol transition of the cysteine side chains near 5.13 μm^{-1} (195 nm) [34].

Binding of group 2B metal ions intensifies the far-ultraviolet (UV) absorption and introduces at the red edge of the absorption spectrum a characteristic shoulder whose position is specific for the metal (Table 2). The difference absorption spectra obtained by subtracting the contribution of the polypeptide chains from that of Cd(II)-metallothionein and Zn(II)-metallothionein reveal broad absorption profiles. They are closely similar in shape and ampli- tude to those of Cd(II) and Zn(II) complexes of 2-mercaptoethanol (Fig. 3), thereby implying their common origin in transitions of the metal-thiolate complexes [8]. By line shape analysis these spectra can be resolved into a minimum of three transitions (Table 2). A similar metal-ion dependency of the position of the lowest energy band is observed in halide complexes of the same metals and is thought to be diagnostic of an electron transfer origin of such bands. According to a semiempirical theory of Jørgensen [35], the frequency σ (μm^{-1}) of these bands is given by the difference of the optical electronegativities of the ligands [$\chi_{opt}(L)$] and the metal [$\chi_{opt}(M)$] linked in the complex according to the equation

$$\sigma = (3.0 \ \mu m^{-1})[\chi_{opt}(L) - \chi_{opt}(M)] \tag{1}$$

Accepting for the thiolate ligand $\chi_{opt}(L) = 2.6$ and for the metal ions the values $\chi_{opt}(M)$ derived from the tetrahedral tetrahalide complexes [1.15 for Zn(II), 1.27 for Cd(II), and 1.50 for Hg(II)], one obtains for the lowest-energy transitions of the metal- thiolate complexes locations which agree well with those found in the spectra of the corresponding metallothioneins and in complexes with 2-mercaptoethanol (Table 2). Hence the lowest-energy band of the metal-thiolate absorption spectrum is identified unambiguously as the first Laporte-allowed electron transfer transition [34]. The close correspondence to the halide reference spectra suggests further- more that the coordination geometry of the metal-thiolate complexes is also tetrahedral (see Sec. 6).

TABLE 2

Location of Metal-Thiolate Transitions in Zn(II)-, Cd(II)-, and Hg(II)-Metallothionein and in Model Complexes with 2-Mercaptoethanol

Metal	Spectral position of metal-induced shoulder in metallothionein [μm^{-1} (nm)]	Contributing bands as suggested by Gaussian analysis		Calculated position of first electron transfer transition [μm^{-1} (nm)]
		Metallothionein [μm^{-1} (nm)]	2-Mercaptoethanol complex [μm^{-1} (nm)]	
Zn(II)	4.55 (220)	4.33 (231) 4.57 (219) 4.87 (205)	4.25 (235) 4.45 (225) 4.80 (208)	4.31 (232)
Cd(II)	4.08 (245)	4.00 (250) 4.33 (231) 4.97 (201)	3.98 (251) 4.30 (232) 4.74 (211)	4.01 (249)
Hg(II)	3.25 (308)	~3.3 (303) ~3.4 (294) ~3.6 (278)	No data available	3.21 (312)

Source: Data reprinted with permission from Ref. 34. © 1981 American Chemical Society.

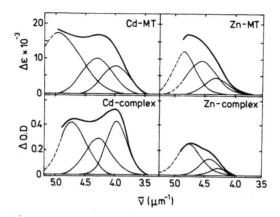

FIG. 3. UV difference absorption spectra of equine Cd(II)- and Zn(II)-metallothionein versus apometallothionein and of Cd(II)- and Zn(II)-complexes with excess 2-mercaptoethanol versus 2-mercapto- ethanol with Gaussian analysis. The molar difference absorptivity, $\Delta\varepsilon$, refers to the concentration of the metal. (Reprinted with per- mission from Ref. 34. © 1981 American Chemical Society.)

The origin of the remaining transitions is not yet clear. Considering the similarity between the closed-shell thiolate and halide anions, it seems reasonable to expect that, in addition to the first electron-transfer transitions which are thought to origi- nate from the highest occupied π molecular orbital (MO) of the com- plex, there is a second one arising from the highest occupied σ MO [34]. However, in light of recent findings that in metallothionein the metal complexes are joined to clusters with some of the cysteine residues serving as threefold-coordinated bridging ligands and others as twofold-coordinated terminal ligands, it is possible that some of the resolved bands can be attributed to corresponding electron trans- fer transitions involving either bridging-sulfur metal MOs or terminal- sulfur metal MOs. To complicate matters further, it is also likely that distortion of the complexes and ligand internal transitions contribute to the difference absorption envelope. Undoubtedly, an unambiguous interpretation of the far-UV spectra requires a still more precise resolution of the contributing bands, both in various metallothioneins and in appropriate model compounds.

4.2. Circular Dichroism of Zn(II)-, Cd(II)-, and Hg(II)-Metallothionein

The chiroptical properties of metallothioneins are also exceptional. Early studies of Cd(II)-metallothionein have shown highly unusual optical rotatory dispersion (ORD) spectra with a large extrinsic Cotton effect associated with the first Cd(II)-thiolate transition near 250 nm [3,36]. Subsequent circular dichroism (CD) measurements on avian [37], rat [10], human [31], and rabbit [32,39] metallo-thionein containing Cd(II), Zn(II), or Hg(II) have revealed CD spectra composed of a number of well-resolved ellipticity bands (Fig. 4, Table 3). However, their general features are quite similar. At the low-energy end of the spectra they display a pair of ellipticity bands with opposite signs.

The molecular mechanism by which optical activity in the metal-thiolate chromophores of metallothionein arises is still largely unknown. While the obvious source of dissymmetry lies in the con-figurational chirality of the amino acid residues, it has been pro-posed that dissymmetric coordination due to distortion of the multi-dentate metal-thiolate complexes is the major contributor to the rotatory strength of the transitions [36]. It has also been suggested that some of the features of the CD spectra might arise from exciton coupling of transitions located in dissymmetrically oriented thiolate ligands [31]. That the generation of rotatory power in metallothio-nein is of complex origin is supported by the recent observation that the Cd(II)-thiolate-associated ellipticity of metallothionein does not increase linearly with cadmium content and that forms of metallo-thionein which contain both Cd(II) and Zn(II) tend to display con-siderably larger Cd(II)-thiolate ellipticity bands than forms con-taining solely Cd(II) [1,37,39]. That the heterogeneity in metal content has a major influence on the spectroscopic features is also indicated by ^{113}Cd NMR data, where in preparations containing both ^{113}Cd(II) and Zn(II) as many as 15 ^{113}Cd resonances were identified as opposed to only 8 ^{113}Cd resonances in preparations containing solely ^{113}Cd(II) [11] (see Sec. 7.1). Analogous effects were

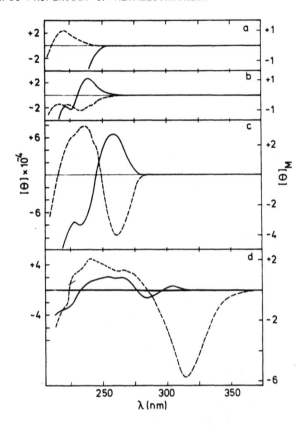

FIG. 4. Circular dichroism (solid line, left scale) and magnetic circular dichroism (dashed line, right scale) spectra of rabbit-liver metallothionein (MT): (a) apo-MT in 0.075 M HCl; (b) Zn(II)-MT; (c) Cd(II)-MT; (d) Hg(II)-MT. Samples (b) to (d) were recorded at neutral pH. The units refer to protein concentration. (From Refs. 32 and 39.)

observed in the ^{111m}Cd PAC spectrum, where samples containing only Cd(II) differed with respect to the NQI parameters from those containing both Cd(II) and Zn(II) [40] (see Sec. 6.2).

Comparison of the CD spectra of metal-containing metallothioneins with the metal-free apometallothionein indicates that below 240 nm, increasingly large ellipticity contributions come from the polypeptide chain. Hence, since metal binding is known to affect the chain conformation (see Sec. 5), it becomes difficult to assess

TABLE 3

Location and Sign of Circular Dichroism (CD) and Magnetic Circular
Dichroism (MCD) Extrema of Apometallothionein (Apo-MT) and of
Zn(II)-, Cd(II)-, and Hg(II)-Metallothionein (MT)[a] Above 210 nm[b]

Apo-MT [39] (nm)		Zn(II)-MT [39] (nm)		Cd(II)-MT [39] (nm)		Hg(II)-MT [32] (nm)	
CD	MCD	CD	MCD	CD	MCD	CD	MCD
(-)222	(+)220	(+)243	(-)243 sh	(+)259	(-)262	(+)305	(-)315
		(-)228	(-)232	(-)233	(+)248 sh	(-)285	(+)265
		(-)223	(-)223	(-)227	(+)237	(+)265 sh	(+)240
					(+)227 sh	(+)253	(+)227 sh
						(-)225 sh	(-)223 sh

[a]All forms contained 7 mol metal per mol protein.
[b]sh, shoulder.

accurately the contribution from metal-thiolate ellipticity in this
region. Nonetheless, difference CD spectra of metallothionein versus
apometallothionein are useful for comparison [31].

4.3. Magnetic Circular Dichroism of Zn(II)-, Cd(II)-, and Hg(II)-Metallothionein

The differentiation of the metal-thiolate transitions from amide
transitions should be easier in magnetic circular dichroism (MCD)
spectra since they are rather insensitive to polypeptide conformation and since the MCD signal is weak for the secondary amide n-π^*
transition at 225 nm [41]. MCD spectra of Zn(II)-, Cd(II)-, and
Hg(II)-metallothionein and of apometallothionein are shown in Fig. 4.
The spectrum of apometallothionein exhibits a positive band at about
220 nm, attributable to the n-π^*amide transition and a negative one
(not shown) at about 200 nm arising from the amide π-π^* transition.
The positions of the MCD bands attributable to the metal-thiolate
transitions are given in Table 3. Some of the bands correspond to
the positions of the resolved Gaussian bands of the difference
absorption spectra (see Sec. 4.1). It is interesting that the
largest MCD bands are associated with the first electron transfer
band, that they are of the same sign, and are markedly increasing
in intensity on going from Zn(II) to Hg(II).

4.4. X-Ray Photoelectron Spectroscopy

One of the most direct means by which bonding of the metal ions to
thiolate groups in metallothionein was established is X-ray photoelectron spectroscopy (ESCA). The method yields information on the charge
distribution in the complex by measuring the electron binding energies of the bonded atoms. Thus in studies on avian and rat metallothioneins, Weser and co-workers [10,42] showed that the $2p_{1/2,3/2}$
core electrons of the cysteine sulfur atoms have binding energies
in the range of those expected for thiolate rather than those

TABLE 4

Core Electron Binding Energies for the S $2p_{1/2,3/2}$
Doublet of Model Compounds and of Different
Metalloforms of Metallothionein (MT)[a]

Compound	Position of maximum (eV)		Half-maximum full width (eV) Vašák et al. [43]
	Weser and Rupp [42]	Vašák et al. [43]	
S_8	163.2	162.9	1.9
Cysteamine-HCl		162.3	1.9
Zn(II)-MT	161.3	161.1	3.1
Cd(II)-MT	161.7	161.7	2.8
Hg(II)-MT	162.7	161.1	3.2
		162.4	

[a]All spectra are referenced to the C 1s binding energy taken as
284.0 eV.

typical for thiols, thereby confirming the earlier suggestion [5,8]
that in native metallothionein all cysteine side chains are deprotonated and participating in metal binding. In addition, these
studies showed that the actual value of the sulfur core binding
energy, E_b, varies somewhat with the type of metal ion bound
(Table 4). Thus measurements on metallothioneins reconstituted
with the various group 2B metal ions revealed that E_b increases on
going from Zn(II) to Cd(II) and Hg(II) by about 1.3 eV (Table 4).
The effect parallels the increased polarization of the sulfur
valence electron shell by metal ions of higher Pauling electronega-
tivity. Not unexpectedly, therefore, the differences in electron
binding energy are also related to the differences in the energy
of the first Laporte-allowed transition of these complexes [34].

        Besides indicating the most probable binding energies for the
core electrons of a given type of atoms, x-ray photoelectron spectra
provide important information on their charge environment. Comparison of the energy profiles of the sulfur $2p_{1/2,3/2}$ electrons in

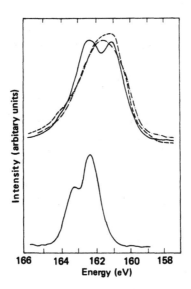

FIG. 5.   S $2p_{1/2,\,3/2}$ electron binding energy profile (ESCA) of
cysteamine-HCl (bottom) and different metalloforms of rabbit-liver
metallothionein (MT) (top):   Zn(II)-MT (---); Cd(II)-MT (-•-);
Hg(II)-MT (——).   For more details, see Table 4.   (From Ref. 43.)

Zn(II)-, Cd(II)-, and Hg(II)-metallothionein showed very substantial
differences from those of the thiol model compound cysteamine hydrochloride or of elemental sulfur (Fig. 5). While these substances
exhibit the expected doublet structure, this feature is not resolved
in the sulfur core electron profiles of the metallothioneins.   In
addition, the unresolved profile is also much broader. Thus the
various metallothioneins display a half-maximum full width of 2.8-
3.2 eV as compared to 1.9 eV in cysteamine hydrochloride and elemental
sulfur (Table 4) [43]. This broadening clearly indicates that metallothionein contains forms of thiolate ligands which differ substantially
from each other in core electron binding energy and hence must be
located in a different charge environment. Such a partition into
distinct forms of thiolate ligands is manifested best in the case of
Hg(II)-metallothionein. It finds a rational explanation in the
recent suggestion that in metallothioneins, the metal ions form

oligonuclear clusters with cysteine residues in which some of the sulfurs are coordinated to one metal ion (terminal ligands) and some to two metal ions (bridging ligands) (see Sec. 7). On electrostatic grounds, it is to be expected that the dual metal coordination reduces the charge density at a bridging sulfur ligand more than does single metal coordination at a terminal ligand [44]. As shown with model complexes containing bridging and terminal chloride ligands, bridging between two metal ions accounts for an increase in the chlorine 2p core electron binding energy by nearly 2 eV [45]. Thus the broadening of the x-ray photoelectron profile of the metallothionein sulfur $2p_{1/2,3/2}$ core electrons offers, in fact, strong support for the postulated metal-thiolate cluster structure.

4.5. NMR Studies

NMR spectroscopy has become a powerful tool in the study of the overall structure of metallothionein and of the organization of its metal-binding sites. Advances in methodology now enable the observation of signals from a variety of nuclei encountered in this protein (e.g., ^{1}H, ^{13}C, ^{113}Cd). In addition, the recent development of two-dimensional ^{1}H NMR spectroscopy has the potential of elucidating the entire spatial structure of this protein.

4.5.1. ^{1}H NMR Spectroscopy

^{1}H NMR spectra of metallothionein and of the apoprotein have been reported for materials from chicken liver [46], from equine, ovine, and human liver, and from equine kidney [47]. The spectra of the apometallothioneins recorded in ^{2}H$_2$O at low pH are of a remarkable simplicity and similarity. In the low-field region, they are all characterized by a complete absence of resonances, reflecting the lack of aromatic amino acids, of histidine, and of unexchanged amide protons. In the high-field region, the distribution of the resonances is closely similar to that of a mixture of the component amino acids, thus permitting their identification (Fig. 6). The complete absence

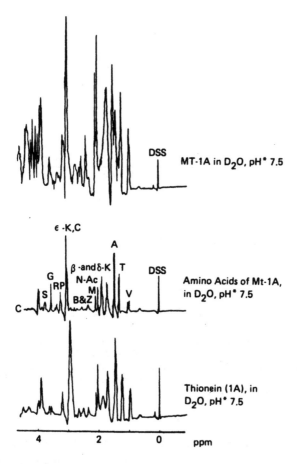

FIG. 6.   270-MHz ^{1}H NMR convolution difference spectra in the high-field region of equine metallothionein-1A in ^{2}H$_2$O. The resonance marked DSS is from the internal standard and the other labels refer to the main amino acid residues contributing to the marked resonances. (From Ref. 94.)

of amide proton resonances indicates that all amide groups have exchanged their protons with deuterons of ^{2}H$_2$O and hence that they must exist in a solvent-exposed conformation. This uniform environment of the secondary amide protons of the peptide backbone is also manifested by the occurrence of a rather narrow signal for the amide protons between 8.5 and 8.1 ppm in ^{1}H$_2$O when water suppression was

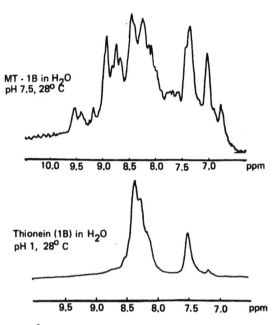

FIG. 7.   270-MHz ^{1}H NMR spectra in the low-field region of equine
metallothionein-1B (MT-1B) and of apometallothionein-1B (thionein-1B)
in ^{1}H$_2$O.   Chemical shift is expressed in ppm downfield from the internal standard DSS.   (Reprinted with permission from Ref. 47. © 1981
American Chemical Society.)

employed [47].   The resonances at 7.51 and 7.18 ppm originate from
the labile side-chain protons of lysine and arginine, respectively
(Fig. 7, bottom).   The temperature dependence of the chemical shifts
of the maximum of the secondary amide resonances (6 ppb/°C) and the
^{2}H/^{1}H exchange rate measured at 5°C in 30 mM HCl is comparable to
that found in random-coil polypeptide chains [48].   Hence, at the
conditions of measurement employed, apometallothionein has a highly
disordered structure, a conclusion supported by independent IR and
CD studies (see Sec. 5).   In contrast to the apoprotein, the metal-
containing forms show NMR features typical of a folded polypeptide
structure.   Thus the ^{1}H NMR spectra of metallothionein recorded in
^{1}H$_2$O (pH 7.5) exhibit in the low-field region, between 6.8 and 9.6
ppm, a broad envelope composed of distinct amide proton resonances

(Fig. 7, top). The various amide protons differ also in the rate of isotopic exchange. Upon dissolution of lyophilized metallothionein in ^2H_2O, the intensity of 12 amide resonances decreases with a rate substantially.lower ($t_{1/2} \approx$ 20-320 min) than that of the other amide hydrogens ($t_{1/2} <$ 2.4 min). Evidence for a partition into more and less exposed amide hydrogens comes also from an appreciable loss of amide proton resonances by saturation transfer when measured in ^1H_2O as well as from temperature dependence studies of the chemical shifts of the amide resonances. Thus measurements of the amide resonances at 15°C in equine metallothionein-1B in ^1H_2O showed that the integrated signal intensity accounts for only about 40% of the value expected for 65 amide protons, suggesting appreciable saturation transfer from ^1H_2O to the exposed amide protons. The chemical shifts of the remaining resonances are much smaller (0.1-4 ppb/°C) than those expected from solvent-exposed amide protons (6-7 ppb/°C), suggesting that the observed protons are to some extent shielded from the solvent [48].

The structural differences between the metal-free and the metal-containing protein are also manifested in the high-field region. However, because of the absence of aromatic residues in the molecule, and hence of a ring current effect, shifts arising from changes in the proton environments are much smaller. Nonetheless, substantial chemical shifts are seen, for instance, at the $-CH_3$ resonances of alanine residues, which in apometallothionein are superimposed at 1.41 ppm and which in metallothionein are distributed between 1.5 and 1.3 ppm [47]. There are also changes that can be traced directly to metal binding. Thus, upon dissociation of the metal from the protein, the $\beta-CH_2$ resonances of Cys are shifted from 3.02 ppm to 2.95 ppm [46]. The binding of metal ions is also characterized by a phase difference of the same signal in the spin-echo spectra (Fig. 8). It reflects a difference in spin-spin coupling of the $\beta-CH_2$ protons in the function of metal binding to the cysteine side chains, thus supporting their participation in metal binding. Another intriguing example is the metal-specific effect on the

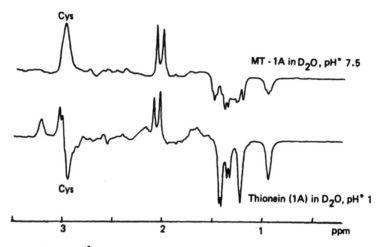

FIG. 8. 270-MHz ^{1}H NMR spin-echo spectra of equine metallothionein-1A (MT-1A) and of apometallothionein-1A (thionein-1A) in ^{2}H$_2$O. The pulse sequence employed was 90°-τ-180°-collect with τ = 71 ms. (Reprinted with permission from Ref. 47. © 1981 American Chemical Society.)

chemical shift of the -CH$_3$ resonances of the sole valine residue (Val-49) in equine metallothionein-1A (Fig. 9). In the solely Zn(II)-containing form, this resonance is located at 0.97 ppm while in forms containing both Cd(II) and Zn(II) there are additional -CH$_3$ resonances at 1.03 and 0.88 ppm whose intensities are proportional to the Cd(II) content. This unusually large perturbation of Val-49 by the type of the metal is probably explained by its location in the sequence between two cysteine residues which have been postulated to serve as the primary chelating site [17]. This striking sensitivity of the ^{1}H NMR spectrum to differences in metal composition, which was also observed in the amide proton region, is an indication for the occurrence of the different metals within the same protein molecule [47].

^{1}H NMR spectroscopy also provides information on protonation equilibria in metallothionein. Thus in equine and ovine metallothionein a resonance arising from an amide proton both in the metal-containing and the metal-free protein was found to titrate reversibly

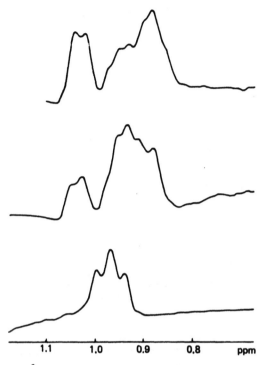

FIG. 9.   270-MHz ^{1}H NMR convolution difference spectra in the upfield region for equine metallothionein-1A in ^2H_2O at pH 7.5.   Effect of metal composition on Val 49 signal.   The relative metal compositions were (top) 55% cadmium and 45% zinc; (middle) 25% cadmium and 75% zinc; and (bottom) 2% cadmium, 2% copper, and 96% zinc.   (Reprinted with permission from Ref. 47.   © American Chemical Society.)

with the protonation of a group of pK_a = 4.0.   Based on the magnitude of the chemical shift (0.24 ppm) of the amide proton resonance, this group was identified as the carboxyl group of the C-terminal alanine residue [47].   Similarly, at high pH an upfield shift of resonances located at 3.0, 1.7, and 1.5 ppm was observed.   These signals have been attributed on the basis of decoupling experiments to the ε-, δ-, β-, and γ-CH_2 groups of lysine residues, respectively.   In the protonated form the position of the ε-CH_2 resonances of lysine of rabbit-liver metallothionein were shown to coincide with that of the β-CH_2 resonances of cysteine, but on deprotonation the lysine

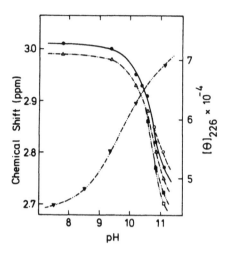

FIG. 10. Effect of pH on lysine NMR resonances and on circular dichroism of rabbit-liver metallothionein. (Left scale) Chemical shift of resolved ε-CH₂ resonances of lysine; (right scale) molar ellipticity (filled triangles) at 226 nm. (From Ref. 49.)

resonances move to high field by about 0.3 ppm and spread into at least five signals (Fig. 10). The pK_a values (in ^2H_2O) characterizing these shifts are close to 10.6. The deprotonation of lysine appears to be linked to a change in protein conformation (see Sec. 5). The spreading out of the resonances implies that upon loss of charge, the lysine side chains are accommodated in distinctly different environments.

4.5.2. ^{13}C NMR Spectroscopy

A ^{13}C NMR spectrum has been reported for a preparation of rabbit-liver metallothionein [50]. Many resonances, including those of carboxyl and carboamide carbons, of the α-carbons of cysteine, of the various methylene carbons of lysine, as well as of 13 different methyl carbons, were surprisingly well resolved.

4.5.3. ^{113}Cd NMR Spectroscopy

The extreme sensitivity of the NMR features of the ^{113}Cd nucleus (12.3% natural abundance) to its coordination environment makes

this isotope a highly suitable natural probe for the study of metal-
ion binding in metallothionein. A natural abundance ^{113}Cd NMR spectrum was first obtained by Otvos and Armitage [50] on rabbit metallothionein containing both Cd(II) and Zn(II). These and similar studies
on ^{113}Cd-enriched rat liver metallothionein by Sadler et al. [51]
were construed as evidence for a variety of different environments
for ^{113}Cd in the preparation examined. The large chemical shifts of
600-700 ppm downfield from $Cd(ClO_4)_2$ suggested, moreover, that Cd(II)
may be bound to more than three sulfur ligands. Using ^{113}Cd-enriched
rabbit metallothionein, Otvos and Armitage [52] subsequently identified about 15 separate ^{113}Cd resonances. By employing ^{113}Cd homonuclear decoupling, they obtained evidence for ^{113}Cd-^{113}Cd spin
coupling and hence for the existence of metal clusters in metallothionein. By analysis of solely ^{113}Cd-containing material, it became
possible to map the organization of these clusters (see Sec. 7.1).

5.   CONFORMATION OF THE POLYPEPTIDE CHAIN IN METALLOTHIONEIN

Although the slow exchange of some of the peptide hydrogens with the
solvent [47,53] and the resistance of metallothionein toward heat
denaturation [7] clearly indicates a stable three-dimensional structure, it has been difficult to obtain a clear picture concerning the
folding of the peptide chain in metallothionein and in apometallothionein. The far-UV CD spectra of both forms display a strong
negative ellipticity band of equal amplitude near 200 nm ($[\theta]_{MRW}$ =
-15,000°) [31] which is known to be typical of random-coil polyamino
acids. This feature remains unchanged upon oxidation of the cysteine
residues to cysteic acid residues, revealing its insensitivity to the
marked structural changes which are expected to accompany the formation of this highly negatively charged sulfonic acid derivative.
The other negative CD band, which in the apoprotein is centered at
about 222.5 nm and which in the metal-containing form is in part
obscured by ellipticity contributions from the metal-thiolate transitions (see Sec. 4.2), suggests the occurrence of some α-helical

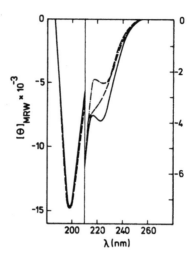

FIG. 11. Circular dichroism spectra of rabbit-liver apometallothionein in 0.02 M HClO$_4$ (——) and in 6 M guanidine-HCl (-·-);
performic acid oxidized apometallothionein in H$_2$O (---). (From
Ref. 54.)

conformation. Judged from the reduction of the amplitude of this
band on exposure to 6 M guanidine-HCl, one can estimate that the
α-helix content of apometallothionein is approximately 5% (Fig. 11).
The remainder of the band which persists under these conditions and
which is also manifested as a shoulder in the oxidized derivative
seems to be a conformation-independent feature. Since a CD spectrum
closely similar to that of apometallothionein has also been observed
with random-coil poly-L-cysteine derivatives [55] but not with
random-coil poly-L-glutamic acid [56] and poly-L-lysine [57], it is
possible that the residual band arises from a perturbation of the
225-nm n-π* amide transition by the neighboring cysteine sulfur atoms.

     The preponderance of a disordered structure in metallothionein
and apometallothionein is also supported by IR spectroscopy [54]. In
this technique, position and shape of the amide I vibration near
1650 cm^{-1} are monitored in a solution of the protein in ^{2}H$_2$O. The
amide I vibration is associated with the C=O stretching mode and is
subject to conformation-dependent exciton-type coupling with neighboring C=O oscillators. Hence the energy and the polarization of

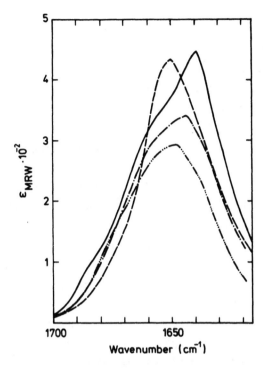

FIG. 12.   IR spectra in the amide I region of rabbit-liver Cd(II),
Zn(II)-metallothionein (-•-), pH 7.5; apometallothionein (-•••-),
pH 1.6; as well as of myoglobin (---) and ribonuclease (——), pH 7.5.
All spectra were taken in ^2H_2O.   (From Ref. 54.)

the bands can yield information on secondary structure folding [58].
For the assessment of peptide chain folding in metallothionein and
apometallothionein, the IR method has the advantage over far-UV CD
measurements that the spectra are not obscured by the metal-thiolate
chromophores.   IR spectra of the two forms of the protein are shown
in Fig. 12.   For comparison, spectra of deuterated reference protein
are also included.   The predominantly α-helix-containing myoglobin
(76%) shows its amide I absorption maximum at 1650 cm^{-1} and the
β-structure-containing ribonuclease (36%) at 1639 cm^{-1}.   The disordered polypeptide conformation in proteins displays a maximum at
about 1645 cm^{-1} and a weak shoulder at 1670 cm^{-1} [59].   Beta structure is characterized by a maximum at 1634 cm^{-1} and by a strong and

a weak shoulder at 1660 and 1680 cm^{-1}, respectively. The maximum of apometallothionein is located at 1647 cm^{-1}, confirming the presence of a substantial proportion of disordered structure. The amide I band of metallothionein has a similar shape but has its maximum shifted to 1643 cm^{-1}, indicating less disordered structure. From comparison of the shape of the spectra with that of the reference proteins, it is also evident that the remaining regular structure must be mainly of the β type. A similar conclusion is reached by a computer analysis of the far-UV CD spectrum of apometallothionein, using a program [60] based on crystallographically defined proteins [59]. Its result indicated 55% disordered structure, 6% α helix, 18% β sheet, and 21% β turn. This is also in reasonably good agreement with the values obtained from the amino acid sequence using the secondary structure prediction method of Chou and Fasman [61]: 48% disordered structure, 10% α helix, 16% β sheet, and 26% β turn. To what extent such quantitative estimates for the secondary structures of apometallothionein also apply to metallothionein remains to be seen. Both the ^{1}H NMR and IR studies suggest that the stabilization by the numerous metal-thiolate cross links is substantial and could impose folding patterns not present in the apoprotein. That differences in overall protein folding between the two forms are relatively minor is implied, however, by the observation that the Stokes radii of 16.1 Å (pH 7.5, 20°C) and of 17.0 Å (pH 2, 20°C) of the native metal-containing and of the metal-free form, respectively, are close [62].

Inseparable from those of metal binding are the effects of electrostatic factors on protein conformation. On complex formation with the apoprotein the bivalent metal ions introduce a total of six negative charges which are not completely balanced by the basic amino acid residues in the polypeptide chain [17]. Thus, depending on the isoprotein type, Cd(II)- or Zn(II)-metallothionein possesses at neutral pH an excess of two or three negative charges (see Sec. 2). A recent study has shown that a further increase in negative charge, as by the loss of positive charges on exposure of metallothionein to pH·11, alters appreciably the far-UV CD in the direction of a loss

of secondary structure (see Sec. 4.5.1). The effects on the CD
spectrum are correlated with the appearance of resonances from
deprotonated lysine residues in the ^{1}H NMR spectrum (Fig. 10)[49].

6. COORDINATION GEOMETRY OF METAL-BINDING SITES IN METALLOTHIONEIN

The simplicity of the far-UV absorption spectra of Zn(II)-, Cd(II)-,
and Hg(II)-metallothionein and their resemblance to those of tetrahedral halide complexes of the same metals (see Sec. 4.1) has suggested that the metal-binding sites of metallothionein are both
chemically and structurally similar and that their coordination
geometry is close to that of a tetrahedron [34]. This arrangement
of the thiolate ligands has also been confirmed by a number of independent spectroscopic studies in which the group 2B metal ions have
been replaced by paramagnetic metal ions whose physical features are
highly sensitive to the mode of coordination.

6.1. Spectroscopic Features of Co(II)- and Ni(II)-Metallothionein

Valuable information on the geometry of metal binding in metallothionein has been obtained from studies on the optical, chiroptical,
and magnetochiroptical properties of derivatives in which the d^{10}
metal ions are replaced by chemically related transition metal ions
such as Co(II) [33,63] and Ni(II) [33]. The preparation of such
derivatives is described in Sec. 3. Co(II)-metallothionein shows a
green color. Ni(II)-metallothionein is yellow. Their spectroscopic
features are summarized in Figs. 13 and 14.

6.1.1. *d-d Absorption Spectra*

The absorption spectrum of Co(II)-metallothionein shows very distinct
features in both the low- and high-energy regions (Fig. 13) [33,63].
In the low-energy region there are broad bands with maxima at 600,

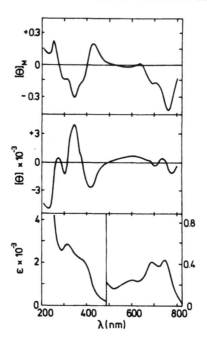

FIG. 13. Magnetic circular dichroism (top), circular dichroism
(middle), and absorption spectra (bottom) of rabbit-liver Co(II)-
metallothionein in 0.05 M Tris-HCl, pH 7.0. The units employed
refer to metal concentration. (Reprinted with permission from
Ref. 33. © 1981 American Chemical Society.)

690, and 743 nm, with molar absorptivities of 245, 420, and 435 M^{-1}/cm,
respectively. These features are due to a d-d transition. They
closely resemble those of inorganic tetrathiolate complexes [64-67]
and of Co(II) derivatives of crystallographically defined metalloproteins with four cysteine ligands ($[Co(Cys)_4]^{2-}$), for example,
Co(II)-rubredoxin [68] and horse-liver alcohol dehydrogenase with
Co(II) bound at the noncatalytic metal-binding site [69]. The
resolved band pattern has been assigned to the spin-allowed
$\nu_3[^4A_2 \rightarrow {}^4T_1(P)]$ transition. The splitting of this transition into
three components with an energy separation of 2175 and 1146 cm^{-1} is
larger than that to be expected from spin-orbit coupling alone.
This is probably due to distortion from T_d symmetry, which results
in removal of the degeneracy of the upper-state $^4T_1(P)$. As expected

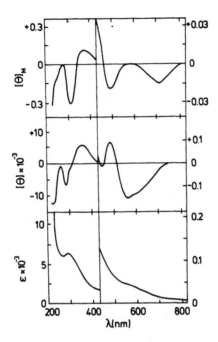

FIG. 14. Magnetic circular dichroism (top), circular dichroism
(middle), and absorption spectra (bottom) of rabbit-liver Ni(II)-
metallothionein in 0.05 M Hepes buffer, pH 7.6. The units employed
refer to metal concentration. (Reprinted with permission from Ref.
33. © 1981 American Chemical Society.)

for Co(II) complexes of T_d symmetry, there is also a spin-allowed
$\nu_2[^4A_2 \rightarrow {}^4T_1(F)]$ transition in the near-infrared region [33,63]. It
manifests itself as a broad band with a maximum at 1275 nm (ε =
165 M^{-1}/cm) and a shoulder at 1150 nm (ε = 155 M^{-1}/cm).

In contrast to Co(II)-metallothionein, the absorption spectrum
of Ni(II)-metallothionein shows in the low-energy region no well-
resolved absorption band (Fig. 14). Instead, there is a gradual
increase in electronic absorption starting at about 1150 nm with
indications of a shoulder near 750 nm. The persistence of absorption bands of similar intensity in this region has been observed in
crystallographically defined model complexes containing a tetrahedral NiS_4 core [65,70]. The absorption is thus taken as evidence
for similar nonplanar Ni(II) coordination in Ni(II)-metallothionein.

The absence of distinct d-d absorption bands is explained in part by
the superimposition of the low-energy edge of the first electron
transfer band centered at about 560 nm (Fig. 14). A similarly poor
resolution of the d-d bands has been noted for the Ni(II) derivative
of azurin, a protein in which Ni(II) is known to be bound tetrahedrally [71].

6.1.2. Magnetic Circular Dichroism Spectra

The strongest evidence for T_d coordination in Co(II)- and Ni(II)-
metallothionein [33,63] comes from their low-energy MCD spectra
(Figs. 13 and 14). Thus Co(II)-metallothionein exhibits an intensive
negative MCD band at 757 nm, a pronounced shoulder at 690 nm, and
both a weak positive and a weak negative band at 628 and 580 nm,
respectively. The same pattern was theoretically predicted and
experimentally observed in a number of inorganic tetrahedral and
pseudotetrahedral Co(II) complexes [72-74] and in Co(II)-substituted
metalloproteins with this coordination geometry [75]. The MCD spectrum of Ni(II)-metallothionein also shows a strong negative ellipticity band at 710 nm which can be assigned to the $\nu_3[^3T_1(F) \rightarrow {}^3T_1(P)]$
d-d transition and which is diagnostic of tetrahedral Ni(II) com-
plexes [76]. In about the same spectral location an absorption band
was recently observed in the Ni(II) derivative of aspartate transcarbamylase (720 nm) [77] and in horse-liver alcohol dehydrogenase
in which the active site Zn(II) was replaced by Ni(II) (680 nm) [78].
In both enzymes the metal ion is bound to ligands arranged in T_d
symmetry.

6.1.3. Electron Transfer Spectra

Since apometallothionein does not absorb at wavelengths above 250 nm
(see Sec. 4.1), it follows that all high-energy absorption bands of
Co(II)- and Ni(II)-metallothionein (Figs. 13 and 14) also arise from
metal binding [33]. Based on current theory, these absorption bands
must be attributed to electron transfer excitations or, below 250 nm,
to ligand internal transitions [79] affected by metal binding. The

same features are typical of other Co(II) and Ni(II) proteins. Thus
the 320-nm band of Co(II)-metallothionein (ε_{Co} = 980 M^{-1}/cm) is also
seen in the Co(II)-substituted forms of rubredoxin [68] and liver
alcohol dehydrogenase [69,80,81]. More details on this high-energy
region are given by the CD and MCD spectra, where seven to eight
bands or shoulders can be resolved (Figs. 13 and 14). The CD profiles of the Co(II) and Ni(II) derivatives are similar, yet they are
appreciably displaced relative to each other on the wavelength scale.
Thus in Co(II)-metallothionein the first electron transfer absorption
band and the accompanying negative ellipticity band are located near
415 nm, while in Ni(II)-metallothionein these two bands occur already
at 560 nm. When the two CD spectra are aligned on an energy scale
such that the first bands of the Co(II) and Ni(II) derivatives coincide, the positions of the next three higher-energy CD bands also
coincide very well (Fig. 15). From this correspondence it is inferred
that the bands arise from homologous electron transfer excitations in
the two metal derivatives and hence that the geometries of the metal
complexes in Co(II)- and Ni(II)-metallothionein are comparable. Tentatively, the first three resolved bands have been assigned to excitations of cysteine-sulfur electrons from two π MOs and one σ MO to
a metal-centered MO [33].

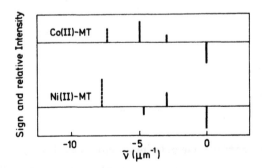

FIG. 15. Bar diagram representation of the relevant part of the
circular dichroism spectra of rabbit-liver Co(II)- and Ni(II)-
metallothionein. The first electron-transfer transition of each
spectrum has been adjusted to 0.0 on the energy scale. (Reprinted
with permission from Ref. 33. © 1981 American Chemical Society.)

6.2. Perturbed Angular Correlation of Gamma-Ray
Spectroscopy of ^{111m}Cd-Metallothionein

Information concerning the coordination geometry of the metal-binding
sites of metallothionein has been obtained recently by the application
of perturbed angular correlation of γ-ray (PAC) spectroscopy [40].
In this nuclear spectroscopic method quadrupole interactions between
electric field gradients of the coordination environment and the
excited state of an appropriate nucleus employed as a spectroscopic
probe are monitored. One of the best suited excited nuclei for such
purposes is the 49-min isomer of ^{111}Cd (i.e., ^{111m}Cd), which is generated by bombardment of ^{108}Pd with 21-MeV alpha particles. This
nucleus decays to the ground state via an intermediate 84-ns state
and the consecutive emission of two gamma quanta, γ_1 and γ_2, in
different directions. The emissions are related by the angular
correlation function $W(\theta,t)$, where θ is the angle between γ_1 and γ_2
and t the delay time of the emission of γ_2 with respect to γ_1 [82].
If this intermediate state interacts with its surrounding (i.e., is
split by a hyperfine interaction), then the angular correlation
function of the probe is perturbed by the nuclear quadrupole interaction (NQI). The result is the development in time of hyperfine
interaction frequencies modulating the amplitude of $W(\theta,t)$. In PAC
measurements such time spectra are determined by measuring with a
four-detector device at fixed 180° and 90° angles the coincident
counting ratios $W_{180°}$ and $W_{90°}$ as a function of delay time. From
the experimental data the NQI parameter sets (ω_i, η_i) are evaluated
according to the equation

$$\frac{W_{180°}(t)}{W_{90°}(t)} = \frac{1 + \Sigma_i \, A_i G(t, \, \omega_i, \, \eta_i)}{1 - 1/2 \, \Sigma_i \, A_i G(t, \, \omega_i, \, \eta_i)} \tag{2}$$

where A_i is the partial unperturbed amplitude for a particular NQI
parameter set and $G(t, \, \omega_i, \, \eta_i)$ the corresponding perturbation function.

PAC spectra of ^{111m}Cd(II)-containing metallothionein are shown
in Fig. 16. ^{111m}Cd(II) suitable for the measurements was prepared

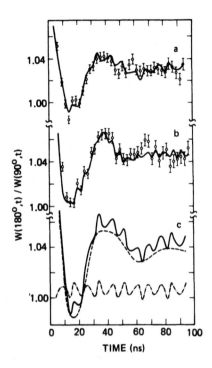

FIG. 16. Perturbed angular correlation of γ-ray (PAC) spectra of rabbit-liver ^{111m}Cd(II)-metallothionein in 65% (weight) sucrose, 0°C. W(180°)/W(90°) is plotted versus delay time. The fully drawn curves represent least squares fits to the spectra. The bars indicate ±1 standard deviation. In all cases the viscosity of the sucrose solution has immobilized the protein within the time scale of the experiment. (a) Fully reconstituted ^{111m}Cd(II)-metallothionein, pH 8.0; (b) 1 equivalent ^{111m}Cd added to (Zn,Cd)-metallothionein, pH 8.0; (c) resolution of the least squares fit of the spectrum in (a) (solid line) yields two NQIs with parameters ω_1 = 116 MHz, η_1 = 0.55 (dashed line), and ω_2 = 579 MHz, η_2 = 0.51 (stippled line). (From Ref. 40.)

by diluting trace amounts (10^{-13} mol) of freshly generated carrier-free ^{111m}Cd(II) with appropriate amounts of "cold" Cd(II). The metal was subsequently incorporated into the protein either uniformly by mixing it with apometallothionein at low pH and adjusting to neutrality (see Sec. 3) or nonuniformly by replacing 1 equivalent of Zn(II) in (Zn,Cd)-metallothionein by direct exchange at neutral pH [40]. The PAC spectra of both samples display the features of

damped oscillating functions (Fig. 16a and b). Resolution of spectrum (a) into frequency components shows that the uniformly labeled
[111m]Cd(II)-metallothionein displays characteristic periodicities
with the frequencies ω_1 of about 120 MHz and ω_2 of about 580 MHz
with relative amplitudes of about 80% and 20%, respectively (Fig.
16c). The occurrence of two very different frequencies is consistent
with the existence of two quite different distorted coordination
geometries for Cd(II) in metallothionein. The strongly damped low-
frequency oscillation is in order of magnitude close to the 65-MHz
frequency observed in horse-liver alcohol dehydrogenase when the
Zn(II) of the structural metal site is substituted by [111m]Cd(II)
[83]. From crystallographic data it is known that at this site four
cysteine thiolate ligands are arranged in a coordination geometry
close to that of a regular tetrahedron. The large relative amplitude
of ω_1 in the PAC spectrum of [111m]Cd(II)-metallothionein suggests that
80% of the metal-binding sites have similar, but weakly distorted T_d
symmetry. By contrast, the undamped 580-MHz frequency ω_2 must arise
from sites of quite different geometric properties. Calculations
based on a point-charge model suggest that a square-planar Cd(II)-
tetrathiolate complex would exhibit a frequency of about 880 MHz.
Alternatively, frequencies of about 400 and 580 MHz could also indicate an extension of the coordination sphere of [111m]Cd(II) to that
of an octahedron, with water or carboxyl ligands, respectively,
serving as the additional ligands. Thus a definitive assignment of
the 580-MHz frequency must wait until additional structural information on the organization of the metal-binding sites becomes available.
Resolution of the PAC spectrum of the nonuniformly labeled, still
some Zn(II)-containing [111m]Cd(II)-metallothionein (Fig. 16b) also
yields two oscillating components. However, their frequencies are
higher (ω_1 = 149 MHz, ω_2 = 714 MHz) than those of the solely Cd(II)-
containing protein (Fig. 16a) [40]. This difference in NQI probably
reflects some geometrical differences in the metal-binding sites
due to the unequalness of some of the sulfur-metal distances in the
two forms examined (see also Sec. 7.1).

6.3. Extended X-Ray Absorption Fine-Structure (EXAFS) Measurements on Zn(II)-Metallothionein

The high x-ray fluxes available from synchrotron radiation have recently made it possible to obtain EXAFS spectra of sheep-liver metallothionein [84]. In this method, x-rays absorbed by the metal atom induce the emission of photoelectrons, which, when backscattered by ligand atoms, modulate the x-ray absorption. Analysis of the observable variations in the x-ray absorption coefficient yields information on the number and type of ligand atoms which surround the absorbing metal atom and on the corresponding interatomic distances. Although the method is applicable to all elements with sufficient absorption cross section, EXAFS has thus far been applied only to the zinc K edge of Zn(II)-metallothionein. The spectrum and its Fourier transform are remarkably simple. They suggest that within the limit of the technique all the zinc sites in the molecule are essentially equivalent and involve the tetrahedral attachment of four sulfur atoms, each with a zinc-sulfur distance of approximately 2.29 Å. No evidence was obtained for the presence of other atoms within a distance of about 4 Å from Zn(II).

7. METAL-THIOLATE CLUSTERS IN METALLOTHIONEIN

Sulfur ligands have a propensity to form polynuclear complexes with many metal ions. In biological systems this tendency is illustrated by the various cluster structures in the iron sulfur proteins, where inorganic sulfide serves as a bridging ligand between iron ions bound to the polypeptide chain through protein thiolate ligands. Metallothionein contains no significant quantities of inorganic sulfur and hence no such mixed structures [8,16]. However, the crowding of thiolate ligands in metallothionein and the unusual stoichiometry of about three thiolate ligands per bivalent metal ion (i.e., $[Me(II)(Cys^-)_3]^-$) typical of all metallothioneins have prompted the suggestion that in this protein the metal complexes

may exist as clusters in which neighboring metal ions are linked
by one or more bridging thiolate ligands [28]. This view has now
received ample support from spectroscopic evidence that in metallothionein most bivalent metal ions are bound in T_d symmetry to four
thiolate ligands (see Secs. 4.1 and 6). This participation of four
sulfur atoms in the coordination of each metal ion requires that of
the 20 cysteine residues, eight provide sulfur bridges between adjacent metal ions. The occurrence of such threefold-coordinated bridging sulfur ligands besides the twofold-coordinated terminal sulfur
ligands is nicely manifested by the broadening of the sulfur core
electron binding energy profile in the x-ray photoelectron spectrum
(see Sec. 4.4). In addition, compelling direct evidence for the
existence of metal-thiolate clusters has been forthcoming from a
number of spectroscopic investigations, among them most prominently
^{113}Cd NMR homonuclear decoupling measurements on ^{113}Cd(II)-metallo-
thionein and from electron spin resonance and magnetic susceptibility
studies on Co(II)-metallothionein.

7.1. Evidence for Clusters from ^{113}Cd NMR Studies

The potential of ^{113}Cd NMR spectroscopy as a highly specific and
natural probe of its environment in metallothionein has been exploited
in an elegant study by Otvos and Armitage [11]. The usefulness of
this method for the study of metallothionein relies both on the sensitivity of the ^{113}Cd chemical shift to subtle differences in coordination which allows the resolution of signals from different sites
(see Sec. 4.5.3) and on the observation that the resonances are split
into multiplets as a result of ^{113}Cd-^{113}Cd interaction. The multiplet
feature offers the possibility to identify the interacting resonances
by homonuclear decoupling studies. Metallothionein suitable for such
studies was isolated from the liver of rabbits subjected to repeated
injections of ^{113}CdCl$_2$. Residual Zn(II) was replaced in the course
of the isolation of the protein by exchange with ^{113}CdCl$_2$ added to
the tissue extract. The solely ^{113}Cd(II)-containing form of metallo-

thionein thus obtained displayed a proton decoupled spectrum with
only eight peaks in the spectral region between 611 and 670 ppm
downfield from the ^{113}Cd resonance of 0.1 M Cd(ClO$_4$)$_2$, as opposed
to the 15 separate resonances counted in the spectrum of a sample
containing both ^{113}Cd(II) and Zn(II) [11]. The occurrence of eight
peaks in the spectrum of a protein with seven metal-binding sites
is not clear. It has been attributed tentatively to some residual
heterogeneity in the metallothionein samples employed in the study.
A spatial relationship between the metal-binding sites was inferred
from selective homonuclear decoupling measurements. In this technique, each of the eight resonances was in turn irradiated by a
selective decoupling pulse and its effect on the multiplet structure
of the other resonances observed. The collapse of a ^{113}Cd multiplet
was taken to indicate spin coupling with the irradiated ^{113}Cd site.

The decoupling analysis reveals that the eight multiplets of
^{113}Cd-metallothionein-2 from rabbit arise from two distinct metal-
thiolate clusters designated A and B. Cluster A, which is thought
to contain four metal ions, gives rise to five signals (Fig. 17 and
Table 5). Two of the signals (7 and 7') possess about half the
intensity of the remaining three signals (1 and 1', 5 and 5', 6 and
6'), suggesting the existence of a variant form of cluster A designated as cluster A'. The latter is thought to occur in about equal
abundance as cluster A. The three remaining signals (2, 3, 4) are
attributed to a three-metal aggregate designated as cluster B. The
rather low intensity of its resonances was attributed tentatively
to a lower metal occupancy of the cluster B site by ^{113}Cd (Table 5).
From spin-coupling information and from optical data [63], Otvos and
Armitage [11] proposed a three-dimensional model for the proposed
clusters of metallothionein. The model that depicts all metal ions
as tetrahedrally linked to four thiolate ligands requires exactly
20 cysteine residues, of which 11 are thought to be located in
cluster A/A', respectively, and 9 in cluster B. The differences in
^{113}Cd chemical shift are explained in part by the unequal coordination environment created by the different number of bridging ligands
at the various sites.

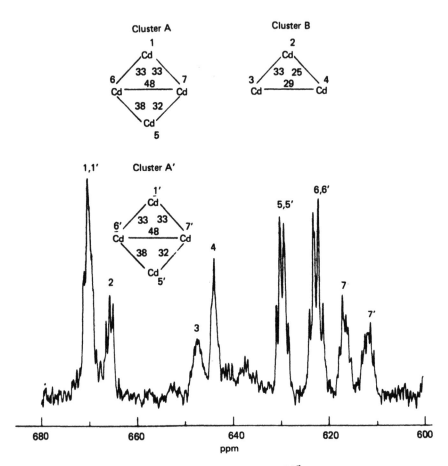

FIG. 17. "Fully relaxed" proton decoupled [113]Cd NMR spectrum of
rabbit-liver Cd(II)-metallothionein-1 (∿8 mM) and schematic repre-
sentations of the metal cluster structures in the protein, established by homonuclear decoupling experiments (see also Table 5).
The spin-coupling connections between adjacent metal ions in the
clusters are indicated by the lines connecting the Cd(II) in the
schematic structures. The number beside each Cd(II) refers to the
corresponding resonance in the [113]Cd NMR spectrum and the numbers
appearing on the lines connecting the metals are measured two-bond
coupling constants (±3 Hz). The cysteine thiolate ligands that
bridge the adjacent metals have been omitted from the drawings for
clarity. (Reproduced by permission from Ref. 11.)

TABLE 5

Chemical Shifts and Integrated Areas of the ^{113}Cd
Resonances in Cd-Metallothionein

Resonance[a]	Chemical shift (ppm)	Relative integrated area[b]
1, 1'	670.4, 670.1	1.03
2	665.1	0.52
3	647.5	0.48
4	643.5	0.65
5, 5'	629.8, 628.8	0.93
6, 6'	622.1, 622.0	1.04
7	615.9	0.48
7'	611.2	0.42
Cluster A[c]	--	3.90
Cluster B[d]	--	1.65
Cluster A plus cluster B	--	5.55

[a]The resonance numbering scheme corresponds to that which appears in Fig. 17.
[b]The areas of the resonances in the Cd-MT-1 spectrum in Fig. 17 are reported relative to the average area of the overlapping resonances 1 and 1', 5 and 5', and 6 and 6'. These three multiplets are each assumed to represent single equivalents of ^{113}Cd^{2+}.
[c]The resonances assigned to the ^{113}Cd^{2+} ions in cluster A are 1, 1', 5, 5', 6, 6', 7, and 7'.
[d]The resonances assigned to the ^{113}Cd^{2+} ions in cluster B are 2, 3, and 4.
Source: Ref. 11.

7.2. Evidence for Clusters from Optical and Magnetic Studies on Co(II)-Metallothionein

Independent evidence for the existence of cluster structures in metallothionein and information concerning their mode of formation comes also from a comparative study of the spectroscopic and magnetic properties of derivatives of metallothionein containing different equivalents of Co(II) [12]. The effect of increasing Co(II)-

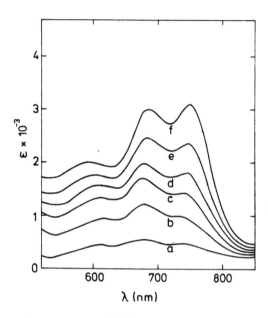

FIG. 18. Absorption spectra of the d-d region of rabbit-liver
Co(II)-metallothionein as a function of Co(II)-to-protein ratio.
Moles of Co(II) per mole of apometallothionein: (a) 1.2; (b) 2.4;
(c) 3.4; (d) 4.6; (e) 5.5; (f) 6.8.   (From Ref. 12.)

to-apometallothionein ratios on the absorption spectrum in the d-d
region is shown in Fig. 18. As discussed in Sec. 6.1, the features
of the protein fully occupied by Co(II) are typical of a distorted
tetrahedral structure. Essentially the same features of the $\nu_3 [^4A_2 \rightarrow {}^4T_1 (P)]$ transition are displayed by the spectra of the incompletely
occupied form of Co(II)-metallothionein, indicating T_d coordination
of the Co(II) sites at all stages of reconstitution. The same conclusion is supported by electron spin resonance (ESR) measurements
at 4 K. However, there are indications of some qualitative spectral
changes attending the filling up of the binding sites with Co(II).
Up to binding of 4 equivalents of Co(II) the d-d profile increases
proportionally without appreciable alteration in shape but upon
further addition of Co(II) there is a progressive blue shift of the
high-energy d-d band from 600 to 590 nm and a change in the relative
intensities of the 690- and 743-nm maxima to yield the spectral

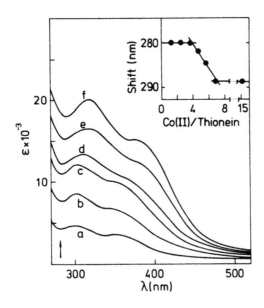

FIG. 19.   Absorption spectra of the electron-transfer region of
rabbit-liver Co(II)-metallothionein as a function of Co(II)-to-
protein ratio.   Moles of Co(II) per mole apometallothionein:
(a) 1.2; (b) 2.4; (c) 3.4; (d) 4.6; (e) 5.5; (f) 6.8.   (Inset)
Dependency of position of the 290-nm minimum (indicated by arrow)
on Co(II)-to-protein ratio.   (From Ref. 12.)

profile of the fully occupied form (Fig. 18).   Analogous spectral

changes upon occupation of metal-binding sites with Co(II) occur in

the region of electron-transfer transitions below 500 nm (Fig. 19).

From MCD and CD measurements of fully reconstituted Co(II)-metallo-

thionein it was concluded in Sec. 6.1.2 that the absorption envelope

contains at least four overlapping electron transfer transitions.

Upon addition of up to 4 equivalents of Co(II) to apometallothionein

there are no qualitative changes discernible.   However, upon further

addition the maximum at 305 nm and the shoulder at 370 nm are shifted

to 320 and 400 nm, respectively.   The spectral changes occurring

above 4 equivalents of Co(II) bound per mole of protein are also

reflected in a red shift of the absorption minimum from 280 nm to

290 nm (Fig. 19, inset).

Differences comparable to those of partially and fully saturated forms of Co(II)-metallothionein also exist between the spectra of crystallographically defined tetrahedral mononuclear and tetranuclear anionic Co(II)-benzenethiolate (SPh) complexes (i.e., $[Co(SPh)_4]^{2-}$ and $[Co_4(SPh)_{10}]^{2-}$, respectively) [64]. The former complex contains only twofold-coordinated or terminal sulfur ligands, while the latter, constituting a thiolate molecular cluster, contains besides four terminal ligands six threefold-coordinated or bridging sulfur ligands. Compared to the mononuclear species, the complex with the cluster structure shows better separation of the bands belonging to the $\nu_3[^4A_2 \rightarrow {}^4T_1(P)]$ transition, a difference comparable to that existing between Co(II)-metallothioneins with low and high Co(II) content, respectively. Similarly, as in the Co(II)-rich metallothionein, the electron-transfer absorption bands of the Co(II)-rich model compound are shifted by about 30 nm toward higher wavelength [12]. This shift is probably a reflection of the relatively stronger polarization to which bridging sulfur ligands are subjected by the coordination to two Co(II).

The observation that qualitative spectral changes occur only after about 4 equivalents of Co(II) are bound suggests that the first four Co(II) bind to separate binding sites, forming tetrahedral complexes with terminal thiolate ligands only, and that subsequently these sites are linked to form Co(II)-thiolate clusters. The same conclusions are supported by ESR measurements of Co(II)-metallothionein [12]. Metallothionein saturated with Co(II) exhibits only a very weak signal (5%) compared to an equivalent concentration of $[CoCl_4]^{2-}$. The spectrum reflects rhombic distortion with g values $g_x \approx 5.9$, $g_y \approx 4.2$, and $g_z \approx 2.0$ (Fig. 20, left). The ESR spectra of Co(II)-metallothionein of different Co(II)-to-apometallothionein ratios display qualitatively the same profile but differ remarkably in intensity. Up to binding of 4 equivalents of Co(II), the intensity of the signal at $g_x \approx 5.9$ increases linearly with Co(II) concentration and is comparable to the signal intensity obtained with the high-spin model complex $[CoCl_4]^{2-}$ (Fig. 20, right). At the highest point

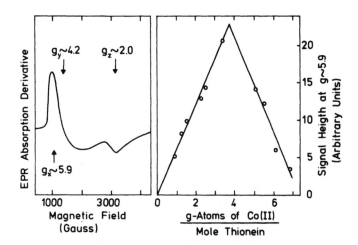

FIG. 20. (Left) Electron spin resonance (ESR) spectrum of rabbit-liver Co(II)$_7$-metallothionein at 4 K; (right) dependency of ESR signal size on Co(II)-to-protein ratio. As a representative measure of the magnitude of the high-spin Co(II) spectrum the amplitude at $g_x \approx 5.9$ is plotted. (From Ref. 12.)

[3.6 Co(II) per apometallothionein] double integration applied to the ESR spectrum of Co(II)-metallothionein revealed 95% of the Co(II) as being ESR detectable. Beyond 4 Co(II) equivalents per apometallothionein, further addition results in a proportionate loss of the ESR signals, yielding on saturation a nearly diamagnetic complex.

The loss of paramagnetism above about 4 Co(II) equivalents per thionein is probably the result of antiferromagnetic coupling of the metals bridged by common thiolate ligands. The intensity drop in the ESR spectra closely parallels the changes observed in the electronic spectra, signaling the transition from magnetically noninteracting tetrahedral Co(II)-tetrathiolate complexes to polynuclear tetrahedral structures. The sharp transition from the nonclustered to the clustered structure indicates preferential formation of the noninteracting complexes.

The reduced paramagnetism of fully Co(II) saturated metallothionein implied by the ESR measurements was confirmed also by magnetic susceptibility measurements using the NMR technique of

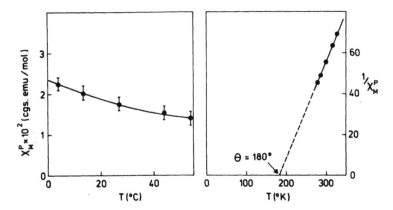

FIG. 21. Temperature dependence of the paramagnetic component of the molar magnetic susceptibility (χ_M^P) of rabbit-liver Co(II)-metallothionein (left) with evidence for Curie-Weiss law behavior (right). cm-g-s electromagnetic units. (From Ref. 12.)

Evans [12]. The paramagnetic component of the molar magnetic susceptibility χ_M^P of a preparation of Co(II)-metallothionein containing 6 Co(II) equivalents is far lower than expected from the total metal content (Fig. 21, left). Moreover, its temperature dependence measured in the temperature range accessible by the solution NMR technique and the temperature stability of the protein does not obey the Curie law. Instead, the change with temperature follows a Curie-Weiss law, $\chi_M^P = C/(T - \theta)$, with a Curie constant C of 2.1 and a Weiss temperature θ of about +180 K (Fig. 21, right). The effective magnetic moment μ_{eff} calculated from χ_M^P ranges from 6.97 to 6.15 Bohr magneton (B.M.) per molecule of Co(II)-metallothionein in the temperature range between 277 and 327 K, respectively. If all metal centers contribute equally to the measured magnetic moment, one obtains values ranging from 2.64 to 2.48 B.M. per Co(II) ion bound to metallothionein. These figures are much lower than the spin-only value of high-spin Co(II) (3.86 B.M.), giving strong support to partial spin canceling by antiferromagnetic coupling of neighboring Co(II) centers in the nearly fully occupied Co(II)-metallothionein. The occurrence of such antiferromagnetic coupling is an unambiguous indication of the

existence of cluster structures in metallothionein. The magnetic
interactions between neighboring paramagnetic high-spin Co(II) are
probably mediated by a superexchange interaction via the bridging
thiolate ligands. This mechanism of magnetic coupling presupposes
the overlap of the s and p orbitals of the bridging ligand with the
d orbitals of the metal. The observation that the temperature
dependence of the magnetic susceptibility does not show the negative
Weiss temperature typical of antiferromagnetically coupled systems,
but a strongly positive value (+180 K) reflects another special
feature of this system. Positive Weiss temperatures are usually
encountered with ferromagnetic substances. Thus this behavior of
Co(II)-metallothionein may imply the existence of some ferromagnetic
type of coupling besides the predominant antiferromagnetic interaction of the Co(II)-thiolate centers. Weak ferromagnetism occurs
in a number of solid-state antiferromagnetically coupled systems,
but it has to our knowledge not been reported for a biological compound. As in some antiferromagnets, ferromagnetic behavior could
arise from noncanceling interactions between sublattices. It is
conceivable that in Co(II)-metallothionein some ferromagnetism could
arise from some cluster-cluster interactions provided that the distance separating them is not too large. Alternatively, it could be
the result of a nonlinear spatial arrangement of the metal centers
within the clusters [85]. However, any more detailed interpretation
of these data must await extension of the magnetic susceptibility
measurements of Co(II)-metallothionein to the low-temperature range.

7.3. Cluster Models

The fixed sulfur-to-metal ratio and the rules of stereochemistry
limit the types of potential cluster structures in metallothionein
rather severely. Thus a comparison with model compounds in conjunction with the body of spectroscopic information at hand may allow
the elaboration of realistic cluster models. One of the few defined

compounds whose crystallographic structure is known and whose spectroscopic properties closely resemble metallothionein are the complexes
of Zn(II) and Co(II) with benzenethiol studied by Dance [64, 86]. In
appropriate solvents these substances form one-dimensional polymers
in which groups of four T_d coordinated bivalent metal ions [Me(II)]
are joined via six bridging benzenethiolate ligands (S) to form regularly structured Me $(II)_4S_6$ adamantane clusters. This well-known
10-vertex structure forms a regular cage and has the overall symmetry
of a tetrahedron. It may be compared to the eight-vertex cubane
cluster known to occur in 4Fe ferredoxins and in Mo-Fe-S clusters
(Fig. 22) [87]. The same adamantane structure has also been found
in a variety of other complexes of group 2B and transition metals
with sulfur ligands, and it is also present in nonmolecular form in
zinc blende (ZnS) [86]. The clusters proposed by Otvos and Armitage
[11] on the basis of ^{113}Cd NMR homonuclear decoupling studies on
metallothionein (see Sec. 7.1) resemble incomplete adamantane structures and may thus represent the first example of their occurrence
in a biological system. However, alternative cluster models should
not be ruled out either. Thus a geometric arrangement in which in

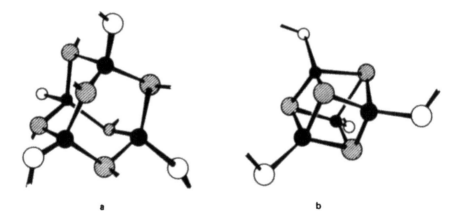

a   b

FIG. 22.   Adamantane-type (a) metal-thiolate cluster proposed for
metallothionein and cubane-type (b) cluster of clostridial 4Fe ferredoxins (adapted by permission from Ref. 87). Solid circles represent
the metal, open circles terminal sulfur ligands, and hatched circles
the bridging sulfur ligands.

analogy to the 2Fe ferredoxins two neighboring T_d Me(II) complexes
are joined edge to edge by two bridging thiolate ligands rather than
only by one as in the adamantane structure is sterically also possible. Based on both structure types and on combinations of them, a
variety of clusters can be constructed which yield the 20:7 ligand-
to-metal stoichiometry found experimentally. Thus unless completely
unambiguous spectroscopic information becomes available, a complete
elucidation of the detailed organization of the clusters in metallothionein must await the outcome of crystallographic studies on the
spatial structure of this protein.

8. STABILITY OF METAL-PROTEIN COMPLEXES
IN METALLOTHIONEIN

The chromophoric features of the metal-thiolate complexes in metallothionein provide a convenient means to assess the stability of the
metal-protein complex and the modes of metal binding and metal
release from the protein. In early studies it was found that the
Cd(II)-thiolate electron-transfer absorption band near 250 nm remained unchanged between pH 11 and 4.5 but that below this value
the absorbance rapidly decreased with the concomitant release of
Cd(II) from the protein [8]. The point of half-maximum absorbance
corresponding to the dissociation of half the metal ions was close
to pH 3 and was noticed to be dependent on protein and metal-ion
concentration. The narrow pH range over which Cd(II)-thiolate
absorption was lost suggested that more than one proton was required
to displace a Cd(II) from its binding site. Graphical analysis
showed that the titration curve could be fit to a model in which the
metal was assumed to be bound to equivalent and independent binding
sites where each site is occupied either by one Cd(II) or three
protons. This 1:3 stoichiometry was also confirmed by measuring
the average number of protons liberated on addition of Cd(II) to
apometallothionein [8]. Based on this model of competition of

protons and Cd(II) for binding sites containing three negatively
charged ligands $(L^-)_3$, an equilibrium constant

$$\beta_{Cd} = \frac{[(L^-)_3 Cd(II)][H^+]^3}{[(LH)_3][Cd(II)]} = 2.1 \cdot 10^{-5} \, M^2 \tag{3}$$

was calculated from the spectrophotometric titration data [8]. An
analogous pH-titration curve is obtained with the primarily Zn(II)-
containing equine-liver metallothionein [16] except that in accordance with the weaker binding of Zn(II) the loss of the metal occurs
at higher pH (Fig. 23 top, right) [62]. The displacement of Zn(II)
by protons is conveniently monitored by following the decrease in
the zinc-thiolate absorption near 220 nm (Fig. 23 top, left). The
difference absorbance at 220 nm of zinc-metallothionein versus apometallothionein remains unchanged between pH 11 and 6.5 but decreases
rapidly below to reach, at about pH 4.4, the point of half-maximum
absorbance. The second, less pronounced titration step, at about
pH 3, is accounted for by the loss of 5% residual Cd(II) present in
this preparation. The similarity of the pH titration curve of
Zn(II)-metallothionein to that of Cd(II)-metallothionein is another
indication that the two group 2B metal ions are bound in a similar
fashion to the protein. The course of the titration curve is again
consistent with the competition of three protons with one Zn(II)
for each binding site. The corresponding equilibrium constant has
the value

$$\beta_{Zn} = \frac{[(L^-)_3 Zn(II)][H^+]^3}{[(LH)_3][Zn(II)]} = 2.1 \cdot 10^{-9} \, M^2 \tag{4}$$

Comparison of β_{Cd} with β_{Zn} indicates that on average, Cd(II) is
bound about 10,000 times more firmly than Zn(II). A value of the
same order of magnitude (i.e., a 3000-fold difference) was found by
competitive displacement of Cd(II) by Zn(II) [8].

Estimates of the actual stability constants of the complexes
of Cd(II) and Zn(II) with the binding sites $(L^-)_3$ have been made
previously from the equilibrium constants above and from an assumed
value, $K_a = 10^{-9}$ M, for the acid dissociation constant of the thiol

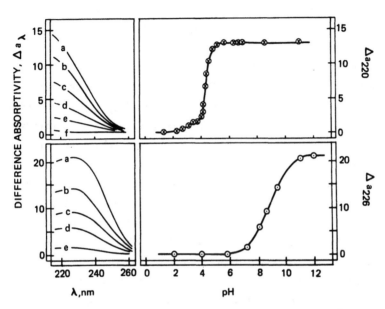

FIG. 23. Spectrophotometric titration of equine-liver Zn(II)-
metallothionein and of apometallothionein. (Top, left) Family of
zinc thiolate difference absorption spectra obtained by subtracting
spectrum of apometallothionein (pH <2) from spectra of Zn(II)-
metallothionein recorded at pH (a) 6.0, (b) 4.6, (c) 4.4, (d) 4.2,
(e) 3.9, and (f) 1.4. (Top, right) Plot of difference absorptivity
Δa_{220} (ml/mg cm) versus pH. Measurements were made in a 1-cm-path-
length cell on solutions containing $1.3 \cdot 10^{-5}$ M equine-liver
($Zn_{5.7}$, $Cd_{0.3}$) metallothionein and 0.1 M NaClO$_4$. pH adjustment
was made with HClO$_4$. (Bottom, left) Family of difference absorption spectra obtained by subtracting the low pH spectrum of apometallothionein (pH <2) from spectra of apometallothionein recorded
at pH (a) 11.0, (b) 9.5, (c) 8.8, (d) 8.0, and (e) 7.4. (Bottom,
right) Plot of difference absorptivity Δa_{226} (ml/mg cm) versus pH.
Measurements were made under exclusion of O$_2$ in a 0.02-cm cell on
solutions containing $5 \cdot 10^{-4}$ M equine-liver apometallothionein
and 0.1 M NaClO$_4$. pH was adjusted by mixing the apoprotein with
an equal volume of an appropriate 0.2 M sodium phosphate-0.2 M
sodium pyrophosphate buffer. (From Ref. 62.)

groups of apometallothionein [91]. That this figure was realistic
is now also documented by direct spectrophotometric titration of
thiol dissociation in apometallothionein (Fig. 23 bottom, right).
Free thiolate ligands exhibit a characteristic strong absorption
band which in simple thiols is located between 230 and 240 nm [88,
89]. In apometallothionein this band is manifested most prominently
in the difference spectrum of the fully deprotonated (pH >11) versus
the fully protonated form (pH <2). It has a difference maximum at
226 nm with a molar difference absorptivity, $\Delta\varepsilon_{thiolate}$, of about
6000 (Fig. 23 bottom, left). When apometallothionein is titrated
with base, deprotonation of the thiol groups is observed between
pH 7 and 12. The pH value of the inflection point (pK_a) is about
8.9, in good agreement with that expected for model thiols [90].
The Henderson-Hasselbach slope $\Delta pH/\Delta \log$ ([thiolate]/[thiol]) characterizing the deprotonation is, however, less than 1.0, indicating
either differences in the pK_a values of the thiol groups or the
influence of electrostatic effects typical for the titration of
polyelectrolytes [92] or both. The large shift of the midpoint of
the spectrophotometric titration curve from 8.9 in apometallothionein
to about 4.4 in Zn(II)-metallothionein and to about 3.0 in Cd(II)-
metallothionein [8] provides a measure of the affinity of Zn(II) and
Cd(II) for the protein. Thus, on the basis of the model of competition of each Me(II) with three protons, the association constant of
the complex is given by

$$K_{Me} = \frac{[(L^-)_3 Me(II)]}{[(L^-)_3][Me(II)]} = \beta_{Me}\beta_{H_3} \qquad (5)$$

where

$$\beta_{H_3} = \frac{[(LH)_3]}{[(L^-)_3][H^+]^3} = \frac{1}{{}'K_{LH} \quad {}''K_{LH} \quad {}'''K_{LH}} \qquad (6)$$

where ${}'K_{LH}$, ${}''K_{LH}$, and ${}'''K_{LH}$ are the acid dissociation constants of
the three protonated sulfur ligands of the binding site. When making
the assumption that

$$'K_{LH} = ''K_{LH} = '''K_{LH} = K_a = 10^{-8.9} \tag{7}$$

one obtains for

$$K_{Cd} = \frac{2.1 \cdot 10^{-5}}{(10^{-8.9})^3} \approx 1 \cdot 10^{22} \; M^{-1} \tag{8}$$

and for

$$K_{Zn} = \frac{2.1 \cdot 10^{-9}}{(10^{-8.9})^3} \approx 1 \cdot 10^{18} \; M^{-1} \tag{9}$$

These values are in the range of the cumulative association constants expected on the basis of published values for the stepwise association constants of thiol ligands [8,93].

At pH values below the pK_a of the thiol group, the apparent association constants $K_{Me,app}$ of the metal-protein complex is given approximately by

$$K_{Me,app} = \frac{K_{Me}}{\beta_{H_3} \cdot 10^{-3pH}} \tag{10}$$

For

$$\beta_{H_3} = \frac{1}{K_a^3} = 10^{26.7} \tag{11}$$

one obtains

$$K_{Me,app} = \frac{K_{Me}}{10^{26.7-3pH}} \tag{12}$$

By this expression the apparent association constants at pH 7 are about $2 \cdot 10^{16} \; M^{-1}$ for Cd(II) and about $2 \cdot 10^{12} \; M^{-1}$ for Zn(II).

The empirical model of binding of group 2B metal ions to apo-metallothionein employed above is consistent with the stoichiometry of about three cysteinyl residues per bivalent metal ion (Σ Cd + Zn) found in all well-characterized forms of this protein [17]. Hence it was proposed that $[Me(II)(Cys^-)_3]^-$ forms the basic metal-binding unit in metallothionein [5]. These negatively charged units are

the principal determinants of the overall negative charge of the
metal-containing proteins. However, these nominal stoichiometric
units were not meant to indicate the existence of rather unusual
three-coordinated metal complexes [17]. In fact, 'as pointed out first
by Weser and Rupp [28] and Webb [14], the 3:1 stoichiometry is entirely
consistent with the now well-established tetrahedral coordination of
bivalent metal ions to four thiolate ligands, one of which is serving
as bridging ligand between adjacent metal centers in the cluster.

The identification of thiolate clusters in metallothionein
(see Sec. 7) renders somewhat doubtful the simplifying assumption
[8] that each metal ion is bound equally firmly to the protein and
that the metal sites in their reactivity are independent of one
another. In fact, preliminary studies indicate that metal binding
to some sites is cooperative, that the buildup of Zn(II) and Cd(II)
complexes is accompanied by qualitative changes in the absorptive
and chiroptical properties of the complex and that there are at
least two classes of metal-binding sites [62]. Nonetheless, the
principal features of the above metal-binding model can be retained.
Adaptation of the reaction equilibrium formulated for binding of the
metal to independent sites to the formation of clusters yields

$$[(LH)_n] + m[Me(II)] \rightleftharpoons [(L^-)_n Me(II)_m] + n[H^+] \qquad (13)$$

where n is the number of thiolate ligands and m the number of metal
ions in the relevant cluster, and where the equilibrium constant is
given by

$$\beta_{cluster} = \frac{[(L^-)_n Me(II)_m][H^+]^n}{[(LH)_n][Me(II)]^m} \qquad (14)$$

By following the formation of the Cd(II) and Zn(II) complexes with
apometallothionein, either at constant pH or at constant metal-ion
concentration, it is possible to obtain information on m, n, and
$\beta_{cluster}$. Thus by titration experiments (not shown) in which at
pH 2 increasing concentrations of Cd(II) were added to the apometallothionein, it was found that Cd(II) binds pairwise (m = 2) to the apoprotein, presumably by forming a thiolate-bridged binary cluster.
The same mode of cooperative complex formation of a pair of metal

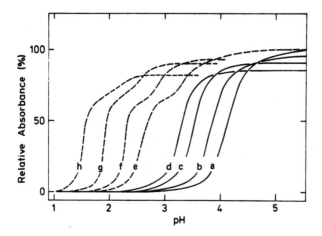

FIG. 24. Effect of metal-ion concentration on spectrophotometric
pH titration of Zn(II)- and Cd(II)-metallothionein from rabbit liver.
Solid lines: Titration of $5 \cdot 10^{-6}$ M Zn(II)-metallothionein, in
0.005 M sodium succinate with HCl in the presence of (a) $1.16 \cdot 10^{-4}$ M;
(b) $1.16 \cdot 10^{-3}$ M; (c) $1.16 \cdot 10^{-2}$ M; and (d) $1.16 \cdot 10^{-1}$ M ZnSO$_4$.
Dashed lines: Titration of $8 \cdot 10^{-6}$ M Cd(II)-metallothionein, in
0.005 M sodium succinate/0.005 M sodium phosphate with HCl, in the
presence of (e) $5 \cdot 10^{-5}$ M; (f) $5 \cdot 10^{-4}$ M; (g) $4.6 \cdot 10^{-3}$ M; and
(h) $4.6 \cdot 10^{-2}$ M CdSO$_4$. In the titrations of Zn(II)-metallothionein,
"% relative absorbance" refers to measurements at 220 nm made relative
to the absorbance measured at neutral pH, in the absence of added
Zn(II). For Cd(II)-metallothionein, "% relative absorbance" refers
to analogous measurements at 250 nm. (From Ref. 62.)

ions (m = 2) with the apoprotein was established for Zn(II) at pH
3.4 [62]. Thus it appears that at least at low pH, pairwise binding
of Cd(II) and Zn(II) is energetically favored over the binding of the
single metal ions.

The effect of pH changes on the formation of the complexes in
the presence of different concentrations of Cd(II) and Zn(II) is
shown in Fig. 24. The results demonstrate nicely that in accordance
with the 1:3 ratio of metal to ligands, a tenfold increase in Me(II)
concentration shifts the titration curve downward by about one-third
of a pH unit. In addition, it is obvious from the shape of the curve
that there is a high- and a low-affinity site for Cd(II). By plot-
ting the pH of the midpoints of the first titration step of each
curve versus m log [Me(II)], where m = 2 (see above), one obtains

for the two families of curves values for n of 5.8 and 6.1 consistent with binding of pairs of Cd(II) and Zn(II), respectively, to
presumably neighboring metal-binding sites with a total of six
thiolate ligands. From the relative displacement of the two titration steps of Cd(II)-metallothionein, it appears that Cd(II) binding
to the second site is about 50 times weaker than to the site dis-
playing cooperative binding. Surprisingly, the data at hand provide
no clear evidence for analogous site-specific differences in binding
of Zn(II).

9. CONCLUSION

The large body of experimental data summarized in this chapter
reflects on the progress made in recent years in the understanding
of the overall structure of metallothionein and of the mode of
metal binding in this unusual protein. Among the most important
results are the unambiguous spectroscopic documentation of tetrahedral metal coordination to four thiolate ligands and the discovery
of discrete metal-thiolate clusters in this protein. These oligonuclear complexes constitute the first example of clusters of group
2B metal ions in a biological system. They differ very clearly both
in composition and structure from those found in the iron-sulfur
proteins and imply by their unique occurrence in metallothionein
that cluster formation with these bivalent metal ions is dependent
on the very special distribution of cysteine residues typical of
this protein. The stringency of the stereochemical requirements
for such bioinorganic structures may, in fact, explain the remarkable preservation of the positions of the cysteine residues in all
metallothioneins throughout vertebrate evolution. In what way these
unique metal-thiolate clusters condition the still elusive function
of metallothionein remains to be established. Clearly, further
studies on the organization of these clusters and on the dynamics
of their formation and degradation are necessary for a proper assessment of the role played by this protein in cellular processes.

ACKNOWLEDGMENT

This work was supported by Swiss National Science Foundation Grant 3.495-0.79.

REFERENCES

1. M. Nordberg and Y. Kojima, in *Metallothionein* (J. H. R. Kägi and M. Nordberg, eds.), Birkhäuser, Basel, 1979, p. 41ff.

2. M. Margoshes and B. L. Vallee, *J. Am. Chem. Soc.*, *79*, 4813 (1957).

3. P. Pulido, J. H. R. Kägi, and B. L. Vallee, *Biochemistry*, *5*, 1768 (1966).

4. R. H. O. Bühler and J. H. R. Kägi, *FEBS Lett.*, *39*, 229 (1974).

5. J. H. R. Kägi, 8th Int. Congr. Biochem. Abstr., Interlaken, September 1970, p. 130ff.

6. M. P. Richards and R. J. Cousins, *Biochem. Biophys. Res. Commun.*, *64*, 1215 (1975).

7. M. Webb, *Biochem. Pharmacol.*, *21*, 2751 (1972).

8. J. H. R. Kägi and B. L. Vallee, *J. Biol. Chem.*, *236*, 2435 (1961).

9. Y. Kojima and J. H. R. Kägi, *Trends Biochem. Sci.*, *3*, 90 (1978).

10. U. Weser, H. Rupp, F. Donay, F. Linnemann, W. Voelter, W. Voetsch, and G. Jung, *Eur. J. Biochem.*, *39*, 127 (1973).

11. J. D. Otvos and I. M. Armitage, *Proc. Natl. Acad. Sci. USA*, *77*, 7094 (1980).

12. M. Vasák and J. H. R. Kägi, *Proc. Natl. Acad. Sci. USA*, *78*, 6709 (1981).

13. J. H. R. Kägi and M. Nordberg, eds., *Metallothionein*, Birkhäuser, Basel, 1979.

14. M. Webb, in *The Chemistry, Biochemistry and Biology of Cadmium*, Vol. 2 (M. Webb, ed.), Elsevier/North-Holland Biomedical Press, Amsterdam, 1979, p. 195ff.

15. K. Lerch, in *Metal Ions in Biological Systems*, Vol. 13 (H. Sigel, ed.), Marcel Dekker, New York, 1981, p. 299ff.

16. J. H. R. Kägi, S. R. Himmelhoch, P. D. Whanger, J. L. Bethune, and B. L. Vallee, *J. Biol. Chem.*, *249*, 3537 (1974).

17. Y. Kojima, C. Berger, B. L. Vallee, and J. H. R. Kägi, *Proc. Natl. Acad. Sci. USA*, *73*, 3413 (1976).

18. I.-Y. Huang, A. Yoshida, H. Tsunoo, and H. Nakajima, *J. Biol. Chem.*, *252*, 8217 (1977).

19. M. Kimura, N. Otaki, and M. Imano, in *Metallothionein* (J. H. R. Kägi and M. Nordberg, eds.), Birkhäuser, Basel, 1979, p. 163ff.

20. D. M. Durnam, F. Perrin, F. Gannon, and R. D. Palmiter, *Proc. Natl. Acad. Sci. USA*, *77*, 6511 (1980).

21. M. M. Kissling and J. H. R. Kägi, in *Metallothionein* (J. H. R. Kägi and M. Nordberg, eds.), Birkhäuser, Basel, 1979, p. 145ff.

22. Y. Kojima, C. Berger, and J. H. R. Kägi, in *Metallothionein* (J. H. R. Kägi and M. Nordberg, eds.), Birkhäuser, Basel, 1979, p. 153ff.

23. J. L. Bethune, A. J. Budreau, J. H. R. Kägi, and B. L. Vallee, in *Metallothionein* (J. H. R. Kägi and M. Nordberg, eds.), Birkhäuser, Basel, 1979, p. 207ff.

24. G. F. Nordberg, M. Nordberg, M. Piscator, and O. Vesterberg, *Biochem. J.*, *126*, 491 (1972).

25. M. G. Cherian, *Biochem. Biophys. Res. Commun.*, *61*, 920 (1974).

26. K. T. Suzuki, *Arch. Environ. Contam. Toxicol.*, *8*, 255 (1979).

27. A. J. Zelazowski and J. A. Szymanska, *Biol. Trace Elem. Res.*, *2*, 137 (1980).

28. U. Weser and H. Rupp, in *Metallothionein* (J. H. R. Kägi and M. Nordberg, eds.), Birkhäuser, Basel, 1979, p. 221ff.

29. J. Shapiro, A. G. Morell, and I. H. Scheinberg, *J. Clin. Invest.*, *40*, 1081 (1961).

30. G. F. Johnson, A. G. Morell, R. J. Stockert, and I. Sternlieb, *Hepatology*, *1*, 243 (1981).

31. R. H. O. Bühler and J. H. R. Kägi, in *Metallothionein* (J. H. R. Kägi and M. Nordberg, eds.), Birkhäuser, Basel, 1979, p. 211ff.

32. W. Bernhard and J. H. R. Kägi, in preparation.

33. M. Vašák, J. H. R. Kägi, B. Holmquist, and B. L. Vallee, *Biochemistry*, *20*, 6659 (1981).

34. M. Vašák, J. H. R. Kägi, and H. A. O. Hill, *Biochemistry*, *20*, 2852 (1981).

35. C. K. Jørgensen, *Prog. Inorg. Chem.*, *12*, 101 (1970).

36. D. D. Ulmer, J. H. R. Kägi, and B. L. Vallee, *Biochem. Biophys. Res. Commun.*, *8*, 327 (1962).

37. H. Rupp and U. Weser, *Biochim. Biophys. Acta*, *533*, 209 (1978).

38. J. H. R. Kägi, Y. Kojima, M. M. Kissling, and K. Lerch, *Sulphur in Biology*, Ciba Found. Symp., *72*, 223 (1980).

39. M. Vašák and J. H. R. Kägi, in preparation.

40. M. Vašák and R. Bauer, *J. Am. Chem. Soc.*, *104*, 3236 (1982).

41. G. Barth, W. Voelter, E. Bunnenberg, and C. Djerassi, *J. Am. Chem. Soc.*, *94*, 1293 (1972).

42. U. Weser and H. Rupp, in *The Chemistry, Biochemistry and Biology of Cadmium* (M. Webb, ed.), Elsevier/North-Holland Biomedical Press, Amsterdam, 1979, p. 267ff.

43. M. Vašák, P. Oelhafen, J. Krieg, and J. H. R. Kägi, in preparation.

44. C. Nordling, *Angew. Chem.*, *84*, 144 (1972).

45. A. D. Hamer and R. A. Walton, *Inorg. Chem.*, *13*, 1446 (1974).

46. H. Rupp, W. Voelter, and U. Weser, *FEBS Lett.*, *40*, 176 (1974).

47. M. Vašák, A. Galdes, H. A. O. Hill, J. H. R. Kägi, I. Bremner, and B. W. Young, *Biochemistry*, *19*, 416 (1980).

48. K. O. Kopple, A. Go, R. H. Logan, Jr., and J. Savrda, *J. Am. Chem. Soc.*, *94*, 973 (1972

49. M. Vašák, C. E. McCleland, H. A. O. Hill, and J. H. R. Kägi, submitted for publication.

50. J. D. Otvos and I. M. Armitage, in *Metallothionein* (J. H. R. Kägi and M. Nordberg, eds.), Birkhäuser, Basel, 1979, p. 249ff.

51. P. J. Sadler, A. Bakka, and P. J. Beynon, *FEBS Lett.*, *94*, 315 (1978).

52. J. D. Otvos and I. M. Armitage, *J. Am. Chem. Soc.*, *101*, 7734 (1979).

53. D. D. Ulmer and B. L. Vallee, *Adv. Chem. Ser.*, *100*, 187 (1971).

54. D. Gilg and J. H. R. Kägi, in preparation.

55. S. Ikeda, A. Fukutome, T. Imae, and T. Yoshida, *Biopolymers*, *18*, 335 (1979).

56. W. C. Johnson, Jr., and I. Tinoco, Jr., *J. Am. Chem. Soc.*, *94*, 4389 (1972).

57. N. Greenfield and G. D. Fasman, *Biochemistry*, *8*, 4108 (1969).

58. J. A. Schellman and C. Schellman, in *The Proteins*, Vol. 2, 2nd ed. (H. Neurath, ed.), Academic Press, New York, 1964, p. 1ff.

59. K. Eckert, R. Grosse, J. Malur, and K. R. H. Repke, *Biopolymers*, *16*, 2549 (1977).

60. S. W. Provencher and J. Glöckner, *Biochemistry*, *20*, 33 (1981).

61. P. Y. Chou and G. D. Fasman, *Biochemistry*, *13*, 211 and 222 (1971).

62. J. H. R. Kägi, unpublished observation.

63. M. Vašák, J. Am. Chem. Soc., 102, 3953 (1980).

64. I. G. Dance, J. Am. Chem. Soc., 101, 6264 (1979).

65. A. Davison and E. S. Switkes, Inorg. Chem., 10, 837 (1971).

66. D. R. McMillin, R. A. Holwerda, and H. B. Gray, Proc. Natl. Acad. Sci. USA, 71, 1339 (1974).

67. R. W. Lane, J. A. Ibers, R. B. Frankel, G. C. Papaefthymiou, and R. H. Holm, J. Am. Chem. Soc., 99, 84 (1977).

68. S. W. May and J.-Y. Kuo, Biochemistry, 17, 3333 (1978).

69. W. Maret, I. Andersson, H. Dietrich, H. Schneider-Bernlöhr, R. Einarsson, and M. Zeppezauer, Eur. J. Biochem., 98, 501 (1979).

70. M. R. Churchill, J. Cooke, J. P. Fennessey, and J. Wormald, Inorg. Chem., 10, 1031 (1971).

71. D. L. Tennent and D. R. McMillin, J. Am. Chem. Soc., 101, 2307 (1979).

72. S. H. Lin and H. Eyring, J. Chem. Phys., 42, 1780 (1965).

73. R. G. Denning, J. Chem. Phys., 45, 1307 (1966).

74. P. J. Stephens, J. Chem. Phys., 43, 4444 (1965).

75. B. Holmquist, T. A. Kaden, and B. L. Vallee, Biochemistry, 14, 1454 (1975).

76. A. B. P. Lever, in Inorganic Electronic Spectroscopy, Elsevier, Amsterdam, 1968, p. 224ff.

77. R. S. Johnson and H. K. Schachman, Proc. Natl. Acad. Sci. USA, 77, 1995 (1980).

78. H. Dietrich, W. Maret, H. Kozlowski, and M. Zeppezauer, J. Inorg. Biochem., 14, 297 (1981).

79. D. Mastropaolo, J. A. Thich, J. A. Potenza, and H. J. Schugar, J. Am. Chem. Soc., 99, 424 (1977).

80. D. E. Drum and B. L. Vallee, Biochem. Biophys. Res. Commun., 41, 33 (1970).

81. A. J. Sytkowski and B. L. Vallee, Proc. Natl. Acad. Sci. USA, 73, 344 (1976).

82. H. Frauenfelder and R. M. Steffen, in Alpha-, Beta- and Gamma-Ray Spectroscopy, Vol. 2 (K. Siegbahn, ed.), North-Holland, Amsterdam, 1965, p. 997ff.

83. I. Andersson, R. Bauer, and I. Demeter, Inorg. Chim. Acta, 67, 53 (1982).

84. C. D. Garner, S. S. Hasnain, I. Bremner, and J. Bordas, Post-FEBS Int. Meet. Metallothionein, Aberdeen, April 1981.

85. J. B. Goodenough, in Interscience Monographs on Chemistry, Vol. 1 (F. A. Cotton, ed.), Interscience, New York, 1963.

86. I. G. Dance, *J. Am. Chem. Soc.*, *102*, 3445 (1980).

87. E. T. Adman, L. C. Sieker, and L. H. Jensen, *J. Biol. Chem.*, *248*, 3987 (1973).

88. R. E. Benesch and R. Benesch, *J. Am. Chem. Soc.*, *77*, 5877 (1955).

89. J. W. Donovan, in *Physical Principles and Techniques of Protein Chemistry*, Part A (S. J. Leach, ed.), Academic Press, New York, 1969, p. 101ff.

90. P. C. Jocelyn, in *Biochemistry of the SH Group*, Academic Press, New York, 1972, p. 52.

91. J. H. R. Kägi and B. L. Vallee, *J. Biol. Chem.*, *235*, 3460 (1960).

92. J. T. Edsall and J. Wyman, in *Biophysical Chemistry*, Vol. 1, Academic Press, New York, 1958, p. 487ff.

93. F. R. N. Gurd and P. E. Wilcox, in *Advances in Protein Chemistry* (M. L. Anson, K. Bailey, and J. T. Edsall, eds.), Academic Press, New York, 1956, p. 311ff.

94. A. Galdes, H. A. O. Hill, J. H. R. Kägi, M. Vašák, I. Bremner, and B. W. Young, in *Metallothionein* (J. H. R. Kägi and M. Nordberg, eds.), Birkhäuser, Basel, 1979, p. 241ff.

Chapter 7

INTERACTION OF ZINC WITH ERYTHROCYTES

Joseph M. Rifkind
Laboratory of Cellular and Molecular Biology
National Institute on Aging
Gerontology Research Center
Baltimore, Maryland

1. INTRODUCTION

Zinc is, next to iron, the most common trace element in living
organisms, and it is an integral part of over 20 essential metalloenzymes. For a long time the role of zinc, with respect to the
erythrocyte, was thought to be limited to that of carbonic anhydrase
[1,2], which is a zinc metalloenzyme responsible for the rapid
equilibrium in blood between carbonic acid and carbon dioxide.

$$CO_2 + H_2O \rightleftharpoons H_2CO_3 \tag{1}$$

It accounts for the large majority of the erythrocyte zinc and plays
an important role in the transport of CO_2 from the tissues to the
lungs.

An additional role for zinc was found in the late 1960s as
part of superoxide dismutase [3,4]. This protein found in the
erythrocyte, previously known as erythrocuprein [5], was shown to
contain two atoms of zinc in addition to two atoms of copper [4].
It catalyzes the reaction

$$2O_2^- + 2H^+ \longrightarrow O_2 + H_2O_2 \tag{2}$$

and is thought to play a role in protecting erythrocytes as well as
other tissues from damage by superoxide ions [6].

More recently there has been considerable interest in the role
of zinc with respect to the enzyme δ-aminolevulinic acid dehydratase
(ALAD) [7-9]. This enzyme catalyzes the condensation of two molecules
of δ-aminolevulinic acid to form porphobilinogen in the early stages
of heme biosynthesis. It is thought to be a zinc enzyme [7], although
it has recently been shown that zinc is not absolutely required for
activity [9].

The decreased activity of this enzyme has been used as a measure of lead toxicity [10-12]. The inhibition of this enzyme by lead
has been used to at least partially explain the anemia as well as
increased accumulation and excretion of heme intermediates resulting
from lead toxicity. Zinc has been shown to reverse the lead inactivation of ALAD [12,13], and it has, in fact, been shown, at least in

one study [14], that the administration of excess zinc could prevent
the clinical symptoms of lead poisoning.

In more recent years, interest in other effects of zinc on
erythrocytes has been generated by a finding that sickle-cell anemia
frequently correlates with a decrease in the erythrocyte concentration of zinc [15], and the demonstration that zinc affects the oxygenation of hemoglobin [16-18]. These observations have led to the
proposed use of zinc for the treatment of sickle-cell anemia [19],
and the importance of delineating and understanding the effect on
the erythrocyte of zinc not tightly associated with metalloenzymes.

By the addition of excessive zinc it has been demonstrated
that zinc can affect the erythrocyte on many different levels. It
has been shown that zinc binds to the outside surface of the erythrocyte membrane [20]. Under proper conditions high concentrations of
zinc can be transported across the membrane [20]. Inside the cell,
in addition to being associated with specific metalloenzymes, zinc
binds to the inner leaflet of the membrane [21], hemoglobin [18],
other proteins [9], and small molecules [22].

This chapter does not discuss the zinc metalloenzymes found in
the erythrocyte, but instead is devoted to the other frequently quite
strong interactions of zinc with the erythrocyte, which are augmented
by the addition of excessive zinc. Emphasis is placed particularly
on those interactions involving the membrane and the hemoglobin.
These interactions are described in detail, and the significance of
these interactions is explored.

2. INTERACTION OF ZINC WITH ERYTHROCYTE MEMBRANES

Zinc interactions with membranes have been implicated in the function
of a number of different membrane systems [23,24], including mast
cells [25], platelets [23,26], macrophages [23,27], and brain micro-
tubules [28]. Appreciable concentrations of zinc have also been
found associated with membranes of sperm cells [29,30], white blood
cells [31], skeletal muscle [32], brain [33], intestine [34], liver

lysosomes [35], kidney [36], in the electric organ of electrophorous electricus [37], and plants, particularly the leaves of *Silene cucubatus* [38].

It has, in fact, been suggested that zinc may be an integral part of some of these membranes and responsible for their structural integrity [39]. It has also recently been suggested that the clinical manifestations of zinc deficiency [24], as well as the pharmocological effects of zinc [40], may be related primarily to interactions of zinc with membranes.

2.1. Stabilization of Erythrocyte Membranes by Zinc

The stabilizing effect of zinc on erythrocyte membranes, as indicated by effects on hemolysis, have been studied extensively. It was found that dietary supplementation of zinc decreases the osmotic fragility of cells [41,42], and that erythrocytes from zinc-deficient animals are more susceptible to osmotic stress [43]. In one study it has been reported that the addition of zinc in vitro to human erythrocytes decreases the osmotic fragility [44]. However, several other studies have failed to detect a significant decrease in osmotic fragility when zinc is added to erythrocytes in vitro [41,45].

It has recently been proposed that the use of the rate of osmotic lysis is a more direct measure of the stability of the membrane to osmotic stress than the pseudoequilibrium osmotic fragility, which is also dependent on the shape and surface-to-volume ratio of the cell [46,47]. Nevertheless, no detectable effect of zinc on the rate of osmotic lysis was observed [21].

The apparent discrepancy between the in vivo results, which indicate that zinc increases osmotic stability, and the bulk of the in vitro results, which indicate that zinc has no affect on osmotic stability, has been explained [43] by the reported inverse relationship between plasma zinc and copper levels [48-50]. Copper is known in vitro to produce lipid peroxidation [51] and increase osmotic hemolysis [43,52]. The in vivo effects of zinc on osmotic lysis can therefore be interpreted as an effect of copper.

Although zinc may not directly reduce the osmotic fragility of
erythrocytes, it has been shown to decrease the hemolysis produced
by a number of different agents. Thus Cu(II)-induced hemolysis [43]
was shown to be decreased by the addition of zinc. Interestingly,
the stabilization by zinc did not produce an appreciable decrease
in the copper-induced lipid peroxidation, even though it has been
reported that zinc protects a variety of membrane systems against
peroxidative damage by various agents [41,53-57].

Zinc has also been shown to protect against hemolysis (Fig. 1)
induced by various bacterial toxins [58-60], complement [61-63], and
other lytic agents [58,59]. It was subsequently shown that the
apparent protection against lysis by vibriolysin [59], Triton X-100
[59,60], saponin, and lysolecithin [60] was due to precipitation of
hemoglobin liberated from the cells [59] as well as in the cell [60].
We have also found that at high enough concentrations of zinc it is
essentially impossible to lyse the erythrocyte [64].

Zinc does, however, directly protect against lysis by other
systems. In the case of staphylococcal toxin and streptolysin, it
was shown to prevent the binding of the lysins to the membrane [60],
while in the case of *Clostridium perfringens* alpha toxin and per-
fringolysin as well as complement-induced lysis [63], steps involving
disruption of the membrane subsequent to the binding are implicated.
The difference between these two cases is shown in Fig. 1 by the fact
that in some cases removal of zinc and toxin after 30 min prevents
lysis (A-D), while in other cases the lysis takes place when both are
removed but not when both are present (E, F).

2.2. Interaction of Zinc with the Outside Surface of the Membrane

Zinc, like other divalent metal ions, with the exception of the
alkaline earth ions, produces agglutination of washed human erythrocytes [20]. The agglutination reaction is due at least in part to
the binding of metal ions to negatively charged groups on the surface
of the membrane, such as the carboxyl groups of neuraminic acid and

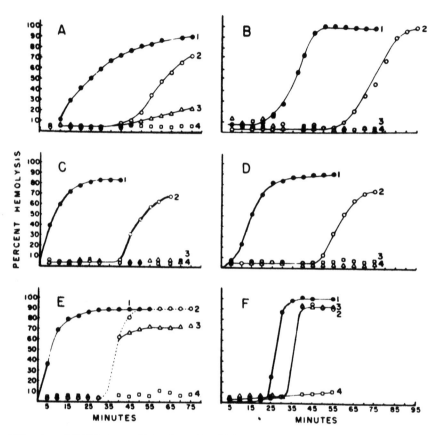

FIG. 1. Inhibition of lysis of rabbit erythrocytes by zinc acetate:
(A) staphylococcal alpha toxin; (B) staphylococcal beta toxin; (C)
streptolysin-O; (D) streptolysin-S; (E) perfringolysin-O; (F) *Clostridium perfringens* alpha toxin. In each figure, (1) toxin-induced
lysis without added zinc; (2) zinc only present first 30 min; (3)
toxin and zinc only present first 30 min; (4) zinc and toxin present
all the time. (From Ref. 60.)

other negatively charged groups on membrane proteins and phospholipids. It has, however, been shown [65] that the order of effectiveness of different metal ions in promoting agglutination is different
from the order of metal ions, which produce charge reversal. This
phenomenon, together with the persistence of agglutination during
vigorous shaking, has been used to suggest that the metal ions involved in agglutination bind quite strongly to the membrane and are
perhaps involved in cross-linking reactions [20].

Studies with liposomes indicate that zinc does bind to phospholipids in liposomes [66,67] and presumably in cellular membranes.
Chvapil et al. [39] have, in fact, found that most of the zinc found
associated with erythrocyte ghosts stays together with the phospho-
lipid fraction when lipids are extracted by organic solvents.

We have found [21] that when erythrocytes are suspended in an
isotonic zinc solution, a concentration of zinc equal to ~15% of the
hemoglobin binds to the erythrocyte in the relatively short time
necessary to spin down the erythrocytes. The large bulk of this
zinc is bound to the outside surface of the membrane, as indicated
by the rapid removal of the large majority of this zinc by EDTA.
Pronase treatment, which degrades some of the proteins on the outside
surface of the membrane, produces a 40% decrease in this binding,
indicating that an appreciable amount of the binding to the outside
surface involves proteins.

In addition to agglutination, another effect that is thought
to be caused by the binding of zinc to the outside surface of the
membrane is an increase in sodium permeability. By comparing the
effect on sodium permeability of different amino and sulfhydryl
reagents [68] it has been proposed that the effect of zinc is due to
the binding of zinc to external amino groups in the sodium channel.

2.3. Transport of Zinc Across the Erythrocyte Membrane

It has been shown that zinc can penetrate the erythrocyte membrane
[69-71]. The process seems to be passive [69] and is not coupled

to the metabolism of the cell [71]. In studies with resealed ghosts, a relatively small enhancement of the zinc influx was observed in the presence of ATP [72]. However, the similarity of the enhancement for ATP, GTP, and CTP suggested that the difference could be explained by the binding of zinc to the nucleotide triphosphate inside the ghosts.

The transport of zinc across the erythrocyte membrane does, however, seem to depend on the free zinc concentration in the zinc incubation medium. Zinc uptake was thus found to be inhibited by nonpenetrating complexing agents such as EDTA [20]. Certain amino acids, particularly histidine, which can complex zinc, also inhibit zinc uptake [20].

Other metal ions, such as Co^{2+}, Mn^{2+}, Cd^{2+}, and Cu^{2+}, were found to have no effect on Zn^{2+} uptake. Fe^{2+} was, however, found to be inhibitory. Even though Cu^{2+} was found to have no effect on zinc uptake by intact cells, an enhanced uptake was found with resealed ghosts [72].

Plasma (Fig. 2) and serum have been shown to decrease dramatically the uptake of zinc by erythrocytes [20,21,27,70,73,74]. Comparing the uptake of zinc in plasma and different buffers, Kruckeberg and co-workers [73,74] found that the enhanced uptake relative to plasma in a modified Krebs phosphate-free bicarbonate buffer (KB) is eliminated when human serum albumin (HSA) is added to KB (Fig. 2). On the basis of these results it was concluded that the decrease uptake is due to zinc binding to HSA. These same investigators found no pH effect on zinc uptake in KB, but a 70% increase in plasma or KB-HSA when the pH was raised from 7.6 to 8.3. A very dramatic temperature dependence [74] of zinc uptake was also found (Fig. 3), with minimal uptake at 5°C but dramatic increases at higher temperatures. The temperature dependence was also quite different in plasma and KB buffer (Fig. 3).

A similar temperature dependence was also observed for the uptake of zinc by resealed ghosts [72], where the small uptake at 16°C was assumed to be due to absorption to the outside of the membrane.

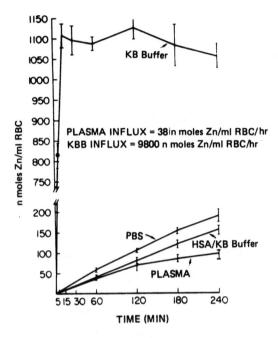

FIG. 2. Uptake of zinc by erythrocytes in different media. (From Ref. 74.)

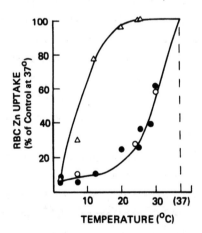

FIG. 3. Temperature variation of uptake of zinc by erythrocytes in different media: △, KB buffer; ○, HSA/KB buffer; ●, plasma. (From Ref. 74.)

Kruckeberg et al. [73] have also found that zinc uptake in plasma is enhanced by unsaturated fatty acids and very appreciably inhibited by the drug Atebrin (quinacrine). Since they do not find these effects without HSA, they believe that these are effects on the interaction of zinc with HSA and not direct effects on the membrane. A decrease in the uptake of zinc [75] reported for trinitrocresol (TNC) in the absence of plasma or HSA suggests a direct effect of TNC on membrane transport of zinc.

The uptake of zinc into the erythrocyte, under favorable conditions, can be very rapid (Fig. 2). Kruckeberg and co-workers [73,74] have calculated that in KB the initial rate of zinc uptake at 37°C, pH 7.4, is 9800 nmol zinc per milliliter of red blood cells (RBC) per hour. The quantity of zinc which can be taken up by the erythrocyte is also very large. At nonsaturating zinc concentrations when corrections are made for the small amounts of hemolysis, >99% of zinc is found associated with the erythrocyte [64]. The uptake of zinc in the erythrocyte was actually found to reach saturating levels higher than that of intracellular hemoglobin [64].

The efflux of zinc from the erythrocytes in all cases is much slower than the influx. This slow efflux and the large concentration difference observed between the inside and outside of the erythrocyte indicate that zinc associated with the erythrocyte is bound relatively tightly to the erythrocyte membrane or intracellular molecules (see Sec. 2.6). Unlike the erythrocyte, other types of cells, such as platelets [26], HeLa cells [76], and sperm cells [77], which also take up zinc lose zinc relatively quickly when suspended in zinc-free medium.

The mechanism for this rapid uptake of zinc is not known. This uptake is much more rapid than the passive transport of calcium, when the calcium pump is arrested [78]. Furthermore, zinc uptake is inhibited by TNC [75], which accelerates Ca^{2+} uptake, indicating a different pathway for Ca^{2+} and Zn^{2+} influx.

The rate of transport of zinc is actually comparable with that obtained by incorporating into the membrane an artificial carrier such as the ionophore A23187, which has been used as a calcium

ionophore, but can function as a zinc ionophore also [79]. This
could suggest that the erythrocyte membrane might contain an integral
zinc carrier. It is perhaps of interest in this respect that zinc
has been found associated with many of the membrane proteins after
purification [39]. We have found, in a preliminary experiment, that
there is a low-molecular-weight membrane protein of <5000 daltons
with a relatively high affinity for Zn^{2+} [80], which can be extracted
from erythrocyte ghosts at low ionic strength.

   Considerably more work is required to determine whether one of
these proteins, which bind zinc, is perhaps able to function as a
Zn^{2+} carrier in the erythrocyte, and thereby facilitate rapid transport of Zn^{2+} and perhaps certain other divalent metal ions across
the membrane.

   Possibilities that the species being transferred is not the
divalent cation Zn^{2+} but the monovalent cations, where zinc is
associated with Cl^- or OH^- ($ZnCl^+$, $ZnOH^+$) must also be considered.
Such possibilities have been considered in other systems [78] and
may permit zinc to utilize the erythrocyte Na^+ or K^+ channels.
Since ATP depletion does not have much of an effect on zinc uptake
[69-72], utilization of the Na^+ or K^+ pumps to transport Zn^{2+} seems
to be ruled out.

   Divalent ions and even monovalent ions probably cannot be
transported directly through lipid bilayers at a significant rate
[81]. It has, however, been recently reported that the lipid
bilayer permeability of $HgCl_2$ is about $1.3 \cdot 10^{-2}$ cm/s, a rate that
is 20-fold higher than the permeability of H_2O through the bilayer
[81]. Since zinc is known to form relatively stable Cl^- complexes
[82], transport of the neutral $ZnCl_2$ through the bilayer can also
be considered [81] a possible pathway for the transport of zinc
through the membrane.

2.4. Effects of Zinc on Sickle-Cell Membranes

Sickle-cell anemia is associated with a modified hemoglobin molecule
Hb-Hb-SS(β-6 Glu $\rightarrow$ Val), which tends to polymerize when deoxygenated

[83,84]. This polymerization causes deformation and sickling of the erythrocytes [85], which can lead to occlusion of small vessels and difficulty in transporting oxygen. It has, further, been shown that the sickling process produces certain modifications in the membrane which can result in cells that retain their abnormal shape even after oxygenation [85-87]. It is thought that these irreversibly sickled cells (ISCs) are responsible for many of the clinical manifestations of sickle-cell anemia [88,89].

It has been reported that the administration of zinc in vivo decreases the number of ISCs from 28.0% before treatment to 18.6% during treatment [90]. Since the ISCs have damaged membranes, these results would seem to be due to the interaction of zinc with the membrane.

It has also been reported that zinc at concentrations as low as 0.3 mM in plasma very significantly improved the ability of partially deoxygenated sickle cells at an oxygen pressure of 15 mm Hg to go through the small pores of Nuclepore filters [91]. This effect of zinc was interpreted as an effect on the membrane and not the hemoglobin, because the zinc concentration was much less than that of hemoglobin. The effect of zinc on the filterability would thus seem to indicate that while the polymerization of deoxygenated Hb-SS may be necessary for the loss in deformability and decreased filterability, which is associated with the sickling of the cell, membrane modifications which can be inhibited to a large extent by zinc are also required for the change in deformability.

Arnone and Williams [92], on the basis of their x-ray evidence for a zinc-binding site on deoxyhemoglobin that links tetramers (see Sec. 3.2), have suggested a possible mechanism for zinc at relatively low concentrations to inhibit hemoglobin polymerization. In that case, it is possible that the effect of zinc on the filterability is due to an effect on hemoglobin and not the membrane. This possibility can be resolved by directly investigating the effect of zinc on the polymerization of deoxygenated Hb-SS, an experiment that has not yet been reported.

An effect of zinc on the morphology of resealed one-time ghosts prepared from sickle red blood cells was also observed. It was found that a large percentage of the control ghosts formed by lysis in 3 mM NaCl and resealed in 10 mM Tris-HCl (pH 7.4) were echinocytes and most of the others were smooth (folded). Lysis in 1.5 mM $ZnSO_4$ very significantly decreased the number of echinocytes and produced an appreciable number of a third type of cell, which is referred to as flat because of its two-dimensional appearance. The appearance of a unique type of morphology for zinc cells seems to require direct interaction of zinc with the membranes.

2.5. Zinc Inhibition of Calcium-Induced Membrane Damage

Increases in intracellular calcium have been shown to produce a number of alterations in the erythrocytes [47]. These include decrease in mean cell volume [93,94], increase in mean cell hemo-globin concentration [93,94], loss of water and potassium [95], marked changes in morphology from diconcave disks to echinocytes [96,97], rapid hydrolysis of intracellular ATP [94], decrease in cell deformability, and stiffening of the erythrocyte membrane [98,99]. It has also been shown that elevated calcium concentrations are found in sickle cells and particularly in ISCs. It has therefore been proposed that the membrane damage associated with sickling is due to this elevated calcium concentration [94,100,101].

Eaton et al. [94] have directly shown that some of the damage produced in erythrocytes by incubating normal and sickle cells with calcium and the ionophore A23187 to increase intracellular calcium can be diminished or inhibited by simultaneously adding zinc. They have thus found that the increase in mean cell hemoglobin concentra-tion (i.e., the cell shrinkage), as well as depletion of ATP for Hb-SS cells, is very significantly diminished in the presence of 0.2 mM zinc. Surprisingly, with normal cells the effect of calcium is similar, but zinc counteracts this effect to a much lesser extent.

The reversal of cell shrinkage is opposite to the earlier report
[102] that both zinc and calcium produce shrinkage of normal erythrocyte ghosts. As indicated by the difference between normal and Hb-SS
cells, the effect of zinc on the shrinkage of the cell may depend on
the exact state of the membrane.

On the basis of these results, it has been proposed [94] that
the effect of zinc on the number of ISCs in patients with sickle-cell
anemia, the filterability at low oxygen pressure of sickle cells [91],
and the decrease in the number of echinocytes found on ghosts prepared
from sickle cells [75] can be interpreted in terms of an antagonism
between the effect of calcium and zinc, whereby zinc prevents the
damage produced by calcium [75,91].

Other antagonistic effects of calcium and zinc have been
reported [103]. Calcium has been shown to promote binding of hemoglobin to the inside of the red cell membrane, while zinc inhibits
this binding and counteracts the effect of calcium [102]. Intracellular calcium has been shown to induce limited proteolysis of one
of the spectrin bands [104] and at somewhat higher concentrations
the cross linking of membrane proteins [47,104,105]. We have shown
that the proteolysis as detected by SDS-PAGE is prevented by the
addition of zinc [21]. The cross linking of membrane protein has
been attributed to Ca^{2+} activation of a transglutaminase present in
the cell [104]. It has been reported that zinc inhibits the erythrocyte transglutaminase [106] and is a potent inhibitor of transami-
dases obtained from other tissues [107].

The intracellular level of Ca^{2+} is kept at its usual low level
by an ATP-dependent Ca^{2+} pump [108,109], which utilizes a Ca^{2+}-Mg^{2+}
ATPase [110] activated by intracellular Ca^{2+}. It has been shown
(Fig. 4) that intracellular Zn^{2+} will inhibit this calcium-stimulated
ATPase activity [111]. Zinc has also been shown to inhibit the phosphorylation of bands 2 and 3, found when erythrocyte proteins are separated by SDS-PAGE, as well as a phosphorylated band which does not
show up by Commassie blue staining, which lies between band 2 and
band 3 and has been designated band 2x [111]. Band 2x has been

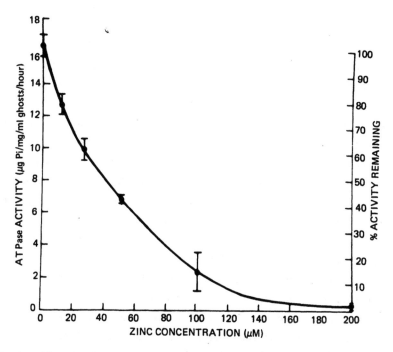

FIG. 4.   Zinc inhibition of ATPase activity of erythrocyte ghosts.
Assay mixtures contained 3 mM ouabain, 3 mM ATP, 1 mM $MgCl_2$, 50 mM
NaCl, 120 mM KCl, and 50 µM $CaCl_2$ and were run at 37°C (pH 7.2).
(From Ref. 121.)

shown to exhibit properties which are consistent with its being
associated with the Ca-Mg ATPase pump [112,113].   Interestingly,
the inhibition of phosphorylation of band 2x requires only 50 mM
zinc, which is an order of magnitude lower than the zinc concentrations required to produce the same inhibition of band 2 and
band 3 phosphorylation [111].   Zinc has also been shown to inhibit
Mg ATPase, which is found on the inner surface of the membrane [114].

On the basis of these results it has been suggested [111] that
the mechanism for the reversal of many of the calcium effects by
zinc may actually be due to a turning off of ATPase (Fig. 4) and
decreased utilization of ATP which results in an increase of the
intracellular ATP level.   According to this proposal, instead of

zinc lowering the intracellular calcium concentration, zinc would be
expected to increase this concentration, because of the inhibition
of the pump. The detrimental effects of calcium which zinc inhibits
are thus perhaps indirectly due to the decrease in intracellular ATP.
Consistent with this expectation it was shown that the intracellular
level of calcium can actually be increased by incubating erythrocytes
with zinc [111].

It has recently been proposed [103] that many of the opposing
effects of calcium and zinc on the erythrocyte may involve calmodulin,
which is a 17,000-dalton protein found in erythrocytes and many other
cells [115-117]. Calmodulin is activated by binding calcium, which
alters the calmodulin conformation and enables it to stimulate the
activity of many enzymes and cellular functions.

In the erythrocyte it has been shown that calmodulin activates
the Ca-Mg ATPase associated with the calcium pump [118]. Evidence
has also been presented for a role of calmodulin in the calcium-
induced retention of hemoglobin by erythrocyte ghosts [119] and in
the calcium-induced shrinkage of erythrocytes [120].

Zinc has been shown to inhibit the calmodulin-stimulated
activity of the Ca-Mg ATPase [121], the calmodulin-stimulated hemoglobin retention [103], as well as several other calmodulin-stimulated
enzyme activates in other cells [103,122].

On the basis of these results it has been suggested [103] that
the major effect of zinc in erythrocyte membranes as well as other
cellular membranes involves inhibition of the calmodulin stimulatory
effect on cellular processes.

2.6. Binding of Zinc to Erythrocyte Membranes

Although there have, as indicated above, been rather extensive
studies on the effects of zinc on the erythrocyte membrane, a
detailed understanding of how zinc produces its effects requires
knowledge of how and where the zinc is actually bound to erythrocyte membrane.

Brewer and Oelshlegel [91] studied the zinc and calcium associated with erythrocyte ghosts by various methods and found that zinc
did bind to ghosts. Furthermore, they demonstrated that zinc produced
a relatively large decrease in the amount of calcium bound, although
not much of an effect of calcium was observed on the zinc bound. On
the basis of these results, it has been suggested that the effect of
zinc on calcium binding may provide a mechanism for the reversal of
many calcium effects on erythrocyte membranes in the presence of zinc.

Chvapil et al. [39] recently showed that when different lipids
and protein fractions of a completely disrupted erythrocyte ghost are
separated, zinc is associated with all fractions of the membrane.
Most of the zinc was, however, associated with phospholipids and the
second fraction of SDS dissolved proteins to elute off a G-200 column.
These results do not necessarily reflect the binding sites of zinc
in an intact membrane, although they do indicate that zinc binds to
the membrane.

In an attempt to understanding something about the binding of
zinc to erythrocyte ghosts, we have compared (Fig. 5) the binding of
zinc and calcium to white hemoglobin-free erythrocyte ghosts [21].
From these results it can be seen that zinc does have an appreciably
higher affinity than calcium for erythrocyte ghosts and that it seems
to involve a greater number of sites. The greater binding of zinc
than calcium explains why calcium has not been found to affect the
zinc binding even though zinc did decrease the calcium binding.

In an attempt to delineate the binding sites further, ghost
proteins were incubated with 0.2 mM EDTA pH 8.0 at 37°C, which is
known to extract 25% of the proteins, including most of the spectrin
(bands 1 and 2) and actin (band 5) [123]. These were separated by
centrifugation and concentrated by ultrafiltration. It was found
that zinc still bound to the pellet, although an appreciable amount
of zinc was bound to the low ionic strength extract. By gel filtration and fractionation using filters of different molecular weight
cutoffs, it was shown [80] that almost all the binding in the low-
ionic-strength extract is to proteins with a molecular weight of

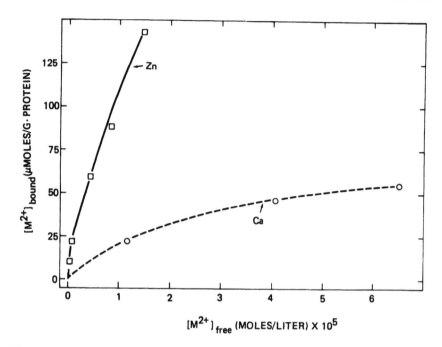

FIG. 5.   Binding of zinc and calcium to erythrocyte ghosts.

>100,000, presumably spectrin. Figure 6 shows that while zinc binds with a high association constant in the region of 10^6, calcium has essentially no affinity to the spectrin preparation. The Scatchard plot for binding of zinc to ghosts indicates that there are at least two classes of zinc sites. On the other hand, the Scatchard plot for binding of zinc to spectrin indicates one type of site. Interestingly, the spectrin binding corresponds to a very large number of zinc-binding sites per spectrin molecule (i.e., to more than 50 zinc-binding sites per spectrin dimer).

It is not certain whether these sites are accessible in the ghosts or become exposed when the spectrin is extracted from the ghosts. The high-affinity class of sites found in the ghosts do, however, have a similar affinity to that of these spectrin sites, consistent with the possibility that these spectrin sites, at least some of them, may be present on intact ghosts.

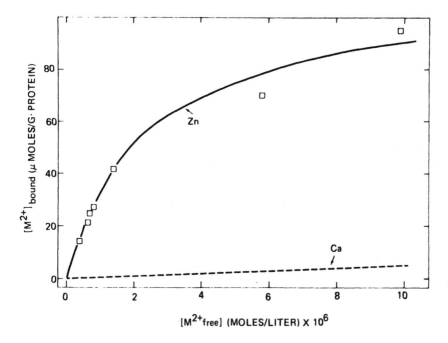

FIG. 6.   Binding of zinc and calcium to partially purified spectrin.

    The role of EDTA for the extraction of spectrin [123] does
perhaps suggest that the spectrin is held to the membrane by metal-
ion interactions, which are consistent with the binding of metal
ions such as zinc to spectrin.
    Not enough is known about this interaction of zinc with the
spectrin to even suggest how this may alter the membrane, although
it is interesting that zinc effects on spectrin phosphorylation have
been reported [111]. Even though calcium does not seem to bind to
these zinc sites, it is possible that the antagonistic effects of
zinc and calcium do not involve binding to the same sites. In fact,
the calmodulin mechanism proposed by Brewer [103] also does not
involve direct competition of zinc for calcium, since it has been
reported that unlike calcium, zinc does not bind directly to free
calmodulin [124]. It is therefore necessary to consider the spectrin
binding in attempting to explain the numerous effects of zinc on
erythrocyte membranes.

3. INTERACTION OF ZINC WITH HEMOGLOBIN

3.1. Effect of Zinc on the Oxygenation of Hemoglobin

An effect on the oxygenation of hemoglobin by zinc was originally reported by Oelshlegel et al. [16]. These investigators found a 9% decrease in P_{50} (i.e., an increase in the oxygen affinity) when incubating normal human erythrocytes as well as erythrocytes of patients with sickle-cell disease in 1.5×10^{-3} M $ZnCl_2$ for $\sim$2 h (Table 1).

In a subsequent paper [17] this same group investigated the effect of zinc on partially purified human hemoglobin preparations and found a somewhat larger 24% decrease in P_{50} from 25.5 to 19.5 mmHg at a zinc/hemoglobin ratio of 0.41 mol zinc per mol hemoglobin tetramer. They also demonstrated that zinc does interact directly with the hemoglobin.

Our studies on the effect of zinc on hemoglobin began in an attempt to explain copper oxidation of hemoglobin [125]. Copper is

TABLE 1

Effect of Zinc on the Oxygenation of Hemoglobin

Preparation	Zinc	$(P_{50})_0/(P_{50})_{Zn}$	Reference
Normal human blood	$1.5 \cdot 10^{-3}$ M	1.1	16
Sickle-cell blood	$1.5 \cdot 10^{-3}$ M	1.1	16
Washed human erythrocytes	$1.5 \cdot 10^{-3}$ M	1.3	17
Washed human erythrocytes	Zinc/heme = 0.25	1.6	64
Purified horse hemoglobin	Zinc/heme = 0.50	2.8	129
Purified human hemoglobin	Zinc/heme = 0.52	3.7	18
Purified human hemoglobin + mM 2,3-DPG	Zinc/heme = 0.52	1.8	18

apparently the only metal ion that rapidly oxidizes hemoglobin [126, 127]. Although most other metal ions had no effect on copper oxidation, it was found that zinc decreased the oxidation of hemoglobin by copper (Fig. 7). A subsequent comparison of the binding of different metal ions to horse hemoglobin [125] indicated that copper and zinc had a similar high affinity for hemoglobin, which was very much larger than that of cadmium, nickel, cobalt, and manganese.

We had previously found that zinc binding to myoglobin increases the oxygen affinity of myoglobin [128]. We therefore investigated the effect of zinc on the oxygen affinity of purified horse hemoglobin [129] and found a much larger effect than that reported by Oelshlegel et al. [16,17] for human hemoglobin (Table 1).

Subsequent studies on purified human hemoglobin [18] stripped of the 2,3-DPG cofactor [130-132] showed a dramatic increase in oxygen affinity (Fig. 8) even greater than that found for horse hemoglobin (Table 1).

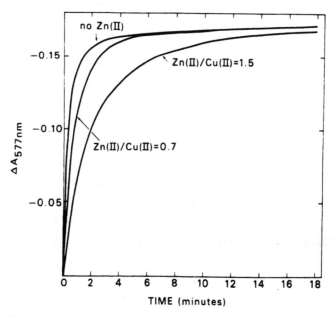

FIG. 7.   Inhibition of copper oxidation of horse hemoglobin by zinc.

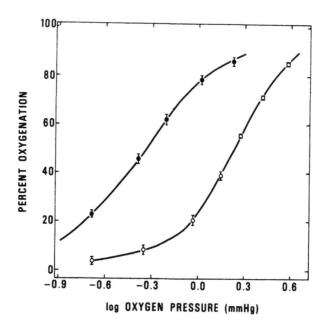

FIG. 8.   Effect of zinc on the oxygenation of stripped human hemoglobin in 0.02 M bistris pH 7.4 at 25°C: ○, no zinc added; ●, zinc/
heme molar ratio = 0.52.   (Reprinted with permission from Ref. 18.
© 1977 American Chemical Society.)

More recent studies by Gilman and Brewer [133] using purified
human hemoglobin preparations obtained changes in oxygen affinity as
great or even greater than those reported by Rifkind and Heim [18].
Increases in oxygen affinity produced by zinc have also been reported
for cow and chicken hemoglobins [134].

In an attempt to explain the effect of zinc on hemoglobin,
oxygenation studies have been performed under various conditions and
with various modified hemoglobins.

2,3-DPG has been shown to bind preferentially to deoxyhemoglobin [130,131] in the crevice between the β subunits [135] and to
play a major role in lowering the oxygen affinity of human hemoglobin.

Experiments were performed at different 2,3-DPG concentrations
[16] to determine whether the initial observations on whole cells
[16] could be explained by zinc binding to 2,3-DPG or zinc blocking

the 2,3-DPG hemoglobin site. These experiments showed that the same
zinc effect on the oxygen affinity was observed irrespective of the
2,3-DPG concentration, and the zinc effect is not simply due to the
displacement of 2,3-DPG. This conclusion was also confirmed by the
reported finding that horse [129] and cow [134] hemoglobin, which do
not interact with 2,3-DPG, are also affected by zinc. Furthermore,
human hemoglobin, which has all the 2,3-DPG removed [18,132], is
also affected by zinc.

   Rifkind and Heim [18] did, however, find that the addition of
2,3-DPG to stripped human hemoglobin does decrease the effect of
zinc on the oxygen affinity (Table 1). In fact, this effect was
proposed as part of the explanation [18] for the much greater effect
of zinc on stripped human hemoglobin than for whole cells or hemo-
lysates [1,2,11,15-18,136]. These results indicate that the increase
in oxygen affinity produced by zinc and the decrease in oxygen affinity produced by 2,3-DPG coupled even though the effect of zinc involves an oxygenation linked interaction with zinc, not just a displacement of 2,3-DPG by zinc.

   The oxygen affinity of hemoglobin is also very sensitive to
pH, a phenomenon known as the Bohr effect. However, all studies thus
far [17,129,133] indicate that the relative effect of zinc is independent of pH. This observation seems to be true, even though the
binding of zinc liberates protons [133] and is dependent on pH [18].

   The reaction of the cysteine β-93 residue with sulfhydryl
reagents is one of the most extensively studied modifications of
hemoglobin [125,137,138]. The sulfhydryl group on this residue is
the only reactive sulfhydryl on hemoglobin and can be specifically
and stoichiometrically blocked by many sulfhydryl reagents [134].

   The reaction of the cysteine β-93 residue with sulfhydryl
reagents has also been shown to produce a relatively large increase
in the oxygen affinity [139,140]. Studies on the effect of zinc on
human hemoglobin reacted with NEM [129,133] at the β-93 cysteine
residue show that the effect of zinc is markedly decreased.

   The early studies on the effect of zinc on intact cells were
done under conditions where it was not clear exactly how much zinc

was being taken up by the erythrocytes. Recent studies indicate
that under proper conditions it is possible to incorporate into
erythrocytes very high concentrations of zinc, which can even be
greater than that of the hemoglobin [64] (see Sec. 2.3).

Since purified hemoglobin studies [18,129] indicate that there
is a very appreciable zinc-induced increase in oxygen affinity (much
larger than those reported by Oelshlegel et al. [16]), even in the
presence of an excess of 2,3-DPG, it was thought that it was worth
repeating the studies on the oxygenation of whole cells, where the
zinc/hemoglobin ratios inside the cells were subsequently measured.
These results [64] indicate that even with whole cells zinc can
produce a considerable increase in oxygen affinity (Fig. 9), which
is almost as great as that produced by purified hemoglobin in the
presence of excess 2,3-DPG (Table 1).

The studies on hemoglobin solutions indicate that zinc decreases
the cooperativity (i.e., the value of the Hill parameter Zn) at intermediate concentrations of zinc, but that the value of n returns to
that found in the absence of zinc at higher zinc concentrations [17,
133,134]. This was explained [134] by the presence of a mixture of
hemoglobins with low and high affinity at intermediate zinc concen-

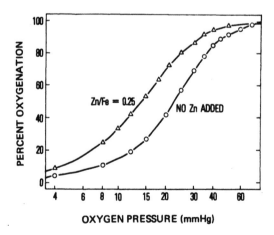

FIG. 9.   Effect of zinc on the oxygenation of intact cells at 37°C
and 40 mmHg CO_2: O, no zinc added; Δ, an intracellular zinc/heme
molar ratio = 0.25.

trations. On the other hand, the whole-cell studies display some
rather unusual changes in shape as the zinc concentrations are
increased. At relatively low zinc concentrations (Fig. 9), an
increase in oxygen affinity is observed with a small decrease in the
sharpness of the curve, similar to that observed for hemoglobin solutions. At higher zinc concentrations, as shown by the comparison of
the oxygenation curve for an intracellular zinc/heme ratio of 0.25
and 0.60 (Fig. 10), there is no further increase in oxygen affinity
(in fact, the P_{50} increases slightly), but the curve becomes very
asymmetric and becomes very sharp at the upper end of the oxygenation curve. A Hill plot for the zinc/heme ratio of 0.6 shows the
n value increasing from the usual value of 3 to 10 near the top of
the oxygenation curve. At still higher zinc concentrations, the
binding curve becomes hyperbolic and it takes a relatively long time
to reach equilibrium.

Considering the inability to lyse the cells at high zinc concentrations (see Sec. 2.1) and the ability of zinc to precipitate
hemoglobin [59,60,129], the results at high zinc concentrations
probably correspond to oxygenation of at least partially precipi-
tated hemoglobin.

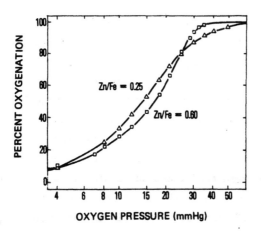

FIG. 10. Effect of zinc on the oxygenation of intact cells at 37°C
and 40 mmHg CO_2: Δ, an intracellular zinc/heme molar ratio = 0.25;
O, an intracellular zinc/heme molar ratio = 0.60.

The intermediate very sharp and asymmetric curves are, however, of considerable interest and seem to reflect linkage phenomena present in the cell, which are not observed at least to the same extent in solution. Possible explanations of these results are currently being investigated.

3.2. Binding of Zinc to Hemoglobin

The binding of zinc to hemoglobin was first detected by Cann [141], who found that zinc produces denaturation of hemoglobin at relatively high concentrations. However, the effect in the oxygen affinity [18, 133] occurs at lower zinc concentrations and does not seem to involve any denaturation. Oelshlegel et al. [17] proposed that the effect of zinc on the oxygenation (see Sec. 3.1) also involved the binding of zinc to hemoglobin. This conclusion was based on their observation that zinc added to hemoglobin chromatographs together with hemoglobin. Subsequently, the binding of zinc to hemoglobin has been carefully measured by various different techniques and by different investigators [18,133,142]. Evidence for the binding of zinc to hemoglobin has also been obtained by a number of other investigators [22,92,143-145].

On the basis of these results and comparisons of binding and oxygenation results [18,133], it has been unambiguously shown that the effect of zinc on the oxygenation of hemoglobin is associated with the binding of two zinc molecules per tetramer with a high association constant to oxyhemoglobin in the range 10^7-10^6 at neutral pH (Fig. 11). Furthermore, there is a lower zinc affinity for deoxyhemoglobin [133,146] than the liganded hemoglobins (oxyhemoglobin or carboxyhemoglobin) (Table 2). These results confirm the expected relationship between the zinc-induced stabilization of the liganded conformation as implied by the increased oxygen affinity and the higher affinity of zinc for hemoglobin in the liganded conformation [147].

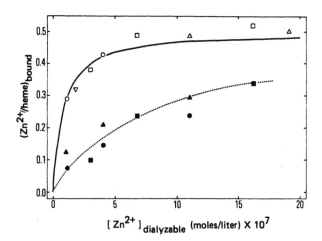

FIG. 11. Binding of zinc to human hemoglobin (open symbols) and the
effect of blocking the β-93 sulfhydryl groups (solid symbols). The
lines are calculated assuming two identical noninteracting zinc
binding sites per tetramer. The various open symbols correspond to
different experiments. ●, ■, two different experiments where the
sulfhydryl groups are blocked by NEM; ▲, the sulfhydryl groups are
blocked by BME. (Reprinted with permission from Ref. 18. © 1977
American Chemical Society.)

The magnitudes of the association constants indicate that zinc
binding to hemoglobin must involve the simultaneous coordination with
several amino acids to form a chelate. In order to explain the mechanism for the rather dramatic increase in oxygen affinity produced by
zinc, it is necessary to try to locate the zinc binding sites. Arnone
and Williams have located by x-ray methods a zinc-binding site on
human deoxyhemoglobin [92]. This site is located at the interface
between tetramers in the region where histidine 116, histidine 117,
and glutamate 26 on the β_1 chain of one tetramer are in close proximity to lysine 16 and glutamate 116 on the α_2 chain of a neighboring
tetramer. It has been proposed that this binding may be related to
the antisickling effect of zinc (see Sec. 2.4) by destabilizing the
tetramer-tetramer interaction responsible for sickling.
    The x-ray site in deoxyhemoglobin may correspond to the highest-
affinity site in concentrated solutions of deoxyhemoglobin or in

TABLE 2

Zinc-Binding Constants

Hemoglobin	Deoxy $(M^{-1} \cdot 10^{-5})$	CO $(M^{-1} \cdot 10^{-5})$	Co/Deoxy
Human	1.8 ± 0.6	10.3 ± 0.2	5.7
Horse	0.9 ± 0.4	7.8 ± 1.1	8.7
Human-NEM	1.0 ± 0.4	0.6 ± 0.1	0.6
Little Rock	0.8	0.2	0.3

cells, but is probably not the oxygen-linked binding site, at least in more dilute solutions. There is thus far no evidence for the association of hemoglobin tetramers associated with the oxygen-linked sites. In fact, direct evidence has been presented that zinc binding actually enhances the dissociation into dimers [134,144]. Furthermore, Rifkind and Heim [18] found evidence to rule out the involvement of histidines 116 and 117 on the β-chain in the binding of zinc to oxyhemoglobin. They thus found that human HbA and HbA_2 have similar stability constants (Fig. 12), even though the δ chains on HbA_2, which replace the β chains of hemoglobin A, has histidine β-116 replaced by arginine and histidine β-117 replaced by asparagine.

Nevertheless, this site found in deoxyhemoglobin must be considered at the high concentrations of hemoglobin found in erythrocytes. It may, in fact, partially explain the changes in shape found for the oxygenation curve when relatively high concentrations of zinc are incorporated into erythrocytes (Fig. 10) [64].

In the absence of x-ray data on zinc binding to liganded hemoglobin, the approach used to obtain information regarding the site of zinc binding has been to study the binding of zinc to various animal hemoglobins, minor fractions of hemoglobin, modified hemoglobins, and abnormal hemoglobins [18,133,134].

A very similar affinity of zinc for hemoglobin has been obtained for human, horse, bovine, sheep, and rabbit hemoglobins [18]. The binding site must therefore involve amino acids common to all these hemoglobins.

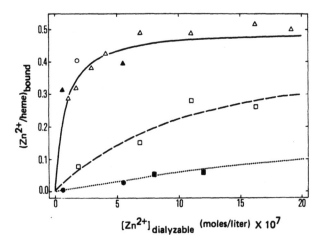

FIG. 12. Binding of zinc to various human hemoglobins. The lines
are calculated assuming two identical noninteracting zinc binding
sites: △,▲, human hemoglobin A; ○, human hemoglobin A_2; □, human
hemoglobin F; ●, purified hemoglobin Abruzzo; ■, purified hemoglobin
Little Rock. (Reprinted with permission from Ref. 18. © 1977
American Chemical Society.)

As indicated above, it has been shown that blocking the β-93
sulfhydryl group, which is present in the hemoglobin of all the
animal species studied, decreases the effect of zinc on the oxygenation of hemoglobin. Zinc-binding studies with the sulfhydryls
blocked show an order of magnitude decrease in the association constant [18] (Fig. 11). The comparison of the binding to carboxyhemoglobin and deoxyhemoglobin [146] (Table 2) further indicates that
this decrease in association constant is primarily for the liganded
hemoglobin, with only a factor of 2 decrease in the constant for
deoxyhemoglobin. The resultant loss of the preferential binding
to liganded hemoglobin explains the elimination of the effect of
zinc on the oxygenation [129,133].

Rifkind and Heim [18] showed that a group with a pK_a ∿6.5 is
involved in the binding of zinc to oxyhemoglobin (Fig. 13). Since
histidines are the only side chains on proteins which generally
have a pK_a in this region, the involvement of histidine in the

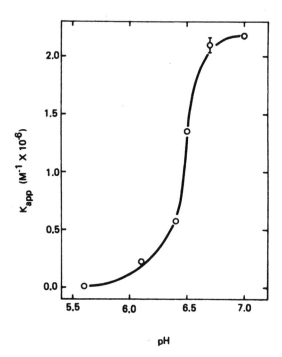

FIG. 13. Effect of pH on the apparent association constant for the
binding of zinc to human hemoglobin. (Reprinted with permission
from Ref. 18. © 1977 American Chemical Society.)

binding is implied. NMR studies [22,148] also indicate that zinc
binding involves histidine residues (Fig. 14).

     As discussed above, binding studies on HbA$_2$ [18] seem to rule
out the involvement of histidine β-116 and β-117, proposed for the
x-ray site on deoxyhemoglobin for binding to liganded hemoglobin.
Gilman and Brewer [133] proposed binding to histidine β-146 on the
basis of an order of magnitude decrease in the affinity of zinc for
human HbCO when the β-146 histidine is removed. Rifkind and Heim
[18] proposed that histidine β-143 is involved in the binding of
zinc to liganded hemoglobin because of a decrease in the association
constant of zinc by two orders of magnitude (Fig. 12) for both hemoglobin Abruzzo [β-143(H21) His → Arg] [149] and hemoglobin Little
Rock [β-143(H21) His → Gln] [150].

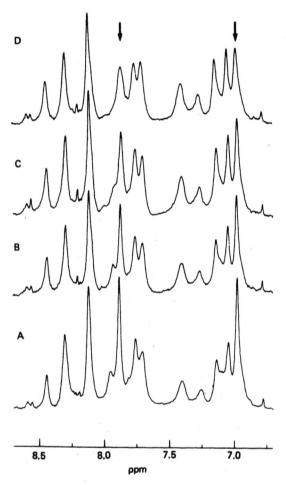

FIG. 14. Histidine region of the ^{1}H NMR spectrum of hemolyzed
erythrocytes at various concentrations of zinc: (A) no zinc;
(B) 1.46 mM; (C) 2.38 mM; (D) 4.55 mM. (From Ref. 22.)

   The lack of a difference between the zinc association constant
of deoxyhemoglobin and carboxyhemoglobin for hemoglobin Little Rock
[146] (Table 2) predicts that the effect of zinc on the oxygen
affinity will not be observed for hemoglobin Little Rock and implicates this amino acid in the oxygen-linked binding site.

The results, on the effect of blocking the β-93 sulfhydryl, removing the β-146 histidine, and substituting β-143 histidine for other amino acids, might suggest formation of a complex involving these three amino acids, all of which are located in the same general region of the molecule on the proximal side of the heme (Fig. 15). An investigation of a model of hemoglobin [18], however, indicates that these three residues do not seem to be located close enough

A Beta Subunit

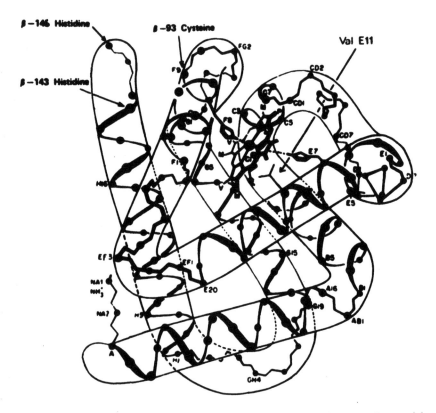

FIG. 15. Diagrammatic sketch showing the course of the polypeptide chain in the β subunit of hemoglobin. The amino acids for which there is evidence for zinc binding are indicated. This figure is a modification of the illustration of the β subunit in Ref. 151. (Reprinted by permission from *Nature*, *228*, 726-734. © 1970 Macmillan Journals Limited.)

together for zinc to coordinate simultaneously to all three residues.
Thus histidine β-143 is directed toward the internal cavity between
the β chains, while cysteine β-93 is closer to the heme pocket and
the $\alpha_1\beta_2$ interface. The distance between the side chains of these
two amino acids is >10 Å, a distance too great for zinc to chelate.
However, it is possible that the strong binding of zinc provides the
necessary free energy to produce the conformational change necessary
to bring these residues together.

In any attempt to relate changes in binding constants of zinc
for different hemoglobins to the location of the zinc-binding site,
it is necessary to consider the possibility that alterations of the
hemoglobin conformation could alter the conformation of hemoglobin
in the region of the zinc-binding site, and thereby the zinc association constant, even though the modification or substitution does
not directly involve amino acids involved in the binding of zinc.
Removal of the β-146 histidine and blocking of the β-sulfhydryl have
thus been shown to produce changes in the conformation of hemoglobin
[151], which could readily produce a change in the zinc association
constant of one order of magnitude, even if zinc is not directly
bound to these residues.

The decrease in the association constant is, however, considerably greater when histidine β-143 is substituted by other amino
acids. Furthermore, the same effect on the stability constant was
found for hemoglobin Little Rock and hemoglobin Abruzzo. Both
arginine and glutamine have a much lower metal-ion affinity than
histidine and would produce a similar decrease of the zinc association constant if they were directly involved in the binding. However, these two amino acids have different charges at neutral pH
and are structurally quite different. Therefore, an indirect effect
of the substitution of histidine for arginine and glutamine in hemoglobins Abruzzo and Little Rock, respectively, would be expected to
have quite a different effect on the association constant. It thus
seems that at least histidine β-143 is one of the amino acids involved in zinc binding.

3.3. Mechanism of Effect of Zinc on Oxygenation

The high affinity of hemoglobin for zinc (see Sec. 3.2) indicates
that the large majority of zinc is binding to hemoglobin in the
concentration range that produces the increase in oxygen affinity.
The difference in affinity between deoxy- and oxy- or carboxyhemoglobin [133,146] further indicates that the binding constant is
greater in the liganded hemoglobin and therefore stabilizes the
liganded conformation of hemoglobin.

Gilman and Brewer [133] and Rifkind and Heim [18] have utilized
their data on the zinc-binding site to propose how zinc stabilizes
the liganded conformation and thereby increases the oxygen affinity.
Gilman and Brewer [133] suggest that zinc destabilizes the deoxy
conformation by binding to histidine β-146, which in deoxyhemoglobin
is involved in an intrachain salt bridge with aspartic acid β-94
[152]. Furthermore, the decrease in the effect of zinc on the oxygen
affinity of hemoglobin with the β-93 sulfhydryls blocked is explained
by the disruption of the salt bridge involving β-146 even in deoxyhemoglobin when the β-93 sulfhydryl is blocked.

Rifkind and Heim [18], having found a particularly large effect
due to substitution for histidine β-143, suggest a mechanism related
to that proposed by Perutz [153] to explain the higher oxygen affinity
of hemoglobin Little Rock (β 143 His → Gln) [150]. This mechanism
proposes a hydrogen bond between the glutamine at β-143 and asparagine
β-139 of the other β chain which would stabilize the liganded conformation. In the same way it was suggested [18] that zinc could coordinate between the asparagine and histidine.

It would, however, seem that such an interaction would decrease
the dissociation constant of tetramers to dimers, which seems to
increase in the presence of zinc [146].

If zinc coordination actually involves both histidine β-146
and β-143, such a mechanism would not seem feasible and instead
destabilization of the β-146 salt bridge may be involved.

It has been shown that under certain conditions zinc causes
dissociation of hemoglobin to dimers [146] and perhaps even to

monomers [134]. Since dimers have a higher oxygen affinity than
tetramers, it is possible that the effect of zinc on the oxygenation
could be explained by dissociation of hemoglobin, without the necessity that zinc bound to the tetramer produce any specific stabilization of the liganded conformation. The ability to observe effects
of zinc on oxygenation that are almost as great as those in dilute
solutions for the high concentration present in erythrocytes seems,
however, to rule out an explanation due solely to dissociation.

4. CONCLUSION

This chapter reviews what is currently known about the interaction
of zinc with hemoglobin and the erythrocyte membranes. The original
impetus for much of this work was the possibility that zinc may be
useful in the treatment for sickle-cell anemia. This expectation
was originally based on the observation that zinc increases the
oxygen affinity of hemoglobin and might inhibit sickling, and subsequently on the improved filterability of sickle cells, which suggests that zinc may inhibit the membrane damage that results from
sickling.

Although zinc has shown some promise for the treatment of
sickle-cell anemia, its usefulness, at this time, is limited by the
inability to elevate plasma zinc to an adequate level. However, the
results on interactions of zinc with erythrocytes as well as other
cells have expanded our understanding of the biological function of
zinc. On the basis of membrane studies, a possible role for zinc
in regulating membrane structure and function is evolving. Furthermore, it is attractive to speculate regarding a possible role for
zinc in fine tuning the oxygen affinity of hemoglobin. This suggestion is based on the similar affinity of zinc for hemoglobin and
spectrin on the cytoplasmic surface of the membrane, which suggests
that zinc may regulate oxygenation by shuttling between hemoglobin
and the membrane.

ABBREVIATIONS

ALAD	δ-aminolevulinic acid dehydratase
ATP	adenosine triphosphate
Bistris	bis(2-hydroxyethyl)imino-tris(hydroxy methyl)-methane
BME	bis(n-maleimidomethyl)ether
CTP	cytosine triphosphate
2,3-DPG	2,3-diphosphoglyceric acid
EDTA	ethylenediaminetetraacetic acid
GTP	guanosine triphosphate
HbA	major hemoglobin component of normal human adult blood
HbA$_2$	minor hemoglobin component of human blood with δ chains instead of β chains
HbF	fetal hemoglobin
HbSS	sickle-cell hemoglobin
HSA	human serum albumin
ISC	irreversibly sickled cell
KB	modified Krebs buffer without phosphate
KB-HSA	modified Krebs buffer plus human serum albumin
NEM	N-ethylmaleimide
NMR	nuclear magnetic resonance
SDS-PAGE	sodium dodecyl sulfate polyacrylamide gel electrophoresis
TNC	trinitrocresol

REFERENCES

1. D. Keilin and T. Mann, *Nature (Lond.)*, *144*, 42 (1939).

2. S. Lindskog, L. E. Henderson, K. K. Kannan, A. Liljas, P. O. Nyman, and B. Strandberg, in *The Enzymes*, 3rd ed., Vol. 5 (P. D. Boyer, ed.), Academic Press, New York, 1971, p. 587ff.

3. J. M. McCord and I. Fridovich, *J. Biol. Chem.*, *244*, 6049 (1969).

4. J. Bannister, W. Bannister, and E. Wood, *Eur. J. Biochem.*, *18*, 178 (1971).

5. H. Markowitz, G. E. Cartwright, and M. M. Wintrobe, *J. Biol. Chem.*, *234*, 40 (1959).

6. J. M. McCord, B. B. Keele, Jr., and I. Fridovich, *Proc. Natl. Acad. Sci. USA*, *68*, 1024 (1971).

7. A. Cheh and J. B. Neilands, *Biochem. Biophys. Res. Commun.*, *55*, 1060 (1973).

8. P. M. Anderson and R. J. Desnick, *J. Biol. Chem.*, *254*, 6924 (1979).

9. I. Tsukamoto, T. Yoshinaga, and S. Sano, *Biochim. Biophys. Acta*, *570*, 167 (1979).

10. H. C. Lichtman and F. Feldman, *J. Clin. Invest.*, *42*, 830 (1963).

11. K. Nakao, O. Wada, and Y. Yano, *Clin. Chim. Acta*, *19*, 319 (1968).

12. Y. Mauras and P. Allain, *Enzyme*, *24*, 181 (1979).

13. M. Abdulla and S. Svensson, *Scand. J. Clin. Lab. Invest.*, *39*, 31 (1979).

14. R. A. Willoughby, E. McDonald, B. J. Sherry, and G. Brown, *Can. J. Comp. Med.*, *36*, 348 (1972).

15. A. S. Prasad, A. Abbasi, and J. Ortega, in *Zinc Metabolism: Current Aspects in Health and Disease* (G. J. Brewer and A. S. Prasad, eds.), Alan R. Liss, New York, 1977, p. 211ff.

16. F. J. Oelshlegel, Jr., G. J. Brewer, A. S. Prasad, C. Knutsen, and E. B. Schoomaker, *Biochem. Biophys. Res. Commun.*, *53*, 560 (1973).

17. F. J. Oelshlegel, Jr., G. J. Brewer, C. Knutsen, A. S. Prasad, and E. B. Schoomaker, *Arch. Biochem. Biophys.*, *163*, 742 (1974).

18. J. M. Rifkind and J. M. Heim, *Biochemistry*, *16*, 4438 (1977).

19. G. J. Brewer, E. B. Schoomaker, D. A. Leichtman, W. C. Kruckeberg, L. F. Brewer, and N. Meyers, in *Zinc Metabolism: Current Aspects in Health and Disease* (G. J. Brewer and A. S. Prasad, eds.), Alan R. Liss, New York, 1977, p. 241ff.

20. H. Passow, in *Effects of Metals on Cells, Subcellular Elements and Macromolecules* (J. Maniloff, J. R. Coleman, and M. W. Miller, eds.), Charles C Thomas, Springfield, Ill., 1970, p. 291ff.

21. K. Araki and J. M. Rifkind, Abstr., XIth Int. Congr. Biochem., Toronto, July 1979.

22. D. L. Rabenstein and A. A. Isab, *FEBS Lett.*, *121*, 61 (1980).

23. M. Chvapil, *Med. Clin. N. Am.*, *60*, 799 (1976).

24. W. J. Bettger and B. L. O'Dell, *Life Sci.*, *28*, 1425 (1981).

25. W. Kazimierczak and C. Maslinski, *Agents Actions*, *4*, 1 (1974).

26. M. Chvapil, P. L. Weldy, L. Stankova, D. S. Clark, and C. F. Zukoski, *Life Sci.*, *16*, 561 (1975).

27. L. Karl, M. Chvapil, and C. F. Zukoski, *Proc. Soc. Exp. Biol. Med.*, *142*, 1123 (1973).

28. V. J. Nicholson and H. Veldstra, *FEBS Lett.*, *23*, 309 (1972).

29. A. W. Martin, C. Lutwak-Mann, J. E. A. McIntosh, and T. Mann, *Comp. Biochem. Physiol.*, *45A*, 227 (1973).

30. B. Bacetti, V. Pallini, and A. G. Burrini, *J. Ultrastruct. Res.*, *54*, 261 (1976).

31. M. Chvapil, *Life Sci.*, *13*, 1041 (1973).

32. R. G. Cassens, W. G. Hoekstra, E. C. Faltin, and E. J. Briskey, *Am. J. Physiol.*, *212*, 688 (1967).

33. K. S. Rajan, R. W. Colburn, and J. M. Davis, *Life Sci.*, *18*, 423 (1976).

34. W. W. Saylor and R. M. Leach, *J. Nutr.*, *110*, 448 (1980).

35. J. C. Ludwig and M. Chvapil, *J. Nutr.*, *110*, 945 (1980).

36. S. Hakomori, T. Ishimoda, N. Kawauchi, and K. Nakamura, *J. Biochem.*, *54*, 458 (1963).

37. D. M. Gettelfinger and G. F. Siegel, *J. Neurochem.*, *31*, 1231 (1978).

38. W. Ernst and H. Weinert, *Z. Pflanzenphysiol.*, *66*, 258 (1972).

39. M. Chvapil, D. Montgomery, J. C. Ludwig, and C. F. Zukoski, *Proc. Soc. Exp. Biol. Med.*, *162*, 480 (1979).

40. G. J. Brewer, in *Metal Ions in Biological Systems*, Vol. 14 (H. Sigel, ed.), Marcel Dekker, New York, 1981.

41. M. Chvapil, Y. M. Peng, A. L. Aronson, and C. F. Zukoski, *J. Nutr.*, *104*, 434 (1974).

42. C. T. Settlemire and G. Matrone, *J. Nutr.*, *92*, 159 (1967).

43. W. J. Bettger, T. J. Fish, and B. L. O'Dell, *Proc. Soc. Exp. Biol. Med.*, *158*, 279 (1978).

44. I.-A. Kabat, J. Niedworok, and J. Blaszczyk, *Zentralbl. Bakteriol. Hyg. I. Abt. Org. B.*, *166*, 375 (1978).

45. K. Weismann and H. I. Mikkelsen, *Arch. Dermatol. Res.*, 269 (1980).

46. K. Araki and J. M. Rifkind, *J. Gerontol.*, *35*, 499 (1980).

47. K. Araki and J. M. Rifkind, *Biochim. Biophys. Acta*, *645*, 81 (1981).

48. G. J. Brewer, in *Zinc and Copper in Medicine* (Z. A. Karcioglu and R. M. Sarper, eds.), Charles C Thomas, Springfield, Ill., 1980, p. 347ff.

49. G. L. Fisher, *Sci. Total Environ.*, *4*, 373 (1975).

50. A. S. Prasad, G. J. Brewer, E. B. Schoomaker, and P. Rabbani, *JAMA, 240*, 2166 (1978).

51. E. D. Wills, *Biochim. Biophys. Acta, 98*, 238 (1965).

52. K. F. Adams, G. Johnson, Jr., K. Hornowski, and T. Lineberger, *Biochim. Biophys. Acta, 550*, 279 (1979).

53. M. W. Radomski and J. D. Wood, *Aerosp. Med., 41*, 1382 (1970).

54. M. Chvapil, J. Ryan, and C. Zukoski, *Proc. Soc. Exp. Biol. Med., 141*, 150 (1972).

55. M. Chvapil, J. Ryan, S. Elias, and Y. Peng, *Exp. Mol. Pathol., 19*, 186 (1973).

56. R. L. Willson, *Ciba Found. Symp., 51*, 331 (1977).

57. C. J. Pfeiffer and C. H. Cho, *Res. Commun. Chem. Pathol. Pharmacol., 27*, 587 (1980).

58. L. S. Avigad and A. W. Bernheim, *Infect. Immun., 13*, 1378 (1976).

59. Y. Takeda, Y. Ogiso, and T. Miwatani, *Infect. Immun., 17*, 239 (1977).

60. L. S. Avigad and A. W. Bernheim, *Infect. Immun., 19*, 1101 (1978).

61. D. W. Montgomery, L. K. Don, C. F. Zukoski, and M. Chvapil, *Proc. Soc. Exp. Biol. Med., 145*, 263 (1974).

62. K. Yamamoto and M. Takahashi, *Int. Arch. Allergy Appl. Immunol., 48*, 653 (1975).

63. M. D. P. Boyle, J. J. Langone, and T. Borsos, *J. Immunol., 122*, 1209 (1979).

64. J. M. Rifkind and R. Anthenelli, unpublished results.

65. A. D. Bangham, B. A. Pethica, and G. V. F. Seaman, *Biochem. J., 69*, 12 (1958).

66. A. McLaughlin, C. Greathwohl, and S. McLaughlin, *Biochim. Biophys. Acta, 513*, 338 (1971).

67. J. Sunamoto, M. Shironita, and N. Kawauchi, *Bull. Chem. Soc. Jpn., 53*, 2778 (1980).

68. Y. Castranova and P. R. Miles, *J. Membr. Biol., 33*, 263 (1977).

69. K. Sivarama, L. Sastry, L. Viswanathan, A. Ramaiah, and P. S. Sarma, *Biochem. J., 74*, 561 (1960).

70. J. E. Ockunewick, S. E. Herrick, and T. G. Hennessey, *J. Cell. Comp. Physiol., 63*, 333 (1964).

71. W. Fuchswans and H. Springer-Lederer, in *Erythrocytes, Thrombocytes, Leukocytes* (E. Gerlach, K. Moser, E. Deutsch, and W. Wilmanns, eds.), Georg Thieme, Stuttgart, 1973, p. 121ff.

72. G. Schmetterer, *Z. Naturforsch, 33*, 210 (1978).

73. W. Kruckeberg, C. A. Knutsen, and G. J. Brewer, in *Zinc Metabolism: Current Aspects in Health and Disease* (G. J. Brewer and A. S. Prasad, eds.), Alan R. Liss, New York, 1977, p. 259ff.

74. W. C. Kruckeberg and G. J. Brewer, *Med. Biol.*, *56*, 5 (1978).

75. W. C. Kruckeberg, F. J. Oelshlegel, Jr., S. H. Shore, P. E. Smouse, and G. J. Brewer, *Res. Exp. Med.*, *170*, 149 (1977).

76. R. P. Cox, *Mol. Pharmacol.*, *4*, 510 (1968).

77. J. Frieberg and O. Nilsson, *J. Med. Sci.*, *79*, 63 (1974).

78. G. Gardos, I. Szasz, B. Sarkadi, and J. Szebeni, in *Membrane Transport in Erythrocytes; Relations Between Function and Molecular Structure* (U. V. Lassen, H. H. Ussing, and J. O. Wieth, eds.), Munksgaard, Copenhagen, 1980, p. 163.

79. D. R. Pfeiffer, R. W. Taylor, and H. A. Lardy, *Ann. N.Y. Acad. Sci.*, *307*, 402 (1978).

80. J. M. Rifkind and E. Hayes, unpublished results.

81. J. Gutknecht, *J. Membr. Biol.*, *61*, 61 (1981).

82. R. M. Smith and A. E. Martell, in *Critical Stability Constants*, Vol. 4, Plenum Press, New York, 1976.

83. J. F. Bertles, R. Rabinowitz, and J. Dobler, *Science*, *169*, 375 (1970).

84. J. T. Finch, M. F. Perutz, J. F. Bertles, and J. Dobler, *Proc. Natl. Acad. Sci. USA*, *70*, 718 (1973).

85. J. Dobler and J. F. Bertles, *J. Exp. Med.*, *127*, 711 (1968).

86. J. F. Bertles and J. Dobler, *Blood*, *33*, 884 (1969).

87. S. Chien, S. Usami, and J. F. Bertles, *J. Clin. Invest.*, *49*, 623 (1970).

88. G. R. Sergeant, B. E. Sergeant, and P. J. Condon, *JAMA*, *219*, 1428 (1972).

89. G. R. Sergeant, *Br. J. Haematol.*, *19*, 635 (1970).

90. G. J. Brewer, L. F. Brewer, and A. S. Prasad, *J. Lab. Clin. Med.*, *90*, 549 (1977).

91. G. J. Brewer and F. J. Oelshlegel, Jr., *Biochem. Biophys. Res. Commun.*, *58*, 854 (1974).

92. A. Arnone and D. Williams, in *Zinc Metabolism: Current Aspects in Health and Disease* (G. J. Brewer and A. S. Prasad, eds.), Alan R. Liss, New York, 1977, p. 317ff.

93. J. Palek, W. A. Curby, and F. J. Lionetti, *Am. J. Physiol.*, *220*, 19 (1971).

94. J. W. Eaton, E. Berger, J. G. White, and H. S. Jacob, in *Zinc Metabolism: Current Aspects in Health and Disease* (G. J. Brewer and A. S. Prasad, eds.), Alan R. Liss, New York, 1977, p. 275ff.

95. M. J. Dunn, *Biochim. Biophys. Acta, 352,* 97 (1974).

96. R. I. Weed and B. Chailley, in *Erythrocytes, Thrombocytes, Leukocytes* (E. Gerlach, K. Moser, E. Deutsch, and W. Wilmanns, eds.), Georg Thieme, Stuttgart, 1973, p. 45ff.

97. J. G. White, *Am. J. Pathol., 77,* 507 (1974).

98. J. G. White, *Semin. Hematol., 13,* 121 (1976).

99. R. I. Weed, P. L. LaCelle, and E. W. Merrill, *J. Clin. Invest., 48,* 795 (1969).

100. J. W. Eaton, T. D. Skelton, H. S. Swofford, C. E. Koplin, and H. S. Jacob, *Nature (Lond.), 246,* 105 (1973).

101. J. W. Eaton, E. Berger, J. G. White, and H. S. Jacob, U.S. Dept. of Health, Education and Welfare, Publ. No. (NIH) 76-1007, 1976, p. 327ff.

102. S. Dash, G. J. Brewer, and F. J. Oelshlegel, Jr., *Nature (Lond.), 250,* 251 (1974).

103. G. J. Brewer, *Am. J. Hematol., 8,* 231 (1980).

104. D. R. Anderson, J. L. Davis, and K. L. Carraway, *J. Biol. Chem., 252,* 6617 (1977).

105. K. L. Carraway, R. B. Triplett, and D. R. Anderson, *Biochim. Biophys. Acta, 379,* 571 (1975).

106. G. E. Siefring, Jr., and L. Lorand, in *Erythrocyte Membranes* (W. C. Kruckeberg, J. W. Eaton, and G. J. Brewer, eds.), Alan R. Liss, New York, 1978, p. 25ff.

107. W. C. Kruckeberg, in *Erythrocyte Membranes* (W. C. Kruckeberg, J. W. Eaton, and G. J. Brewer, eds.), Alan R. Liss, New York, 1978, p. 35.

108. K. S. Lee and B. C. Shin, *J. Gen. Physiol., 54,* 713 (1969).

109. H. J. Schatzmann, *Experientia, 22,* 364 (1966).

110. Y. N. Cha, B. C. Shin, and K. S. Lee, *J. Gen. Physiol., 57,* 202 (1971).

111. W. C. Kruckeberg, R. Bargal, and G. J. Brewer, in *Erythrocyte Membranes* (W. C. Kruckeberg, J. W. Eaton, and G. J. Brewer, eds.), Alan R. Liss, New York, 1978, p. 139ff.

112. P. A. Knauf, F. Proverbio, and J. F. Hoffman, *J. Gen. Physiol., 63,* 324 (1974).

113. S. Katz and R. Blostein, *Biochim. Biophys. Acta, 389,* 314 (1975).

114. M. D. White and R. B. Ralston, *Biochim. Biophys. Acta, 599,* 569 (1980).

115. W. Y. Cheung, T. J. Lynch, and R. W. Wallace, in *Advances in Cyclic Nucleotide Research* (W. J. George and L. F. Ignarro, eds.), Raven Press, New York, 1978, p. 233ff.

116. W. Y. Cheung, *Biochem. Biophys. Res. Commun.*, *38*, 533 (1970).

117. S. Kakiuchi, R. Yamazaki, and H. Nakagima, *Proc. Jpn. Acad.*, *46*, 587 (1970).

118. G. H. Bond and D. L. Clough, *Biochim. Biophys. Acta*, *323*, 592 (1973).

119. J. C. Aster and G. J. Brewer, *Clin. Res.*, *27*, 636A (1979).

120. G. J. Brewer and J. C. Aster, *Blood*, *54*, 24A (1979).

121. G. J. Brewer, J. C. Aster, C. A. Knutsen, and W. C. Kruckeberg, *Am. J. Hematol.*, *7*, 53 (1979).

122. T. E. Donnelly, *Biochim. Biophys. Acta*, *522*, 151 (1978).

123. G. Fairbanks, T. L. Steck, and D. F. H. Wallach, *Biochemistry*, *10*, 2606 (1971).

124. Y. M. Lin, Y. P. Liu, and W. Y. Cheung, *J. Biol. Chem.*, *249*, 4943 (1974).

125. J. M. Rifkind, in *Metal Ions in Biological Systems*, Vol. 12 (H. Sigel, ed.), Marcel Dekker, New York, 1981, p. 191ff.

126. J. M. Rifkind, *Biochemistry*, *13*, 2475 (1974).

127. J. M. Rifkind, L. D. Lauer, S. C. Chiang, and N. C. Li, *Biochemistry*, *15*, 5337 (1976).

128. J. M. Rifkind, M. H. Keyes, and R. Lumry, *Biochemistry*, *16*, 5564 (1977).

129. J. M. Rifkind, unpublished results.

130. R. Benesch and R. E. Benesch, *Biochem. Biophys. Res. Commun.*, *26*, 162 (1967).

131. A. Chanutin and R. R. Curnish, *Arch. Biochem. Biophys.*, *121*, 96 (1967).

132. M. Berman, R. Benesch, and R. E. Benesch, *Arch. Biochem. Biophys.*, *145*, 236 (1971).

133. J. G. Gilman and G. J. Brewer, *Biochem. J.*, *169*, 625 (1978).

134. J. G. Gilman, F. J. Oelshlegel, Jr., and G. J. Brewer, in *Erythrocyte Structure and Function* (G. J. Brewer, ed.), Alan R. Liss, New York, 1975, p. 85ff.

135. A. Arnone, *Nature (Lond.)*, *237*, 146 (1972).

136. A.-I. Kabat, J. Niedworok, J. Kedziora, J. Blaszczyk, and G. Bartosz, *Zentralbl. Bakteriol. Hyg. I. Abt. Org. B*, *169*, 436 (1979).

137. J. M. Rifkind, in *Inorganic Biochemistry* (G. L. Eichhorn, ed.), Elsevier, Amsterdam, 1973, p. 832ff.

138. J. M. Rifkind, *Biochim. Biophys. Acta*, *273*, 30 (1972).

139. J. F. Taylor, E. Antonini, M. Brunori, and J. Wyman, *J. Biol. Chem.*, *241*, 241 (1966).

140. K. Moffat, S. R. Simon, and W. A. Konigsberg, *J. Mol. Biol.*, *58*, 89 (1971).

141. J. R. Cann, *Biochemistry*, *3*, 903 (1964).

142. I. I. Wu, M. L. Borke, and N. C. Li, *J. Inorg. Nucl. Chem.*, *40*, 745 (1978).

143. R. K. Gupta, J. L. Benovic, and Z. B. Rose, *J. Biol. Chem.*, *253*, 6165 (1978).

144. R. D. Gray, *J. Biol. Chem.*, *255*, 1812 (1980).

145. G. Amiconi, L. Civalleri, S. G. Condor, F. Ascoli, R. Santucci, and E. Antonini, *Hemoglobin*, *5*, 231 (1981).

146. J. M. Rifkind and J. M. Heim, unpublished results.

147. J. Wyman, Jr., *Adv. Protein Chem.*, *4*, 407 (1948).

148. C. Ho, in *Erythrocyte Structure and Function* (G. J. Brewer, ed.), Alan R. Liss, New York, 1975, p. 103.

149. C. Bonaventura, J. Bonaventura, G. Amiconi, L. Tentori, M. Brunori, and E. Antonini, *J. Biol. Chem.*, *250*, 6273 (1975).

150. P. A. Bromberg, J. O. Alben, G. H. Bare, S. P. Balcerzak, R. T. Jones, B. Brimhall, and F. Padilla, *Nature New Biol.*, *243*, 177 (1973).

151. M. F. Perutz, *Nature (Lond.)*, *228*, 726 (1970).

152. M. F. Perutz, *Nature (Lond.)*, *228*, 735 (1970).

153. M. F. Perutz, *Nature New Biol.*, *243*, 180 (1973).

Chapter 8

ZINC ABSORPTION AND EXCRETION
IN RELATION TO NUTRITION

Manfred Kirchgessner and Edgar Weigand
Institut für Ernährungsphysiologie der Technischen
Universität München
Freising-Weihenstephan
Federal Republic of Germany

1. INTRODUCTION

Zinc as an essential trace element has many physiological functions in both humans and animals (see Chap. 9) and must accordingly be supplied in adequate amounts at all stages of life. How much zinc is available at the site of its metabolism in tissues and organs of the mammalian body will depend not only on alimentary supply, which may greatly vary with different dietaries, but also on the percentage absorbed by the digestive tract and the excretion of this element via various routes. Therefore, it is the intention of this chapter to review relevant aspects of zinc absorption and excretion from a nutritional point of view and to relate their significance for the dietary zinc requirement in some concluding remarks.

2. ABSORPTION

2.1. Site of Absorption

According to a number of studies on the laboratory rat [1], zinc can be absorbed throughout the total length of the small intestine, whereas stomach and large intestine contribute very little to the absorption of this element [2-4]. Whether significant amounts of zinc might be absorbed from the human stomach is not clear. Animal studies employing various techniques have yielded discrepant results as to the small intestinal region having maximum absorptive capacity [1,5]. Figure 1, showing the relative absorptive efficiency of jejunal and ileal segments, suggests that the major site of absorption might shift with the zinc status. Dietary zinc depletion stimulated zinc uptake and transfer by everted sacs most markedly in the distal jejunum and ileum [6]. Considering the length and surface area of the various segments of the small bowel, the transit time of the digesta, and the endogenous secretion of zinc (see Sec. 3.1), most zinc may be absorbed from the jejunum. In healthy humans fed mixed meals, intraluminal quantities of zinc recovered at different sites were highest at the distal duodenum and substantially declined

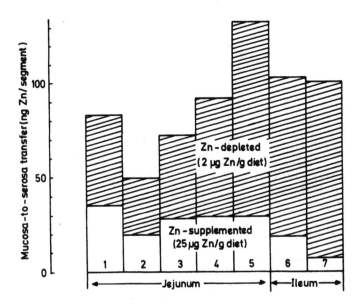

FIG. 1. Mucosa-to-serosa transfer of zinc by everted jejunal and
ileal segments of equal length from zinc-depleted versus zinc-
supplemented rats. (After data from Ref. 6.)

in the jejunum, especially in the proximal portion [7]. Another
study on human subjects with and without jejunoileostomy also suggests the jejunum as the major site of intestinal zinc absorption
[8].

2.2. Regulation of Absorption

Numerous studies provide firm evidence that intestinal zinc absorption varies in homeostatic response to an altered zinc supply status
of the body and that the intestine itself is a major site for this
control. Accordingly, much recent research has been devoted to the
mechanism(s) regulating the absorptive process.

Intestinal zinc absorption involves an active, energy-requiring
step [9]. Metabolic inhibitors such as 2,4-dinitrophenol (2,4-DNP),
iodoacetate, or ouabain substantially lower the flux of zinc across
everted intestinal segments [9-11]. Table 1 shows that the inhibitory

TABLE 1

Effect of Dietary Zinc Supply and 2,4-Dinitrophenol (2,4-DNP)
$(5 \cdot 10^{-5}$ M) on the Bidirectional Transfer
of Zinc Across Everted Jejunal Sacs

Dietary pretreatment	Transfer (ng zinc/sac in 60 min)			
	Mucosa → control	Serosa + 2,4-DNP	Serosa → control	Mucosa + 2,4-DNP
Zinc-deficient rats (2.4 ppm zinc in dry diet	63.1	23.4	10.4	11.1
Pair-fed rats (110 ppm zinc in dry diet)	11.7	8.5	15.6	9.1

Source: Data from Ref. 12.

effect of 2,4-DNP on zinc transfer from the mucosal to the serosal
side of jejunal sacs from rats is most marked in the case of zinc-
depleted animals. Lack of oxygen, metabolizable hexose, or sodium
in the incubating medium also diminishes the in vitro absorption of
zinc [9].

With respect to the dependence on zinc concentration, intestinal absorption displays saturation kinetics indicating a carrier-
mediated transport mechanism [4,13]. Figure 2 shows, after the
Lineweaver-Burk approach, a double-reciprocal plot of the daily
rates of true zinc absorption versus intake according to data of an
in vivo study with young rats. This graph yields a Michaelis-Menten-
type constant K_a of 1.31 mg of zinc per day. This value, reflecting
zinc intake at half-maximum absorption rate, corresponds to approximately 115 ppm of zinc in the dry diet fed to these animals. This
is a supply level substantially higher than the estimated dietary
zinc requirement of rats growing at a very rapid rate [15,16].
Accordingly, it is indicated that under normal nutritional circumstances, zinc absorption remains far below the saturation limit of
the physiological absorptive capacity.

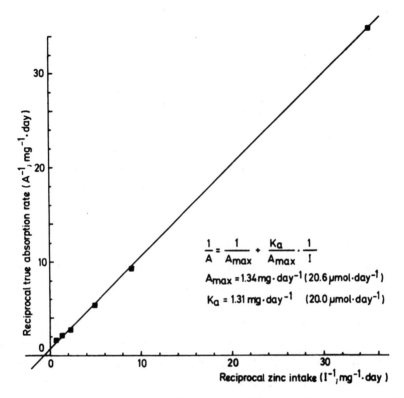

FIG. 2. Double-reciprocal plot of true absorption rate versus zinc
intake in young rats fed a basal diet containing graded levels of
zinc ranging from 5.6 up to 141 ppm zinc. (After data from Ref. 14.)

Zinc absorption may be seen to take place in at least three
stages [11]: (1) uptake from the lumen by the brush border of the
epithelial cells lining the small intestine, (2) transport through
the absorptive cells, and (3) discharge across the basolateral cell
membrane to the vascular compartment. Numerous studies indicate
that control is exerted at one (or more) of these steps and that
subcellular changes occur in homeostatic response to zinc status.
There is little evidence to assume that the absorptive process
might be regulated by the secretion of specific zinc-binding ligands
into the intestinal lumen. This does, however, by no means exclude

that such ligands would be important for the absorptive process. In fact, zinc has the potential to form complexes with many different components of both dietary and endogenous origin under the prevailing pH conditions in the intestinal lumen. Accordingly, both high- and low-molecular-weight ligands ingested with the food or generated during gastrointestinal digestion could materially affect zinc absorption, especially so by either facilitating or impairing mucosal zinc uptake. For example, in vitro and in situ studies [10,11,17,18] have regularly demonstrated that cysteine and histidine greatly enhance intestinal zinc uptake and transport when present in high molar excess. These two amino acids form zinc complexes of much higher stability than others [19]. In vivo, however, the formation and effective stability of such complexes will depend not only on the particular pH and zinc:ligand ratio, but also on the relative abundance of other cations and other organic constituents of high chelate-forming potential. In this context it seems highly questionable that particular amino acids or other low-molecular-weight compounds reported to improve intestinal zinc absorption, for instance, picolinic acid [20] or citrate [13,21,22], would have a specific functional role in the absorptive process. Forth and Rummel [23] postulate that trace element absorption from complexes of very high stability depends on the absorbability of the particular complex. Accordingly, liver copper accretion from various copper amino acid complexes added to the diet was related to the absorption characteristics of the particular amino acid [24].

The ingestion of strong chelating agents such as ethylenediaminetetraacetic acid (EDTA) might shunt zinc away from the normal absorption route. In this regard in vitro work [10,11] has shown that mucosal zinc uptake from EDTA-containing media is greatly reduced, especially in the case of intestines from zinc-depleted animals, whereas transmural zinc transport to the serosal side is actually enhanced and, with respect to its magnitude, about related to the transfer of the ligand. Also, 2,4-DNP addition did not affect mucosal zinc uptake in the presence of EDTA and even increased the mucosa-to-serosa transfer of both zinc and EDTA. The therapeutic

effect of halogenated hydroxyquinolines in patients suffering from
acrodermatitis enteropathica [25,26] might similarly be explained
by facilitating zinc absorption via a pathway different from the
normal carrier-mediated process.

The mucosal cell lining would seem to take a precedential
position in controlling zinc absorption. Everted intestinal sacs
prepared from zinc-depleted rats regularly display a much greater
uptake and mucosa-to-serosa transport of zinc than control segments
from animals fed zinc-supplemented diets [10,11,27]; see also Fig. 1
and Table 1. This increased absorptive capacity seems to be rather
specific for zinc, because dietary zinc depletion does not enhance
absorption of other cations, such as calcium, iron, or manganese
[28-30]. The absorption of copper, however, which shows a mutual
interaction with zinc (see Ref. 31), is markedly increased after
zinc depletion [28,30].

As to the zinc concentration of the intestinal wall, little
difference is indicated between zinc-depleted and zinc-supplemented
rats [13,32]. However, sudden drastic zinc loading either orally
[33] or parenterally [13,34] leads to a marked transient increase
in the zinc content of intestinal tissue. Menard and co-workers
[33] reported a concomitant elevation of metallothionein-bound zinc
in mucosal cytosol, whereas zinc associated with high-molecular-
weight fractions did not significantly change after dietary zinc
supplementation. Metallothionein does not seem to be a regular
major zinc-binding protein of enterocytes; however, its synthesis
in the small intestine and other tissues is rapidly induced by oral
or parenteral zinc loading [35,36]. The de novo synthesis of intestinal zinc-metallothionein seems to attain its maximum rate around
6 h after zinc administration [33,37]. This coincides approximately
with the time lapse for oral [33] or parenteral zinc dosing [34,37]
to bring about a notable decrease in zinc absorption. Thus an inverse relationship is indicated between zinc absorption and the
mucosal zinc-thionein content [35,36]. In the case of a gradual
elevation of the zinc status of depleted rats by prolonged intra-

venous zinc infusion, as many as 94 h were needed to induce a significant reduction in the intestinal zinc absorption [38].

These studies demonstrate that intestinal zinc absorption can
respond rather quickly to fluctuations in the zinc supply status by
short-term changes in subcellular components. Mucosal metallothionein
has been suggested to serve as a temporary zinc storage protein from
where the metal could subsequently either be passed on to the vascular
compartment or returned to the intestinal lumen [35]. Further work
will help to understand the function of this and other zinc-responsive
mucosal constituents involved in the cellular events controlling zinc
absorption and excretion. It has recently been realized that previous
findings on the presence of significant amounts of low-molecular-
weight zinc-binding ligands in the enterocyte may have resulted from
the degradation of metallothionein or other macromolecular zinc-
binding proteins during tissue preparation and cell fractionation
[39,40] or from procedural artifacts [40]. Accordingly, most recent
investigations in different laboratories [18,33,40] found very little
zinc associated with low-molecular-weight fractions in mucosal homoge-
nates of the rat small intestine. As already pointed out by Cousins
[35], this does not exclude the possible involvement of small zinc-
binding ligands in the absorption process.

Further research is needed to clarify the regulatory steps and
metabolic integration of the zinc-absorbing process from the site of
its uptake at the luminal surface and its conveyance across the
enterocyte to the site of its discharge at the basolateral membrane.
Smith and Cousins [30] suggested from work with a luminal and vascular perfusion system that the rate-limiting step in zinc absorption
might be at the basement membrane. However, an incremental saturation of the zinc-transporting compounds of blood--predominantly
plasma albumin according to recent studies [41,42]--can hardly provide
a plausible explanation for the short-term or long-term reduction in
zinc absorption rates following dietary zinc supplementation. In
zinc-depleted rats, intestinal absorption of this metal continued
to proceed at an elevated rate, although plasma zinc concentrations
had reached extremely high levels due to prior intravenous zinc

injection [13,34]. Also, serum has been shown to have a much higher
zinc-binding capacity than that normally attained under physiological
conditions [43]. These observations again point to the small intestinal wall as the major site for the homeostatic regulation of zinc
absorption.

Acrodermatitis enteropathica (AE), a rare disease inherited as
an autosomal recessive trait in humans [25], has been demonstrated
to result from zinc malabsorption [26,44]. In vitro work with jejunal
biopsy samples from AE patients and healthy persons suggests a faulty
cellular zinc uptake as the primary reason for this abnormality [45].
Apart from the therapeutic value of human milk (see Sec. 3.3) and
halogenated hydroxyquinolines (see earlier in this section), oral
zinc doses in a 10- to 20-fold excess of the normal nutritional
intake also proved to be effective in the clinical remission of this
disorder [44]. This suggests that the normal pathway for zinc absorption is either not totally blocked or that sufficient zinc may cross
the intestinal wall via alternative routes at very high supply levels,
perhaps by transmucosal diffusion. The dose-related course of the
true absorption rate shown in Fig. 6 indicates that passive diffusion
does not materially contribute to zinc absorption under normal nutritional circumstances. In this regard it is also of interest that a
recent report [46] suggested that calcium in high luminal concentration, but still in a much closer molar ratio to zinc than normally
present in adequate dietaries, would depress the passive component
of jejunal zinc absorption in rats.

Recently, Jackson et al. [13] postulated that, apart from the
carrier-mediated transport system, an additional, high-affinity
mechanism would exist in the rat gut, being responsible for a rapid
translocation of zinc from the intestinal lumen to the body tissues
after induction by prior low-zinc nutrition. The experimental basis
for this hypothesis, however, must be regarded as inconclusive,
since numerous studies [6,10,11,27-29,32,47] show that zinc uptake
by the small intestinal wall is indeed greatly enhanced after dietary
zinc depletion, and also during the latter third of gravidity [48].
Furthermore, Jackson et al.'s study [13] did not take into account

the major effect of isotope dilution in the intestinal lumen as it
is evident from other studies [49,50]. Zinc supply status markedly
modifies the bidirectional flux of zinc across the mucosal wall
(Table 1, Fig. 6).

2.3. Influence of Nutrition

How much zinc is absorbed from the digestive tract of the healthy
mammalian body depends greatly on nutrition on the whole. Naturally,
the height of the particular zinc intake level plays a key role.
However, various other dietary constituents, both organic and inorganic in nature, may either ameliorate or impair the extent of
intestinal zinc accretion, mainly by way of influencing zinc binding.

2.3.1. Zinc Supply Status

As already outlined in Sec. 2.2, the body accretes zinc in the small
intestine via a carrier-facilitated saturable mechanism which is
highly responsive to an altered supply status. The quantitative
significance of this variable absorption for the overall zinc homeostasis of the body is, however, not readily evident from studies conducted to elucidate regulation of the absorption process. Similarly,
conventional balance studies measuring merely the difference between
zinc intake and total fecal output do not allow us to evaluate the
actual or true extent of absorption in relation to the metabolic
needs for this trace element, because feces regularly contain zinc
from body pools (i.e., of endogenous origin; see Sec. 3.1). Figure
3 illustrates the quantitative response of percent true and apparent
zinc absorption in relation to a widely varied dietary zinc supply
according to an in vivo study with young laboratory rats [14,51].
In the case of deficient to markedly suboptimum supply, reflected by
stagnant or very poor body growth, true absorption was virtually
complete. This absorptive efficiency, however, steadily declined as
the supply rose. At the highest dietary zinc level, which supplied

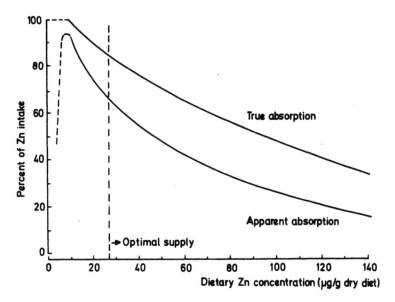

FIG. 3. Efficiency of true and apparent zinc absorption in young
rats in homeostatic response to varied dietary zinc supply. (Modified from Ref. 51.)

this element in substantial excess of the estimated dietary requirement for the rapid growth of these animals [16], the extent of true
absorption was confined to about one-third of intake. It is further
evident from Fig. 3 that true zinc absorption had already fallen
markedly below the 100% limit before intake reached the optimum
supply margin. This points to a beginning saturation of the transport mechanism before the metabolic needs are fully met. It must,
however, be realized that, despite this supply-requirement-related
decline in the absorptive efficiency, the absolute daily rates of
true absorption increased with zinc intake following saturation
kinetics, as is evident from Fig. 6.

     The homeostatic dependence of true zinc absorption on the
supply-requirement status is also reflected in studies with other
species, for instance the lactating dairy cow [52] or even in the
case of humans. Sandström and Cederblad [53] observed a close positive relationship between the intake and the quantity of zinc

absorbed by adult persons fed composite meals containing animal
protein. Conversely, percent true absorption decreased with zinc
intake, as is to be expected according to the relationships shown
in Figs. 3 and 6. Similarly, increased zinc intake in the case of
children [54] and adolescents [55,56] was associated with a correspondingly elevated fecal elimination of the element, reflecting a
lowered apparent and also true absorptive efficiency.

2.3.2. Other Dietary Components

Numerous studies show that zinc utilization from different food
sources and dietaries may vary appreciably. This property is most
commonly referred to as *(bio)availability* in the English literature.
If this term is used, it should, for the sake of clarity, be used to
mean *absorbability*, defining the fractional or percent amount of the
element ingested that is at best truly available for absorption [57,
58]. In this connotation it defines a potential value--that is, a
maximal value reached only if the zinc supply-requirement relationship is appropriately chosen. In studies designed to assess this
relative merit of different zinc sources for human and animal nutrition, absorbability usually is not determined directly because of
the procedural problems involved, but rather indirectly by a diversity
of response criteria, such as growth and zinc concentrations in various tissues, especially in serum and bone. This indirect approach,
however, assumes that these and other parameters correlate closely
with absorbability or, often more so, with the total efficiency of
utilization (i.e., absorbability times the metabolic efficiency [57]).
This, however, is likely to hold true only in the range of deficient
to suboptimum zinc supply, a situation not sufficiently realized in
all bioassays. Figure 3 demonstrates that percent true absorption
of zinc from the very same basal diet may vary over a wide range
depending on dietary zinc supply relative to the animals' particular
requirement. Indirect measures of zinc absorbability or total efficiency may be expected to lose their responsiveness as the zinc
supply moves well into the range of optimum intake, since adequate

amounts of zinc may be absorbed in either case to meet the metabolic
needs.

There is little evidence that the chemical nature of zinc as
naturally present within the food itself or added as supplement would
play a critical role regarding its absorbability under usual dietary
conditions. Nutritional studies support this view [59,60]. The
particular complex or dietary source would be important only if it
would resist degradation when exposed to the physicochemical environment of the gastrointestinal canal. For instance, copper absorbability from diets supplemented with various copper amino acid complexes
showed certain differences related to the configuration and absorption
characteristics of the particular ligand, whereas molecular size or
solubility of the complexes were no traits of major significance [24].

Of primary importance for the absorbability of dietary zinc--
apart from the influence of the supply status (Sec. 2.3.1) and physiological state of the animal or human (Sec. 2.4)--are the possible complex reactions and interactions encountered in the gastrointestinal
lumen between zinc, other metal cations, and the organic compounds of
both exogenous and endogenous origin. Accordingly, major shifts in
the metal-ligand interrelationships or luminal pH may ultimately
alter the fraction of absorbable zinc. Also, because of this multifactorial situation, responses observed in vitro must not necessarily
be relevant in vivo. Examples for the significance of organic food
components and minerals for the absorbability of zinc are reviewed
briefly in the following two subsections.

(a) *Organic components*. Numerous studies indicate that zinc
is more readily available for absorption from animal foods than from
plant products [61-65]. According to absorption studies by Sandström
and co-workers [53,66], this also applies to human nutrition. Phytate
(myoinositol hexaphosphate) is generally considered to be the primary
factor responsible for the poorer absorbability of zinc from plant
proteins and cereal products. It is widely accepted that this strong
chelating agent renders zinc unabsorbable by forming water-insoluble
complexes, accentuated by the presence of calcium, under the pH

conditions prevailing in the small intestinal lumen [67,68]. In
vivo, the absorbability of zinc from a zinc-phytate complex added as
such to a phytate-free diet was found to be just as high as that
from equivalent supplements of zinc chloride, fumarate, or histidine
[60].

Whether or not phytate interferes with zinc absorption depends
greatly on the overall composition of the meals and diets, in particular on the presence of other metal cations, mainly calcium, and
more important, the presence of foodstuffs that generate competing
ligands during their digestion. Accordingly, phytate must be present
in sufficiently high molar excess over zinc before it will commence
to exert its detrimental effect [64,69,70]. For example, Sandström
and co-workers [53,66] reported data showing that in adult persons
zinc absorption from composite meals that were based on whole-meal
bread and about balanced in zinc content markedly improved as the
protein content increased due to the combination with various animal
protein foods. It seems likely that the amino acids from protein
digestion increasingly competed as potential zinc-binding ligands
with the phytate of the whole-meal bread. In vitro work [11,13] has
shown that amino acids, especially histidine and cysteine, improve
intestinal zinc uptake and transmural transport when present in high
molar excess over zinc (see Sec. 2.2). Also, organic acids, at
sufficiently high intake, may improve zinc absorbability. Such a
response was indicated, for instance, in piglets fed diets supplemented with 1 or 2% fumaric acid [71].

In vitro observations [72] on the binding of divalent metal
cations by wheat bread and its major components led to the hypothesis
that, in addition to phytate, the indigestible plant fiber would be
a major determinant of the availability of zinc and other essential
elements in plant foods. Earlier, Becker and Hoekstra [61] had
already reported that cellulose added at graded levels to a semisynthetic diet for rats markedly diminished the intestinal transit
time and absorption of an oral radiozinc dose. This work is supported by several newer investigations showing that cellulose

supplementation of diets containing the same or similar levels of zinc reduces zinc absorption in laboratory rats [73] and in humans [74,75]. Similarly, hemicellulose has been reported to cause a significant increase in fecal zinc loss and ultimately lead to a negative zinc balance in adolescent and adult men [75,76]. Similar observations were also made in another balance study in which fruit and vegetables were fed as fiber sources to men [77]. By contrast, no adverse effect on zinc excretion and balance was evident in the case of dietary pectin supplementation [75,78]. Polyuronic acids (e.g., alginates) are otherwise known to form rather stable complexes with metal ions. There is, however, also a similar number of animal and human nutritional studies that failed to find any significant influence of plant fiber constituents on zinc availability [73,79, 80].

Regarding the effect of the plant cell wall constituents on mineral availability, several mechanisms have been considered, in particular the cation-exchange properties of fibrous material, dilution of the intestinal contents, and the faster passage rate of the digesta past the intestinal absorption sites (see Ref. 81). A central point seems to be the fact that plant fiber constituents, as well as phytate, are not degradable to a major extent by microbial activity before they reach the large intestine of humans and other monogastric species, whereas zinc absorption is essentially limited to the small intestine (see Sec. 2.1). Further research is needed to improve our understanding not only of the physiological effects of plant cell wall constituents, and carbohydrates in general, but also of the nutritional circumstances under which these dietary constituents might materially affect mineral absorption and retention.

Availability of zinc from milk has recently received special attention, in part stimulated by its noted differential value in pediatric nutrition of the AE patient. Since the clinical manifestation of this genetic disorder (see Sec. 2.2) is usually not encountered before the children are weaned to cow's or formula milk, it has been hypothesized that breast milk, as opposed to cow's milk,

must contain some factor that renders zinc more absorbable. Recent
research demonstrating major differences in the molecular association
of zinc in human milk versus milk from other mammals (see Sec. 3.3)
would seem to support this hypothesis. However, this is insufficient
evidence to conclude that the low-molecular-weight zinc complex of
human milk would serve a specific function in zinc absorption of the
neonate or, conversely, that the lack of such a complex would make
cow's milk a poorer zinc source for infants and children, including
the AE patient. It seems teleologically sound if the composition of
the mother's milk is adapted to the digestive functions of the neonate, and conversely. One might therefore basically expect that zinc
of this food source (and others) would be less available for absorption, the less digestible these and other such constituents having a
high zinc content or zinc-binding affinity would be in the human
infant. Accordingly, it may be speculated that cow's milk becomes
a less efficient zinc source for the AE patient not so much because
the molecular association of this metal is very different from breast
milk as that an additional digestive defect, apart from zinc malabsorption, is present, most likely concerning proteins that are absent
or not present in equivalent quantities in human milk. A similar
suggestion had been forwarded previously by Moynahan [25], but has
apparently not been pursued further in newer case studies. At any
event, it seems unlikely that zinc deficiency per se would result in
a substantially impaired nutrient digestibility. Both energy and
protein digestibility differed relatively little between severely
zinc-deficient rats and zinc-supplemented pair-fed control animals
fed a semisynthetic diet with casein as the protein source [82,83].

   (b) Minerals. Several minerals, in particular calcium,
phosphorus, and various trace metals, may markedly lower zinc absorbability by antagonistic interactions in the digestive canal. Of
major importance for the notable occurrence of such interactions are
the molar ratios of the particular reactants involved. Especially,
imbalances in the nutritional supply of minerals bear the enforced
risk that zinc or other essential trace elements are quantitatively

affected in their absorption from the gut. Basically, such imbalances
may result either from an excessive intake of one or more elements or
from the inadequate supply of one element relative to the supply of
others [84]. Within the scope of this chapter, just a few examples
will be depicted to point out the complexity of the situation. More
extensive reviews concerning zinc absorption and metabolism as affected
by interactions with other trace elements have been presented recently
elsewhere [31,84].

It has been known for quite some time from animal experiments
that calcium may markedly impair zinc absorption so that ultimately
zinc deficiency is induced or an existing deficiency accentuated. As
in the case of trace element interactions, this calcium-zinc antagonism
cannot be seen isolated but rather as a multifactorial reaction system.
Since the time when this topic was extensively reviewed by Becker and
Hoekstra [61], 10 years ago, our knowledge on the mode of this interaction has not basically improved. Greater awareness has come, however, that the calcium effect depends critically on overall food composition, in particular the content and type of phosphate (mainly
phytate) and protein (see also Sec. 2.3.2(a)). Human studies confirm
and complement the findings of the numerous animal studies. For
example, Sandström et al. [66] found that supplementation of a meal
based on wholewheat bread with calcium in the form of dairy foods
did not impair but rather ameliorated zinc absorption, since this
composing of the meal not only improved the calcium/phosphorus ratio,
but also resulted in a several-fold increase in the content of protein of high biological value. Another recent study with humans
confirms the significance of nonphytate phosphorus for zinc absorption. A 2.5-fold raise of the daily phosphorus intake above 1 g
daily by supplementing the diet with monohydrogen phosphate brought
about a marked increase in fecal zinc excretion and a negative zinc
balance [85].

There is a mutual interaction between zinc and copper at the
site of absorption [31]. While deficient copper nutrition hardly
affected the intestinal capacity to absorb zinc [28,86], Van Campen

[87] reported that the intraduodenal dosing of copper in 40-fold excess over zinc markedly reduced zinc absorption. This finding was confirmed in a more recent in vivo study [88] showing that the feeding of a diet containing 24 versus 3 ppm of copper caused a 20% reduction in zinc absorption (12 ppm of zinc in the diet). A further increase in copper supply remained without effect. The mode of this interaction is not evident; however, intestinal metallothionein was seemingly not involved [88].

It is of practical nutritional interest that iron supplementation of foods or the application of iron-fortified vitamin-mineral preparations might interfere with normal zinc absorption. Solomons and Jacob [89], measuring the response in plasma zinc concentrations as absorption index after the oral dosing of ferrous and zinc sulfates to human adults fasted overnight, reported a slightly to markedly depressed zinc absorption, when the Fe/Zn ratio was raised from 0 to 1:1 and then up to 3:1. There was no such difference, however, when heme chloride was used as the iron source and Atlantic oysters as the zinc source. This finding again emphasizes the major significance of organic ligands for the absorptive process in the intestine. Other studies in which rats received pasteurized milk with heavy iron fortification provided no evidence that the intestinal capacity for zinc absorption could have been diminished by the high iron intake [90,91].

Findings with respect to the effect of the nutritional iron status on zinc absorption are in part conflicting. Spry and Piper [92] noted an increased zinc retention in iron-deficient rats. It is, however, not evident whether this was due to an enhanced zinc absorption, even though some in vitro and in situ absorption studies [93,94] found that iron depletion would increase the absorption capacity not only of iron but also of zinc among other trace metals. In other investigations intestinal zinc transfer was not enhanced after dietary iron depletion [6,95]. In perfusion experiments on mice fed an iron-deplete diet [94], duodenal zinc uptake and transfer was significantly reduced when either iron, cobalt, or manganese were

also added to the zinc-containing perfusate in a molar excess of
10 over zinc. Also, the enhanced zinc absorption noted after dietary
iron depletion as compared to iron-replete controls was suppressed by
adding these other trace metals to the duodenal perfusate.

There is no evidence that manganese would adversely affect zinc
absorbability under common circumstances. For example, jejunal zinc
uptake and transfer was not significantly different between rats fed
a low-manganese versus an adequate diet [29]. In support of this
finding, zinc contents of various tissues were not or only slightly
reduced [96]. Overall, an equidirectional response is indicated
between zinc and manganese in regard to their retention.

It is to be expected that still other trace metals, especially
at excessively high intakes, will impair zinc accretion of the body
by antagonistic interactions at the site of absorption. For instance,
there is a recent report [97] that implicates tin as a zinc antagonist.
Apparent zinc absorption and zinc contents in tibia and kidneys were
diminished when the diet of growing rats contained about 200 ppm tin,
a level considered as moderate by the authors.

2.4. Influence of the Physiological State

Zinc metabolism is affected by the physiological state of the human
and animal body (age, gravidity, lactation). This also concerns
absorption as the accretory pathway through which sufficient zinc
must enter the central pool for distribution according to the particular metabolic needs at a given time.

2.4.1. Age

According to various animal studies, zinc absorption and retention
are inversely related to age [98]. The absorptive efficiency for
metals during the newborn phase is generally rather high. Figure 4
shows after data by Sherif et al. [48] that the jejunal zinc transfer
capacity of rat pups declined drastically near weaning age down to

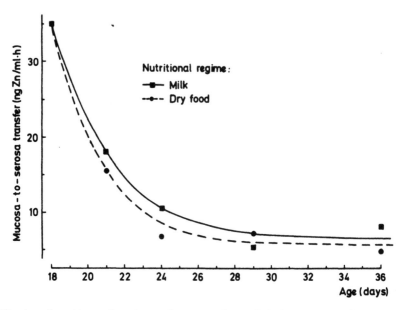

FIG. 4. In vitro zinc transfer capacity of jejunal sacs from weanling
rats allowed to consume either solely mother's milk or dry food beyond
18 days of age. (After data from Ref. 48.)

the level characteristic for adolescent rats. This age-related
response was also evident in littermates restricted to consume merely
mother's milk before and beyond the normal weaning age. Hence this
abrupt fall in the absorptive efficiency cannot be related directly
to the nutritional change from milk to solid food. Concurrently with
the transfer capacity, zinc uptake by the intestinal tissue had also
decreased sharply in both nutritional groups; however, the fall was
somewhat more marked in the case of the early-weaned rats as compared
to their littermates still suckling milk [48].

It may be speculated that the carrier-mediated zinc absorption
mechanism characteristic of the adolescent and adult body (see Sec.
2.2) is not yet or not fully developed in the neonatal rat and that
maturation does not occur until weaning. It has been reported [99]
that 6-day-old rat pups fed ^{65}Zn-labeled, zinc-supplemented milk
(about 76 µg Zn/ml total) retained at least as much of the radio-

isotope as pups fed labeled control milk (3 µg Zn/ml), probably
indicating immaturity of homeostatic control mechanism. In the case
of a functional saturable absorption mechanism, the relationships
shown in Figs. 3 and 6 would have predicted a greatly decreased ^{65}Zn
retention in the pups fed the supplemented milk.

For calcium, evidence has been reported [100] that intestinal
absorption of this element proceeds largely by a passive mechanism
in suckling rats and changes to a carrier-mediated mechanism at the
time of weaning. An abrupt decline in the intestinal absorption
capacity, as shown in Fig. 4 for zinc, has also been established for
several other heavy metals, including, for example, iron [101,102],
manganese [48], and even mainly toxic metals [101,103]. In all cases,
the timing of this response coincided closely with the termination of
the ability of the intestinal mucosa to take up immunoglobulins and
other proteins by pinocytosis, which ends in the rat about 18 days
after birth (see Ref. 104). This suggests that absorption of zinc
and other trace metals occurs mainly by pinocytotic uptake during
the neonatal life of the rat. Species differences are, however,
likely. From a teleological viewpoint it could be argued that the
neonate body does not need its own homeostatic mechanisms regulating
intestinal absorption as long as nutrition is limited to milk from
its own mother, who would take the physiological responsibility to
produce milk with the proper pattern of essential trace elements
matching the needs of the young and, hopefully, to exclude potentially
toxic amounts and elements as early as during passage of the maternal
digestive tract.

Figure 4 and other research indicates a further slight decline
in zinc absorption even after the age of weaning. These findings are
substantiated by the quantitative data in Table 2, showing that in
weaned rats fed the same diet, an age difference of 4 weeks was sufficiently long for percent true zinc absorption to be markedly lower
in the case of the older animals. This difference was attributed to
reflect fully the particular zinc supply status of the two age groups
[98] in agreement with the relationships shown in Figs. 3 and 6. The

TABLE 2

Age-Related Change in Zinc Absorption and Excretion
in Growing Rats Fed the Same Diet

Age (weeks):	5	9
Growth rate (g/day):	5.1	5.2
Food consumption[a] (g/day):	10.2	14.6
True absorption		
µg/day	129	173
% of intake	98	92
Apparent absorption		
µg/day	122	140
% of intake	93	74
Endogenous zinc in feces		
µg/day	7	33
% of intake	5	17

[a]12.9 µg zinc per g dry matter.
Source: Data from Ref. 98.

older rats, growing at the same rate as the younger ones, consumed a
much greater quantity of the diet and hence had a correspondingly
higher zinc intake. Accordingly, the true absorption rates were
actually higher in the older animals. There is no convincing evidence to assume that the body's physiological ability to homeostatically regulate zinc absorption in the adolescent and adult
stage would decrease because of aging. Studies with humans suggest
that zinc status as indicated by the plasma and hair zinc levels can
be maintained until into a late age [105,106]. Taper et al. [107]
also reported that zinc retention was comparable between two groups
of persons with an average age of 68 years despite a threefold
difference in dietary intake of the element.

2.4.2. Pregnancy and Lactation

During gravidity and lactation the nutritional requirement for zinc
as for other minerals and nutrients is greatly increased, mainly
because of the accretion for fetal growth and postnatally for milk

production. Also, additional zinc may be deposited in extrauterine
tissues of the maternal body, for instance in the liver, because of
a pronounced pregnancy anabolism [108]. One possibility to make
available sufficient zinc to meet this greatly augmented metabolic
need is to raise the efficiency of intestinal absorption. In support
of this, in vitro [109] and in situ [110] work demonstrated that the
intestinal capacity of gravid rats to absorb zinc is indeed enhanced,
especially in the latter part of pregnancy, when zinc accretion is
highest, according to the exponential development of the fetuses.
This is illustrated in Fig. 5 showing the mucosa-to-serosa zinc
transfer of gravid and lactating rats compared to nongravid, non-
lactating controls. Also in the case of gravid sows, a markedly
improved apparent absorption and retention of zinc was noted during
the last third of pregnancy [111,112], indicative of a higher extent
of true absorption.

Figure 5 further suggests a sharp decline in the absorptive
capacity of the rat intestine during the perinatal stage, so that
the control level would be reached in an early stage of lactation.

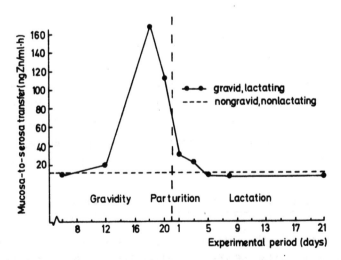

FIG. 5. Change in the in vitro zinc transfer capacity of jejunal
segments during gravidity and lactation. (From Ref. 109.)

In contrast, Davies and Williams [110] reported that zinc absorption rates of lactating rats may yet increase above those of gravid females. This apparent discrepancy might be related to nutritional and physiological differences, especially with regard to zinc supply status and pregnancy anabolism. During lactation voluntary food consumption, and hence zinc intake, is usually greatly elevated. With respect to the physiological and cellular changes leading to the enhanced zinc absorption efficiency during reproduction, further research is needed. It has been noted that the dry matter and protein content of the intestinal tissue is increased during gravidity and lactation [109].

3. EXCRETION

Zinc excretion via the various orifices and from the surface of the mammalian body is affected to a varying extent by nutrition. As to zinc homeostasis, elimination of endogenous zinc through the digestive tract is of quantitative significance for attaining the necessary balance between absorption and metabolic needs of the element.

3.1. Fecal Excretion

Feces contain a highly variable amount of zinc from endogenous pools apart from the unabsorbed fraction of dietary origin. Determination of this endogenous complement of total fecal zinc under normal nutritional circumstances requires suitable isotope procedures, for example, the dynamic radioisotope dilution technique [113,114].

3.1.1. Secretion and Excretion of Endogenous Zinc

Endogenous zinc in the lumen of the gastrointestinal tract originates from the desquamation of the epithelial lining, the discharge of mucins, and digestive secretions such as saliva, gastric and

intestinal juices, bile and pancreatic fluid, and from the transmucosal flux directed from the vascular bed to the luminal pool.
Accordingly, Matseshe et al. [7] reported that appreciably more zinc
passed the distal duodenum than was contained in the meal fed to
human subjects. As to the quantitative contribution of the various
secretions to the luminal pool of endogenous zinc, information is
still fairly limited and in part conflicting.

Various studies with animals [115,116] and humans [117] indicate that pancreas, bile, and duodenal secretion may account overall
for only a minor portion of the total fecal elimination of radiozinc
administered by intravenous injection. In the pancreatic juice, zinc
and protein contents are closely related [116,118,119]. It is not
clear whether the pancreas secretes zinc complexes other than the
procarboxypeptidases, especially in response to varied zinc nutrition.
Lönnerdal et al. [120] reported that zinc in pancreatic juice of rats
was associated only with high-molecular-weight fractions, whereas
bile contained predominantly low-molecular-weight zinc complexes.

Table 3 shows the intestinal uptake and secretion of labeled
zinc injected intravenously into adolescent rats. Within 3 h after
injection, a total of about 16% of the dose was recovered in the
intestinal tract, mainly in the wall. Zinc uptake was highest in
the distal half of the small intestinal wall, whereas cecum and colon
contained a much lower amount. The luminal zinc content increased
from the proximal to the distal segments of the small intestine.
This difference could, at least partially, be due to the peristaltic
transport of luminal contents, but might also indicate that the distal
portions of the small intestine perhaps play a particular role in the
secretion of endogenous zinc when dietary zinc supply is high. In
this context, it is of interest that most of the intestinal zinc-
metallothionein may be present in the jejunum [30]. However, the
possible significance of this zinc-responsive metalloprotein (see
Sec. 2.2) for the intestinal excretion of zinc needs to be established
yet. Similarly, the role of the Paneth cells in zinc homeostasis is
unclear. Elmes [122] reported that these cells in the crypts of the

TABLE 3

Intestinal Uptake and Luminal Secretion of an Intravenous Zinc
Dose (310 µg ^{65}Zn-Labeled Zinc) in Adolescent Rats Allowed
Adequate Zinc Nutrition (124 µg Zinc/g Dry Diet)

Segment	µg zinc/segment 3 h after zinc dosing	
	Wall	Lumen
Segment of small intestine[a]		
1 (proximal)	9.4	0.32
2	8.8	0.34
3	11.2	0.55
4 (distal)	11.1	0.69
Large intestine (wall plus lumen contents)		
Cecum	3.5	
Colon	4.9	

[a]Postmortem, the small intestine was separated into four segments
of equal length.
Source: Data from Ref. 121.

intestinal mucosa become more numerous in distal direction from the
duodenum to the ileum and that zinc loading increases the percentage
of cells containing histochemically detectable zinc. In the case of
zinc-deficient rats, dithizone-reactive Paneth cells were absent in
all regions of the small intestine.

As to the mechanism of zinc secretion by the small intestinal
wall, in vitro work [6,9] has provided evidence that the transport
of this trace element from the serosal to the mucosal surface may
also be an energy-requiring process. The data presented in Table 1
suggest, however, that such a secretory route may not be present or
may be depressed in the case of zinc deficiency. These and other
in vitro findings [11] imply that changes occurring within the
intestinal tissue itself may be of primary importance for regulating
the secretion of endogenous zinc in response to an altered nutritional

zinc status. In situ and in vivo studies [12,38,121] also demonstrated that a greater fraction of radiozinc administered by intravenous injection or infusion is retained in the small intestine of
rats fed a zinc-adequate diet compared to zinc-depleted animals.
Nutritional copper deficiency had no effect on the small intestinal
secretion of endogenous zinc [86].

The zinc supply status may also influence zinc secretion by
other tissues connected with the digestive tract. Reduced zinc concentrations were noted in parotid saliva from human patients with
hypogeusia [123]. According to nutritional studies [124,125], salivary zinc does not readily respond to dietary zinc depletion.
Sullivan et al. [119], however, found a diminished zinc secretion
via bile and pancreatic juice in zinc-depleted baby pigs. This
study further indicates that biliary zinc flow may be comparable
with that from the pancreas regardless of zinc nutrition.

In 1962, Cotzias et al. [126] reported that oral or intraperitoneal zinc administration altered the whole-body retention and
excretion of parenterally dosed radiozinc in mice. Since then
numerous studies with different species have provided firm evidence
that the fecal excretion of this element from endogenous pools is
greatly restricted in response to dietary depletion and, conversely,
is augmented by abundant zinc supply (see Ref. 127). Regarding the
quantitative significance of this response for zinc homeostasis,
see Sec. 3.1.2.

Due to the saturation kinetics of the absorptive process (see
Sec. 2.2), it is to be expected that the amount of unabsorbed zinc
passing along the small intestine greatly increases as the dietary
zinc supply becomes more abundant. A concomitantly greater secretion of endogenous zinc via the various routes will also contribute
materially to this luminal pool. The efficiency of reabsorption of
endogenous zinc may, therefore, be closely related to the absorbability of dietary zinc, especially if extensive "mixing" should
occur, depending on the chemical reactivity of the various compounds.
There is no scientific evidence to suggest that the chemical nature

of certain endogenous zinc compounds could be specifically designed
to either favor or disfavor its reabsorption from the lumen of the
small intestine.

Furthermore, food constituents that change the absorbability
of dietary zinc (see Sec. 2.3.2) may ultimately be expected to
modify the fecal output of endogenous zinc [128]. For example,
Davies and Nightingale [129] found that dietary phytate supplementation increased fecal zinc loss from zinc-deficient rats. This
finding further suggests that, even in the state of a pronounced
deficiency, the intestinal pool of endogenous zinc is much larger
than that indicated by the actual excretion in the feces.

3.1.2. Homeostatic Role of the Endogenous Fecal Zinc

The excretion of endogenous zinc via the digestive tract may vary
within wide limits depending on the nutritional zinc status. In
this regard and in regard to zinc homeostasis of the body, two
nutritional situations are particularly important: the situation of
deficient to suboptimal zinc nutrition and the situation of optimal
supply. Figure 6 illustrates the response of the endogenous fecal
zinc excretion in young rats over an appropriately wide range of
dietary zinc supply. For comparison, true and apparent absorption
rates are also shown. Apparent zinc absorption, reflecting about
overall net retention, steeply increased as intake rose from deficient to marginally adequate amounts and then assumed a definite
plateau regardless of the continued linear increase in zinc intake.
A corresponding response has been noted for the growth rate of these
animals. Fecal zinc was limited to the inevitable endogenous loss
when the dietary supply was deficient to markedly suboptimal. A
similar response of the fecal zinc excretion to the feeding of a
depletion diet is indicated in balance studies with women [130] and
lactating dairy cows [52].

According to Fig. 6, the excretion of endogenous zinc began
to rise markedly above the inevitable fecal loss before zinc intake
reached adequate dietary supply levels, perhaps reflecting an

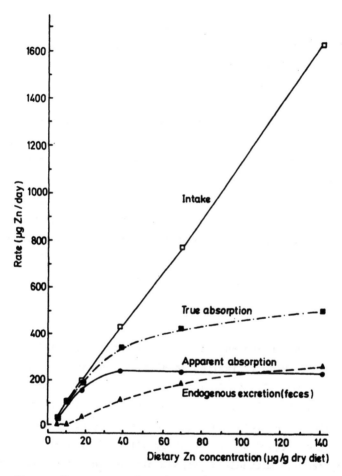

FIG. 6.   True and apparent zinc absorption and fecal excretion of
endogenous zinc in relation to dietary zinc supply and intake in
young rats.   (Modified from Ref. 128.   © 1980 Journal of Nutrition,
American Institute of Nutrition.)

activation of the secretory routes and/or a beginning saturation of
the absorptive process due to an increased intestinal zinc pool.
Over the wide plateau region of apparent absorption, any additional
increase in the true absorption rate was offset by an equivalent
increment of the fecal excretion of zinc from endogenous pools [128].
The data presented in Table 2 show that endogenous fecal zinc excretion increases with the age of adolescent rats in accordance with
the age-related improvement in the zinc supply status (see Sec.
2.4.1).

In summary it may be stated in agreement with the relationships shown in Figs. 3 and 6 that the absorptive efficiency, declin-
ing with increasing dietary zinc supply, is certainly not the sole
but quantitatively the primary factor to limit zinc accretion according to the particular metabolic needs of the body, especially when
intake greatly exceeds requirements [14,128]. The still necessary
"fine control" of retention is brought about by an appropriate variation in the fecal excretion of zinc from endogenous pools [51,58].

3.2. Urinary Excretion

In the healthy human and animal, renal zinc excretion remains fairly
constant despite wide variations in dietary zinc intake. Alimentary
zinc depletion may ultimately lead to a markedly lower urinary zinc
elimination, as is evident from balance studies with young women
[130] and adolescent rats [14,82]. In lactating dairy cows, however,
urinary zinc excretion was not notably affected by dietary zinc
depletion, despite markedly depressed plasma zinc levels and the
manifestation of conspicuous deficiency symptoms [131].

Greger and Snedeker [85] reported that men fed a high-protein
diet excreted significantly more urinary zinc than did men fed a low-
protein diet. In their opinion this response might be related to the
difference in the intake of histidine and cysteine, since these amino
acids, given either orally or parenterally in sufficiently large quantities, greatly augment renal zinc loss [132-134]. Administration

of chelating agents that are eliminated via the kidneys, for example
EDTA, may also substantially increase urinary zinc above normal
levels. Zincuria has been found in a number of diseases, in the
case of tissue injuries, and during starvation (see Ref. 135).

3.3. Excretion and Distribution in Milk

Zinc excretion from the maternal body via the mammary glands depends
on the amount of milk secreted and its zinc concentration. In studies
with lactating goats [136] and dairy cows [137], alimentary zinc deficiency did not adversely affect milk production and its nutrient composition. Zinc concentration, however, greatly varies with species,
stage of lactation, and zinc status apart from great individual vari-
ability. If human milk is assumed to contain, on the average, about
1.5 µg of zinc per ml, cow's milk would have a three-times higher
zinc content. A Finnish longitudinal study on 27 nursing mothers
[138], however, demonstrated a substantial decrease in milk zinc
during lactation. From week 2 to week 31 postpartum, mean zinc concentration fell by a factor of 10, from 4.0 ppm down to 0.44 ppm.
This decline was steepest during the first 8-10 weeks. A marked
decrease in milk zinc concentration during the course of lactation
has also been noted in the case of cows (see Ref. 139), sows [140],
and other species [141]. Colostrum of all species is highly enriched
with zinc, but there is a very rapid decline during the transition
phase to normal milk.
    Zinc nutrition also greatly affects milk zinc concentration.
In dairy cows fed a deficient ration with 6 ppm zinc, mean zinc concentration of milk fell from an initial level of 7.0 ppm down to
2.3 ppm during a 6-week depletion phase [142] and returned to normal
levels during subsequent repletion. Over the range of adequate zinc
nutrition, milk zinc concentration responds relatively little to
major differences in intake [139]. This suggests a homeostatically
controlled zinc secretion into milk. Analogous to these findings
in animal studies, it may be assumed that zinc deficiency would also

lower zinc concentration in milk from lactating women. When a
Finnish study on 15 nursing mothers [143] did not find a significant
relationship between dietary zinc supply and the content of this
trace metal in milk samples, it must be realized that zinc intake
evidently did not include deficient levels.

With respect to zinc distribution in milk, marked species
differences are indicated. In milk from the cow, sheep, goat, dog,
and rat, the major portion of this element is bound to the casein
fraction [141,144]. In human milk, however, containing much less
casein than bovine milk, most zinc is present in the whey [141],
where a portion of about 10% of the total zinc is bound to a low-
molecular-weight complex [145]. A Californian research group [22,
146] recently reported that this low-molecular complex of human milk
represents zinc citrate and that very much less zinc eluted with the
equivalent fraction from cow's milk despite a higher citrate content.

In the case of rat milk, which is essentially free of citrate,
all the zinc was associated with high-molecular-weight proteins [40].
Piletz and Ganschow [147] postulate that bovine milk also contains
a low-molecular-weight zinc-binding complex. It would, however,
normally be aggregated with the casein micelles, whereas there might
be insufficient aggregation in human milk because of a lower content
and size of micelles. Furthermore, Cousins and Smith [145] demonstrated that incubating bovine milk with added Zn^{2+} induces the
association of zinc with low-molecular-weight fractions, although
not to quite the same extent as in the case of human milk. These
authors hold it likely that not just one molecular species but rather
a heterogeneous group of compounds, including free amino acids, could
contribute to the low-molecular-weight zinc-binding complex of human
milk.

In general, zinc distribution in milk may seem to vary between
and within species mainly according to the content of this metal
relative to the presence of various proteins and other organic constituents, all differing in regard to their aggregating and metal-
binding properties. The fat fraction of milk from all species
studied contained very little zinc [141,144].

3.4. Integumental and Menstrual Loss

Zinc lost from the body surface, mainly by epidermal desquamation,
hair growth, and sweating, must be regarded, like urinary zinc, as
an inevitable loss that does not actively partake in maintaining
zinc homeostasis. During zinc deficiency such surface losses may
be somewhat diminished: for instance, by a reduced zinc concentration
in sweat [148]. Overall, integumental zinc loss may reach a sizable
magnitude not to be neglected in quantifying adequate dietary allowance. For example, preadolescent children [149,150] and adult persons
[130,151] have been observed to lose up to approximately 1 mg and more
zinc per day via sweating. This is a quantity exceeding the inevitable endogenous zinc loss in feces and urine.

Menstrual zinc loss amounting to a total of about 0.5 mg per
period [152] is negligibly small compared with the recommended daily
dietary allowance of 15 mg of zinc for adolescent and adult women
[153].

4. SIGNIFICANCE FOR ZINC NUTRITION

The quantitative significance of absorption and excretion for zinc
nutrition may be realized best by a factorial consideration of the
dietary requirement. The necessary dietary allowance, also referred
to as gross requirement, may be presented as the quotient of the net
requirement divided by the efficiency of total utilization [58].

The actual needs of zinc in metabolism for its functions in
body maintenance, growth, and reproduction define the total net
requirement. Quantitatively, the net zinc requirement for maintenance consists of the various endogenous losses from the body surface,
the urogenital tract, and the digestive tract. In the latter case,
however, only the inevitable endogenous loss may be truly considered
as maintenance component.

Endogenous zinc excreted in feces for homeostatic and other
nutritional reasons (see Sec. 3.1) should be treated as part of the

overall true efficiency of utilization according to the factorial
model concept discussed previously in more detail [58,128]. During
the growing and reproductive stages the total net requirement exceeds
the maintenance need according to the accretion of zinc for tissue
deposition, fetal development, or excretion with the milk.

The efficiency of total utilization of zinc and other essential
minerals represents the product of two factors, true absorption and
metabolic efficiency [57]. Obviously, the necessary dietary allowance
to meet the actual metabolic needs must be the higher the less efficient true absorption and true retention are. Total utilization and
its constituent factors may vary greatly, for two principal reasons.
First, dietary zinc may not be fully available for intestinal absorption due to dietary composition (see Sec. 2.3.2). Second, absorption
and fecal excretion of endogenous zinc must be expected to change for
homeostatic reasons (see Secs. 2.3.1 and 3.1.2). In fact, this variable endogenous zinc excretion for nutritional reasons quantitatively
reflects the metabolic efficiency. The experimental determination
and homeostatic dependence of total zinc utilization has been illustrated previously in a model study with fast-growing laboratory rats
[128]. Dietary zinc requirements assessed by the factorial approach
for fast-growing rats [16] and lactating dairy cows [154] showed
close agreement with estimates derived from dose-response relationships [15,139].

An increasing number of balance studies is conducted with
humans in the intention of determining nutritional requirements or
differences in the efficiency of utilization of zinc and other trace
elements. In this regard, it must, however, be emphasized that the
indirect determination of absorption and retention on the basis of
conventional balance studies is a meaningful approach only if metabolic needs and hence actual accretion o: the particular element are
indeed high relative to its dietary supply. Otherwise, the usual
variance of intake and excretion data, especially in the case of
fecal excretion, will a priori prevent the detection of real differences with statistical confidence. Accordingly, absorption and

retention data, as observed in numerous balance studies with humans
ranging in age from neonates to adults, commonly have rather wide
fiducial limits. Hence dietary treatments imposed would have to be
appropriately large to detect changes in zinc absorption and balance
with statistical confidence. Systematic errors are indicated when-
ever zinc balances of children and especially of adults are found to
be significantly or even highly significantly positive. Basically,
one would expect that regressions between zinc intake and net absorption or retention describe a plateau corresponding to the actual zinc
accretion over the range of adequate zinc nutrition (see Fig. 6).
According to Sandstead et al. [155], actual growth-related zinc
accretion may reach daily rates of at most 0.15-0.2 mg in children
at the ages of 1-9 years and 0.65-0.8 mg in adolescents. These
values are relatively small (at most about 2-5%) compared with the
dietary zinc allowance recommended by the U.S. Food and Nutrition
Board [153].

Consequently, specific and sensitive criteria must be available
even more so in human than in animal nutrition for estimating the
dietary requirement and utilization on the basis of dose-response
relationships. Such criteria should still be sufficiently responsive
in the range of suboptimum zinc nutrition resulting either from marginal intake or from limited absorbability. Among the zinc metalloenzymes, the activity of the alkaline phosphatase is a sensitive
parameter readily assessed in blood serum. In model studies with
rats of weanling up to nearly adult age, the response of the activity
of this enzyme in serum to zinc injection proved to be a sensitive
test to indicate states of marginally adequate supply [156,157].
Since the activity of this enzyme is determined before and after
zinc injection in this response technique, the problem of specifying
norm values to indicate adequate zinc nutrition is avoided. Another
equally responsive parameter is the zinc-binding capacity of serum
[43,157]. The particular merit of this latter assay is that it can
be performed in vitro.

REFERENCES

1. F. J. Schwarz and M. Kirchgessner, *Nutr. Metab.*, *18*, 157 (1975).

2. D. R. Van Campen and E. A. Mitchell, *J. Nutr.*, *86*, 120 (1965).

3. A. H. Methfessel and H. Spencer, *J. Appl. Physiol.*, *34*, 58 (1973).

4. N. T. Davies, *Br. J. Nutr.*, *43*, 189 (1980).

5. D. L. Antonson, A. J. Barak, and J. A. Vanderhoof, *J. Nutr.*, *109*, 142 (1979).

6. F. J. Schwarz and M. Kirchgessner, *Z. Tierphysiol. Tierernaehr. Futtermittelkd.*, *34*, 67 (1974).

7. J. W. Matseshe, S. F. Phillips, J.-R. Malagelada, and J. T. McCall, *Am. J. Clin. Nutr.*, *33*, 1946 (1980).

8. K.-E. Andersson, L. Bratt, H. Dencker, and E. Lanner, *Eur. J. Clin. Pharmacol.*, *9*, 423 (1976).

9. S. Kowarski, C. S. Blair-Stanek, and D. Schachter, *Am. J. Physiol.*, *226*, 401 (1974).

10. R. I. Hutagalung, F. J. Schwarz, and M. Kirchgessner, *Arch. Tierernaehr.*, *27*, 347 (1977).

11. F. J. Schwarz and M. Kirchgessner, *Z. Tierphysiol. Tierernaehr. Futtermittelkd.*, *39*, 68 (1977).

12. F. J. Schwarz and M. Kirchgessner, in *Trace Element Metabolism in Man and Animals*, Vol. 3 (M. Kirchgessner, ed.), Arbeitskreis Tierernährungsforschung Weihenstephan, Freising-Weihenstephan, West Germany, 1978, p. 110ff.

13. M. J. Jackson, D. A. Jones, and R. H. T. Edwards, *Br. J. Nutr.*, *46*, 15 (1981).

14. E. Weigand and M. Kirchgessner, *Nutr. Metab.*, *22*, 101 (1978).

15. E. Weigand and M. Kirchgessner, *Z. Tierphysiol. Tierernaehr. Futtermittelkd.*, *39*, 16 (1977).

16. E. Weigand and M. Kirchgessner, *Z. Tierphysiol. Tierernaehr. Futtermittelkd.*, *39*, 84 (1977).

17. F. J. Schwarz and M. Kirchgessner, *Z. Tierphysiol. Tierernaehr. Futtermittelkd.*, *35*, 257 (1975).

18. K. T. Smith, R. J. Cousins, B. L. Silbon, and M. L. Failla, *J. Nutr.*, *108*, 1849 (1978).

19. P. S. Hallman, D. D. Perrin, and A. E. Watt, *Biochem. J.*, *121*, 549 (1971).

20. G. W. Evans and E. C. Johnson, *J. Nutr.*, *110*, 1076 (1980).

21. E. Giroux and N. J. Prakash, *J. Pharm. Sci.*, *66*, 391 (1977).

22. B. Lönnerdal, A. G. Stanislowski, and L. S. Hurley, *J. Inorg. Biochem.*, *12*, 71 (1980).

23. W. Forth and W. Rummel, *Physiol. Rev.*, *53*, 724 (1973).

24. M. Kirchgessner and E. Grassmann, in *Trace Element Metabolism in Animals* (C. F. Mills, ed.), Churchill Livingstone, Edinburgh, 1970, p. 277ff.

25. E. J. Moynahan, *Lancet*, *2*, 399 (1974).

26. K. Weismann, S. Hoe, L. Knudsen, and S. S. Sørensen, *Br. J. Dermatol.*, *101*, 573 (1979).

27. M. Kirchgessner, F. J. Schwarz, and E. Grassmann, *Bioinorg. Chem.*, *2*, 255 (1973).

28. F. J. Schwarz and M. Kirchgessner, *Int. Z. Vitamin- Ernaehrungsforschung*, *44*, 116 (1974).

29. F. J. Schwarz and M. Kirchgessner, *Z. Tierphysiol. Tierernaehr. Futtermittelkd.*, *43*, 272 (1980).

30. K. T. Smith and R. J. Cousins, *J. Nutr.*, *110*, 316 (1980).

31. M. Kirchgessner, F. J. Schwarz, E. Grassmann, and H. Steinhart, in *Copper in the Environment*, Part 2 (J. O. Nriagu, ed.), Wiley, New York, 1979, p. 433ff.

32. F. J. Schwarz and M. Kirchgessner, *Int. Z. Vitamin- Ernaehrungsforschung*, *44*, 258 (1974).

33. M. P. Menard, C. C. McCormick, and R. J. Cousins, *J. Nutr.*, *111*, 1353 (1981).

34. F. J. Schwarz and M. Kirchgessner, *Z. Tierphysiol. Tierernaehr. Futtermittelkd.*, *37*, 31 (1976).

35. R. J. Cousins, *Am. J. Clin. Nutr.*, *32*, 339 (1979).

36. R. J. Cousins, *Nutr. Rev.*, *37*, 97 (1979).

37. B. C. Starcher, J. G. Glauber, and J. G. Madaras, *J. Nutr.*, *110*, 1391 (1980).

38. F. J. Schwarz and M. Kirchgessner, *Res. Exp. Med. (Berl.)*, *170*, 241 (1977).

39. R. J. Cousins, K. T. Smith, M. L. Failla, and L. A. Markowitz, *Life Sci.*, *23*, 1819 (1978).

40. B. Lönnerdal, C. L. Keen, M. V. Sloan, and L. S. Hurley, *J. Nutr.*, *110*, 2414 (1980).

41. K. T. Smith, M. L. Failla, and R. J. Cousins, *Biochem. J.*, *184*, 627 (1978).

42. J. K. Chesters and M. Will, *Br. J. Nutr.*, *46*, 111 (1981).

43. H.-P. Roth and M. Kirchgessner, *Res. Exp. Med. (Berl.)*, *177*, 213 (1980).

44. I. Lombeck, H. G. Schnippering, F. Ritzl, L. E. Feinendegen, and H. J. Bremer, *Lancet*, *1*, 855 (1975).

45. D. J. Atherton, D. P. R. Muller, P. J. Aggett, and J. T. Harries, *Clin. Sci.*, *56*, 505 (1979).

46. N. F. Adham and M. K. Song, *Nutr. Metab.*, *24*, 281 (1980).

47. C. J. Hahn and G. W. Evans, *Am. J. Physiol.*, *228*, 1020 (1975).

48. Y. S. Sherif, F. J. Schwarz, and M. Kirchgessner, *Arch. Tierernaehr.*, *31*, 597 (1981).

49. E. Weigand and M. Kirchgessner, *Nutr. Metab.*, *20*, 307 (1976).

50. G. W. Evans, E. C. Johnson, and P. E. Johnson, *J. Nutr.*, *109*, 1258 (1979).

51. E. Weigand and M. Kirchgessner, in *Trace Element Metabolism in Man and Animals*, Vol. 3 (M. Kirchgessner, ed.), Arbeitskreis Tierernährungsforschung Weihenstephan, Freising-Weihenstephan, West Germany, 1978, p. 106ff.

52. M. Kirchgessner, W. A. Schwarz, and H.-P. Roth, in *Trace Element Metabolism in Man and Animals*, Vol. 3 (M. Kirchgessner, ed.), Arbeitskreis Tierernährungsforschung Weihenstephan, Freising-Weihenstephan, West Germany, 1978, p. 116ff.

53. B. Sandström and Å. Cederblad, *Am. J. Clin. Nutr.*, *33*, 1778 (1980).

54. C. R. Meiners, L. J. Taper, M. K. Korslund, and S. J. Ritchey, *Am. J. Clin. Nutr.*, *30*, 879 (1977).

55. J. L. Greger, R. P. Abernathy, and O. A. Bennett, *Am. J. Clin. Nutr.*, *31*, 112 (1978).

56. J. L. Greger, S. C. Zaikis, R. P. Abernathy, O. A. Bennett, and J. Huffman, *J. Nutr.*, *108*, 1449 (1978).

57. M. Kirchgessner, H. L. Müller, E. Weigand, E. Grassmann, F. J. Schwarz, J. Pallauf, and H.-P. Roth, *Z. Tierphysiol. Tierernaehr. Futtermittelkd.*, *34*, 3 (1974).

58. E. Weigand and M. Kirchgessner, *Z. Tierphysiol. Tierernaehr. Futtermittelkd.*, *39*, 325 (1977).

59. G. S. Ranhotra, R. J. Loewe, and L. V. Puyat, *Cereal Chem.*, *54*, 496 (1977).

60. E. Weigand and M. Kirchgessner, *Z. Tierphysiol. Tierernaehr. Futtermittelkd.*, *42*, 137 (1979).

61. W. M. Becker and W. G. Hoekstra, in *Intestinal Absorption of Metal Ions, Trace Elements and Radionuclides* (S. C. Skoryna and D. Waldron-Edward, eds.), Pergamon Press, Elmsford, N.Y., 1971, p. 229ff.

62. B. L. O'Dell, C. E. Burpo, and J. E. Savage, *J. Nutr.*, *102*, 653 (1972).

63. B. Momčilović, B. Belonje, A. Giroux, and B. G. Shah, *J. Nutr.*, *106*, 913 (1976).

64. N. T. Davies and S. E. Olpin, *Br. J. Nutr.*, *41*, 590 (1979).

65. D. A. Hardie-Muncy and A. I. Rasmussen, *J. Nutr.*, *109*, 321 (1979).

66. B. Sandström, B. Arvidsson, Å. Cederblad, and E. Björn-Rasmussen, *Am. J. Clin. Nutr.*, *33*, 739 (1980).

67. P. Vohra, G. A. Gray, and F. H. Kratzer, *Proc. Soc. Exp. Biol. Med.*, *120*, 447 (1965).

68. D. Oberleas, M. E. Muhrer, and B. L. O'Dell, *J. Nutr.*, *90*, 56 (1966).

69. H.-J. Lantzsch, H. Schenkel, and I. Nickerl, in *Trace Element Metabolism in Man and Animals*, Vol. 3 (M. Kirchgessner, ed.), Arbeitskreis Tierernährungsforschung Weihenstephan, Freising-Weihenstephan, West Germany, 1978, p. 460ff.

70. E. R. Morris and R. Ellis, *J. Nutr.*, *110*, 1037 (1980).

71. M. Kirchgessner and F. X. Roth, *Z. Tierphysiol. Tierernaehr. Futtermittelkd.*, *44*, 239 (1980).

72. J. G. Reinhold, F. Ismail-Beigi, and B. Faraji, *Nutr. Rep. Int.*, *12*, 75 (1975).

73. N. Gruden, M. Buben, and M. Ciganović, *Nutr. Rep. Int.*, *20*, 757 (1979).

74. F. Ismail-Beigi, J. G. Reinhold, B. Faraji, and P. Abadi, *J. Nutr.*, *107*, 510 (1977).

75. L. M. Drews, C. Kies, and H. M. Fox, *Am. J. Clin. Nutr.*, *32*, 1893 (1979).

76. C. Kies, H. M. Fox, and D. Beshgetoor, *Cereal Chem.*, *56*, 133 (1979).

77. J. L. Kelsay, R. A. Jacob, and E. S. Prather, *Am. J. Clin. Nutr.*, *32*, 2307 (1978).

78. K. Y. Lei, M. W. Davis, M. M. Fang, and L. C. Young, *Nutr. Rep. Int.*, *22*, 459 (1980).

79. R. C. Y. Tsai and K. Y. Lei, *J. Nutr.*, *109*, 1117 (1979).

80. H. H. Sandstead, J. M. Muñoz, R. A. Jacob, L. M. Klevay, S. J. Reck, G. M. Logan, F. R. Dintzis, G. E. Inglett, and W. C. Shuey, *Am. J. Clin. Nutr.*, *31*, S 180 (1978).

81. D. Oberleas and B. F. Harland, in *Zinc Metabolism: Current Aspects in Health and Disease* (G. J. Brewer and A. S. Prasad, eds.), Alan R. Liss, New York, 1977, p. 11ff.

82. J. Pallauf and M. Kirchgessner, *Arch. Tierernaehr.*, *26*, 547 (1976).

83. M. Kirchgessner, H.-P. Roth, and E. Weigand, in *Trace Elements in Human Health and Disease* (A. S. Prasad, ed.), Academic Press, New York, 1976, p. 189ff.

84. M. Kirchgessner, A. M. Reichlmayr-Lais, and F. J. Schwarz, in *Proc., 12th Int. Congr. Nutr.*, Alan R. Liss, New York, 1981, p. 189ff.

85. J. L. Greger and S. M. Snedeker, *J. Nutr.*, *110*, 2243 (1980).

86. F. J. Schwarz and M. Kirchgessner, *Z. Tierphysiol. Tierernaehr. Futtermittelkd.*, *41*, 335 (1979).

87. D. R. Van Campen, *J. Nutr.*, *97*, 104 (1969).

88. A. C. Hall, B. W. Young, and I. Bremner, *J. Inorg. Biochem.*, *11*, 57 (1979).

89. N. W. Solomons and R. A. Jacob, *Am. J. Clin. Nutr.*, *34*, 475 (1981).

90. B. Momčilović and D. Kello, *Nutr. Rep. Int.*, *15*, 651 (1977).

91. N. Gruden and B. Momčilović, *Nutr. Rep. Int.*, *19*, 483 (1979).

92. C. J. F. Spry and K. G. Piper, *Br. J. Nutr.*, *23*, 91 (1969).

93. W. Forth, in *Trace Element Metabolism in Animals* (C. F. Mills, ed.), Churchill Livingstone, Edinburgh, 1970, p. 298ff.

94. P. R. Flanagan, J. Haist, and L. S. Valberg, *J. Nutr.*, *110*, 1754 (1980).

95. S. Pollack, J. N. George, R. C. Reba, R. M. Kaufman, and W. H. Crosby, *J. Clin. Invest.*, *44*, 1470 (1965).

96. D. Heiseke and M. Kirchgessner, *Zentralbl. Veterinaermed. Reih. A*, *25*, 307 (1978).

97. J. L. Greger and M. A. Johnson, *Food Cosmet. Toxicol.*, *19*, 163 (1981).

98. E. Weigand and M. Kirchgessner, *Biol. Trace Elem. Res.*, *1*, 347 (1979).

99. B. Momčilović, *Period. Biol.*, *80*, 141 (1978).

100. F. K. Ghishan, J. T. Jenkins, and M. K. Younoszai, *J. Nutr.*, *110*, 1622 (1980).

101. G. B. Forbes and J. C. Reina, *J. Nutr.*, *102*, 647 (1972).

102. N. D. Gallagher, R. Mason, and K. E. Foley, *Gastroenterology*, *64*, 438 (1973).

103. D. M. Taylor, P. H. Bligh, and M. H. Duggan, *Biochem. J.*, *83*, 25 (1962).

104. I. G. Morris, in C. F. Code, ed., *Handbook of Physiology*, Sec. 6: *Alimentary Canal*, Vol. 3: *Intestinal Absorption*, American Physiological Society, Washington, D.C., 1968, p. 1491ff.

105. J. M. Hsu, *World Rev. Nutr. Diet.*, *33*, 42 (1979).

106. S. C. Vir and A. H. G. Love, *Am. J. Clin. Nutr.*, *32*, 1472 (1979).

107. L. J. Taper, D. M. Burke, F. J. DeMicco, and S. J. Ritchey, *Fed. Proc.*, *39*, 430 (1980).

108. M. Kirchgessner, R. Spoerl, and U. A. Schneider, in *Trace Element Metabolism in Man and Animals*, Vol. 3 (M. Kirchgessner, ed.), Arbeitskreis Tierernährungsforschung Weihenstephan, Freising-Weihenstephan, West Germany, 1978, p. 440ff.

109. F. J. Schwarz, M. Kirchgessner, and S. Y. Sherif, *Res. Exp. Med. (Berl.)*, *179*, 35 (1981).

110. N. T. Davies and R. B. Williams, *Br. J. Nutr.*, *38*, 417 (1977).

111. M. Kirchgessner, R. Spoerl, and D. A. Roth-Maier, *Z. Tierphysiol. Tierernaehr. Futtermittelkd.*, *44*, 98 (1980).

112. M. Kirchgessner, F. J. Schwarz, and D. A. Roth-Maier, in *Trace Element Metabolism in Man and Animals*, Vol. 4 (J. McC. Howell, J. M. Gawthorne, and C. L. White, eds.), Australian Academy of Science, Canberra, 1981, p. 85ff; distributed by Springer-Verlag, Heidelberg.

113. E. Weigand and M. Kirchgessner, *Nutr. Metab.*, *20*, 314 (1976).

114. E. Weigand and M. Kirchgessner, *Z. Tierphysiol. Tiereraehr. Futtermittelkd.*, *42*, 44 (1979).

115. M. L. Montgomery, G. E. Sheline, and I. L. Chaikoff, *J. Exp. Med.*, *78*, 151 (1943).

116. J. C. Pekas, *Am. J. Physiol.*, *211*, 407 (1966).

117. E. B. Miller, A. Sorscher, and H. Spencer, *Radiat. Res.*, *22*, 216 (1964).

118. M. Adler, P. Robberecht, M. Mestdagh, M. Cremer, A. Delcourt, and J. Christophe, *Gastroenterol. Clin. Biol.*, *4*, 441 (1980).

119. J. F. Sullivan, R. V. Williams, J. Wisecarver, K. Etzel, M. M. Jetton, and D. F. Magee, *Proc. Soc. Exp. Biol. Med.*, *166*, 39 (1981).

120. B. Lönnerdal, B. O. Schneeman, C. L. Keen, and L. S. Hurley, *Biol. Trace Elem. Res.*, *2*, 149 (1980).

121. F. J. Schwarz and M. Kirchgessner, *Zentralbl. Veterinaermed. Reih. A*, *23*, 836 (1976).

122. E. Elmes, *J. Pathol.*, *118*, 183 (1976).

123. R. I. Henkin, R. E. Lippoldt, J. Bilstad, and H. Edelhoch, *Proc. Natl. Acad. Sci. USA*, *72*, 488 (1975).

124. G. A. Everett and J. Apgar, *J. Nutr.*, *109*, 406 (1979).

125. J. H. Freeland-Graves, P. J. Hendrickson, M. L. Ebangit, and J. Y. Snowden, *Am. J. Clin. Nutr.*, *34*, 312 (1981).

126. G. C. Cotzias, D. C. Borg, and B. Selleck, *Am. J. Physiol.*, *202*, 359 (1962).

127. M. Kirchgessner and F. J. Schwarz, in *Nuclear Techniques in Animal Production and Health,* International Atomic Energy Agency, Vienna, 1976, p. 81ff.

128. E. Weigand and M. Kirchgessner, *J. Nutr.*, *110*, 469 (1980).

129. N. T. Davies and R. Nightingale, *Br. J. Nutr.*, *34*, 243 (1975).

130. F. M. Hess, J. C. King, and S. Margen, *J. Nutr.*, *107*, 1610 (1977).

131. W. A. Schwarz and M. Kirchgessner, *Arch. Tierernaehr.*, *25*, 597 (1975).

132. R. I. Henkin, B. M. Patten, P. D. Re, and D. A. Bronzert, *Arch. Neurol.*, *32*, 745 (1975).

133. R. M. Freeman and P. R. Taylor, *Am. J. Clin. Nutr.*, *30*, 523 (1977).

134. A. A. Yunice, R. W. King, S. Kraikitpanitch, C. C. Haygood, and R. D. Lindeman, *Am. J. Physiol.*, *235*, F40 (1978).

135. H. H. Sandstead, K. P. Vo-Khactu, and N. Solomons, in *Trace Elements in Human Health and Disease*, Vol. 1: *Zinc and Copper* (A. S. Prasad, ed.), Academic Press, New York, 1976, p. 33ff.

136. B. Groppel and A. Hennig, *Arch. Exp. Veterinaermed.*, *25*, 817 (1971).

137. W. A. Schwarz and M. Kirchgessner, *Wirtschaftseigene Futter*, *21*, 169 (1975).

138. E. Vuori and P. Kuitunen, *Acta Paediatr. Scand.*, *68*, 33 (1979).

139. M. Kirchgessner and E. Weigand, *Arch. Tierernaehr.*, *32*, 569 (1982).

140. M. Kirchgessner, D. A. Roth-Maier, and R. Spoerl, *Z. Tierphysiol. Tierernaehr. Futtermittelkd.*, *44*, 233 (1980).

141. B. Lönnerdal, C. L. Keen, and L. S. Hurley, in *Trace Element Metabolism in Man and Animals*, Vol. 4 (J. McC. Howell, J. M. Gawthorne, and C. L. White, eds.), Australian Academy of Science, Canberra, 1981, p. 249ff; distributed by Springer-Verlag, Heidelberg.

142. M. Kirchgessner and W. A. Schwarz, *Zentralbl. Veterinaermed. Reih. A*, *22*, 572 (1975).

143. E. Vuori, S. M. Mäkinen, R. Kara, and P. Kuitunen, *Am. J. Clin. Nutr.*, *33*, 227 (1980).

144. S. Parkash and R. Jenness, *J. Dairy Sci.*, *50*, 127 (1967).

145. R. J. Cousins and K. T. Smith, *Am. J. Clin. Nutr.*, *33*, 1083 (1980).

146. L. S. Hurley, B. Lönnerdal, and A. G. Stanislowski, *Lancet*, *1*, 677 (1979).

147. J. E. Piletz and R. E. Ganschow, *Am. J. Clin. Nutr.*, *32*, 275 (1979).

148. A. S. Prasad, A. R. Schulert, H. H. Sandstead, A. Miale, and Z. Farid, *Lab. Clin. Med.*, *62*, 84 (1963).

149. M. E. Harrison, C. Walls, M. K. Korslund, and S. J. Ritchey, *Am. J. Clin. Nutr.*, *29*, 842 (1976).

150. S. J. Ritchey, M. K. Korslund, L. M. Gilbert, D. C. Fay, and M. F. Robinson, *Am. J. Clin. Nutr.*, *32*, 799 (1979).

151. J. R. Cohn and E. A. Emmet, *Ann. Clin. Lab. Sci.*, *8*, 270 (1978).

152. J. L. Greger and S. Buckley, *Nutr. Rep. Int.*, *16*, 639 (1977).

153. Food and Nutrition Board, *Recommended Dietary Allowances*, 9th ed., National Academy of Sciences, Washington, D.C., 1980.

154. E. Weigand and M. Kirchgessner, *Z. Tierphysiol. Tierernaehr. Futtermittelkd.*, *47*, 1 (1982).

155. H. H. Sandstead, L. M. Klevay, J. M. Muñoz, R. A. Jacob, G. M. Logan, S. J. Reck, F. R. Dintzis, G. E. Inglett, and W. C. Shuey, in *Spurenelemente* (E. Gladtke, G. Heimann, and I. Eckert, eds.), Georg Thieme, Stuttgart, 1979, p. 105ff.

156. H.-P. Roth and M. Kirchgessner, *Zentralbl. Veterinaermed. Reih. A*, *27*, 290 (1980).

157. M. Kirchgessner and H.-P. Roth, in *Trace Element Metabolism in Man and Animals*, Vol. 4 (J. McC. Howell, J. M. Gawthorne, and C. L. White, eds.), Australian Academy of Science, Canberra, 1981, p. 327ff; distributed by Springer-Verlag, Heidelberg.

Chapter 9

NUTRITIONAL INFLUENCE OF ZINC ON THE ACTIVITY OF ENZYMES AND HORMONES

Manfred Kirchgessner and Hans-Peter Roth
Institut für Ernährungsphysiologie der Technischen
Universität München
Freising-Weihenstephan
Federal Republic of Germany

1. INTRODUCTION

Zinc deficiency has been demonstrated in a number of birds and mammals, including human beings. Among the most frequently reported deficiency symptoms are loss of appetite; retardation or even cessation of growth; skin defects, including parakeratotic lesions; hair

loss; impaired wound healing; and defects leading to reproductive
failure. Poor appetite and failure to grow are commonly noted as
the earliest conspicuous responses to zinc deficiency, although not
in adult animals. Little is known about the metabolic defects responsible for the zinc-defiency symptoms, but much scientific effort has
been invested, especially in recent years, to elucidate the biochemical changes in the human or animal body suffering from an inadequate
supply [1-5] of zinc.

2. EFFECTS OF ZINC ON METALLOENZYME ACTIVITY

The essential nature of zinc for the living system is fundamentally
based on its role as an integral part of a number of metalloenzymes
and as a cofactor for regulating the activity of specific zinc-
dependent enzymes. The level of zinc in cells can therefore govern
many metabolic processes, specifically carbohydrate, fat, and protein
metabolism and nucleic acid synthesis or degradation, through the
initiation and/or regulation of the activity of zinc-dependent
enzymes. The decrease in the activity of a particular enzyme in
response to deficient zinc nutrition depends on how tightly the zinc
cation is bound to the protein (thermodynamic stability) or how fast
the rate of exchange of the ligands is (kinetic stability). These
are also the reasons why only a few of the known zinc metalloenzymes
(see Table 1) respond sensitively and rapidly to a deficient zinc
supply.

2.1. Changes in Enzyme Activities

2.1.1. Alkaline Phosphatase

Some metalloenzymes contain concrete metal-binding sites that are
important for catalytic function and essential for structural stability. Thus alkaline phosphatase from *Escherichia coli,* for example,
contains 4 g-atoms of zinc per mol; 2 are essential for catalytic

TABLE 1

Zinc Metalloenzymes in the Animal and Human Body

Enzyme	EC number	Activity change[a]	Examples of occurrence
Alcohol dehydrogenase	1.1.1.1	-	Liver, adipose tissue, lung
Glutamate dehydrogenase	1.4.1.3	0	Liver, cerebrum, kidneys, cardiac muscle, etc.
Malate dehydrogenase	1.1.1.37	0	Brain, adipose tissue, liver, pancreas, etc.
Lactate dehydrogenase	1.1.1.27	0	Brain, adipose tissue, cardiac muscle, kidneys, liver, pancreas, etc.
Glyceraldehydephosphate dehydrogenase	1.2.1.13		Muscle
RNA polymerase	2.7.7.6	-	Liver, kidneys
DNA polymerase	2.7.7.7	-	Liver, kidneys
Alkaline phosphatase	3.1.3.1	-	Bone, mucosa, serum, kidneys
Leucine aminopeptidase	3.4.1.1		Kidneys
Carboxypeptidase A	3.4.2.1	-	Pancreas
Carboxypeptidase B	3.4.2.2	-	Pancreas
Dipeptidase	3.4.3	-	Kidneys
AMP aminohydrolase	3.5.4.6		Muscle
Carbonic anhydrase	4.2.1.1	-	Erythrocytes
δ-Aminolevulinic acid dehydrogenase	4.2.1.24	-	Blood

[a]In zinc deficiency: 0, unchanged in comparison with pair-fed controls, -, reduced in comparison with pair-fed controls.
Source: Ref. 16.

activity [6], while the additional 2 zinc atoms stabilize the protein structure [7]. Accordingly, alkaline phosphatase exhibits quick loss of activity in experimental zinc deficiency. In the serum of rats, Roth and Kirchgessner [8] found the activity of alkaline phosphatase to decrease by as much as 25% after just 2 days of dietary zinc

depletion and by 50% after 4 days. This loss of activity was not
due to reduced food intake; in fact, the activity of this serum
enzyme had already diminished before growth and food intake notice-
ably decreased. Furthermore, there was no difference in the activity
between restrictively fed and ad libitum-fed control groups throughout
the experimental period. Also, the serum alkaline phosphatase was
greatly decreased in zinc-deficient lactating cows that did not
restrict their feed intake [9].

The activity of alkaline phosphatase returned nearly to the
level of that in control animals or in animals having adequate zinc
within 3 days after zinc had been added to the diet or injected [8,
10] (see also Fig. 1). Although preincubation of serum from zinc-
deficient rats with zinc in vitro enhanced the activity of the alkaline phosphatase, the level of control serum was not reached because

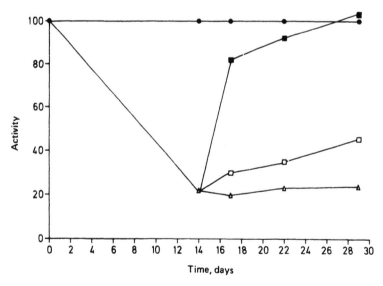

FIG. 1. Activities of alkaline phosphatase in serum of depleted,
repleted, and control rats (pair-fed control rats = 100). ●, Pair-
fed control rats (96 mg zinc per kg diet); △, zinc-deficient rats
(1.2 mg zinc per kg diet); □, zinc-repleted rats (4.5 mg zinc per
kg diet); ■, zinc-repleted rats (12 mg zinc per kg diet). (From
Ref. 16.)

the activity there was also stimulated by zinc addition. This demonstrates that the decrease in alkaline phosphatase activity in zinc-deficient status is the result of lowered enzyme concentration, caused by reduced enzyme synthesis or enhanced degradation [10].

Henkin et al. [11] found the activity of the alkaline phosphatase in the leukocytes of 106 patients with gustatory and olfactory disturbances, a zinc deficiency symptom, to lie significantly below normal values. Similarly, in volunteers in whom a slight zinc deficiency was produced experimentally for the first time with human beings [12], the activity of the alkaline phosphatase in serum slowly decreased after zinc restriction. After zinc supplementation of the diet, all the other dietary constituents remaining the same except for zinc intake, the activity of the alkaline phosphatase doubled within 8 weeks. This also proves that this biochemical change in enzyme activity is to be attributed solely to zinc deficiency. In patients with acrodermatitis enteropathica, a lethal autosomal, recessive disease that is probably due to a disorder in zinc absorption [13], the rate of renal zinc excretion and the activity of the serum alkaline phosphatase were also greatly reduced, in addition to the zinc contents in plasma and skin [14,15]. This indicates a severe state of zinc deficiency. An oral supply of 22-44 mg of zinc twice daily (see Table 2) brought about rapid and complete clinical remission in all four patients within 2 weeks and the return of biochemical indices such as plasma zinc content, renal zinc excretion, and alkaline phosphatase activity of serum to normal values [15].

The clinical effect of zinc therapy was equally fast on skin lesions, diarrhea, lack of appetite, and emotional depression. This immediate clinical response to zinc therapy before the biochemical parameters of the zinc supply status became fully normal shows that the nutritive effect of zinc supplementation is to be attributed to an amelioration of the zinc deficiency and not to a pharmacological effect of excessive zinc [14]. Ten patients with parenteral nutrition developed skin lesions similar to those seen in acrodermatitis enteropathica, and showed low serum zinc levels and reduced activity

TABLE 2

Biochemical Data on Four Patients with Acrodermatitis
Enteropathica Before and After Zinc Therapy

Biochemical index	Before zinc therapy	After zinc therapy
Plasma zinc (µg/100 ml)	25 ± 10	112 ± 28
Renal zinc excretion (µg zinc/24 h)	47 ± 11	714 ± 384
Alkaline phosphatase in serum (IU/liter)	57 ± 14	181 ± 18

Source: Abstracted from data given in Ref. 14.

of the alkaline phosphatase in serum [17]. In Friesian calves with
Adema disease, a malady with symptoms paralleling those of acrodermatitis enteropathica of humans, it was consistently found that the
serum zinc contents were subnormal and the alkaline phosphatase
activity reduced, and that both increased immediately after zinc
therapy [18-20].

In arthritis rheumatica, an increasingly disabling disease of
unknown etiology, there are also indications of deficient zinc status
[21]. After 12 weeks of oral zinc sulfate therapy, the activity of
the alkaline phosphatase in serum had increased significantly, whereby
the serum zinc content was closely correlated with the activity of the
alkaline phosphatase. Recent studies on patients with sickle-cell
anemia showed that zinc plays an important role in this disease and
that some of these patients suffer from zinc deficiency [22-24]; zinc
therapy therefore has proved to be beneficial. In the Middle East,
dietary supplementation with zinc augmented the activity of the alkaline phosphatase in the serum of zinc-deficient persons showing
dwarfism [25].

2.1.2. *Carbonic Anhydrase*

The carbonic anhydrase of erythrocytes also requires zinc for its
physiological functions. The first zinc metalloenzyme was detected by
Keilin and Mann [26,27], it contains 1 g-atom of zinc per mol. In
rats the activity of carbonic anhydrase was reduced by about 20 and
40% within 2 and 4 days, respectively, after feeding them a zinc-
deficient diet [28]. However, at the end of the 30-day experiment,
there was no longer a difference in the activities of the deficient
and the control animals. The number of erythrocytes per cubic milli-
meter of blood had increased in this time by 40%. The body thus
strives to maintain the activity of this essential enzyme by raising
the erythrocyte concentration in the blood. When the activity is
expressed per unit of erythrocytes, a reduction in the activity of
the carbonic anhydrase in the blood can be demonstrated soon after
the start of zinc depletion and later in the stage of extreme zinc
deficiency [29]. Also, in the intestine and stomach of zinc-deficient
rats carbonic anhydrase was reduced by 33 and 47%, respectively [30].
In patients with sickle-cell anemia the activity of the carbonic
anhydrase in the red blood cells correlated closely with the zinc
content [31].

2.1.3. *Carboxypeptidases*

Two additional zinc metalloenzymes important in protein digestion
are the pancreatic carboxypeptidases A and B. Each possesses 1 g-
atom of zinc per mol and has a molecular weight of about 34,000.
In zinc-deficiency studies carboxypeptidase A showed a reduced
activity in the pancreas of rats [32-35] and pigs [36]. The
activity of this enzyme was about 25% lower in the pancreas of rats
after 2 days of zinc depletion [35]. In repletion studies the
activity of carboxypeptidase A returned to nearly normal values
within 3 days after zinc supplementation (Fig. 2). Also in the
case of pancreatic carboxypeptidase B, a reduction of its activity
by roughly 50% was found in zinc-deficient rats compared with pair-
fed and ad libitum-fed control animals [35].

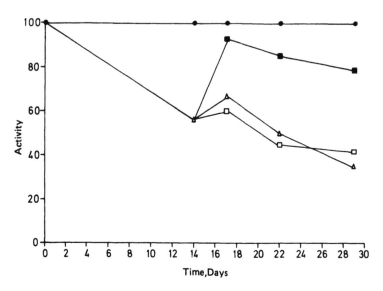

FIG. 2. Activities of pancreatic carboxypeptidase A of depleted,
repleted, and control rats (pair-fed control rats = 100). Symbols
as in Fig. 1. (From Ref. 16.)

2.1.4. Dehydrogenases

The lactate, malate, alcohol, and glutamate dehydrogenases are other
zinc metalloenzymes that differ in their molar contents of zinc.
They show, depending on species and tissue, unchanged or only slightly
reduced activities in response to zinc deficiency [1]. In experimental
zinc deficiency in human beings [12] the activity of the lactate dehydrogenase in the plasma of all subjects correlated with the phases
of zinc restriction and zinc repletion. During zinc restriction
lactate dehydrogenase activity was decreased; during zinc repletion
a significant elevation of the activity was observed.

2.1.5. RNA and DNA Polymerase

In 1959, Wacker and Vallee showed that zinc occurred in several
highly purified preparations of RNA and DNA from very different
sources. On the basis of these observations many experimental
studies were undertaken to investigate whether this element plays

a role in nucleic acid metabolism and in protein synthesis [37].
Several years ago zinc became known as an essential constituent of
the DNA-dependent DNA and RNA polymerase. Using highly purified
preparations of DNA-dependent RNA polymerase from E. coli, Scrutton
et al. [38] determined its zinc content as 2 g-atoms per mol of
enzyme (molecular weight 370,000). A homogeneous DNA polymerase was
isolated by Slater et al. [39] from E. coli and sea urchins and determined to contain 2 and 4 g-atoms of zinc per mol of enzyme. Definite
proof that the DNA polymerase of E. coli is a metalloenzyme was
established by Springgate et al. [40,41]. The RNA-dependent DNA
polymerase, that is, the reverse transcriptase, also is a zinc
metalloenzyme [42,43] and occurs in a number of viruses. These
transcription enzymes have a key position in nucleic acid metabolism
and hence also in protein biosynthesis. Lieberman and Ove [44]
found in in vitro experiments with kidney cells from rats that zinc
ions are necessary for DNA synthesis. Removal of zinc from the
growth medium resulted in inhibition of both DNA synthesis and DNA
polymerase and thymidine kinase activity. Terhune and Sandstead
[45] studied the influence of zinc deficiency on the activity of
the DNA-dependent RNA polymerase in the nuclei of liver cells of
suckling rats born of zinc-deficient mothers. The activity of this
enzyme steadily decreased in the young from the tenth day of life,
demonstrating that zinc is necessary for the activity of the nuclear
DNA-dependent RNA polymerase in mammalian liver. In a similar study
on the brains of prenatal zinc-deficient rats, reduced RNA polymerase
activity was also found, in addition to smaller brain size and diminished DNA synthesis [46]. Similarly, the activity of DNA polymerase
was significantly lower in embryos of zinc-deficient rats than in
those of control animals from day 9 to 12 of pregnancy.

On the basis of these observations it may be concluded that
zinc-deficient organisms show impaired nucleic acid synthesis.
Although the exact mechanism responsible for the defective protein
synthesis during zinc deficiency cannot yet be stated with certainty,
it appears likely that it also results from impaired nucleic acid
synthesis.

2.1.6. Ribonuclease

Fernandez-Madrid et al. [47] and Somers and Underwood [48] expressed
the opinion that Zn^{2+} can inhibit ribonuclease, whereby RNA degradation would be lowered. Also, enhanced activity of this enzyme was
observed in zinc-deficient tissues of experimental animals [49].
Hence the zinc content in cells regulates the activity of ribonuclease, so that the rate of RNA catabolism also depends on the
tissue zinc concentration. Thus the testes of zinc-deficient rats
contained less zinc, RNA, DNA, and protein, but at the same time
they exhibited an elevated ribonuclease activity [48]. The primary
defect is also the increased ribonuclease activity, which leads to
increased protein catabolism. In similar experiments [49], the
activities of ribonuclease and deoxyribonuclease were studied in
the testes, kidneys, bones, and thymus of zinc-deficient rats.
Whereas the deoxyribonuclease activity did not differ between the
zinc-deficient and pair-fed control rats, the ribonuclease activity
was elevated in the zinc-deficient tissues. The authors believe
that the increased ribonuclease activity is responsible for the
lowered RNA/DNA ratio observed in previous investigations with zinc-
deficient pigs [36], as well as for the lowered protein synthesis
and the growth depressions observed regularly in many species,
including human beings.

   In the plasma of patients with sickle-cell anemia a reduced
zinc concentration and an elevated ribonuclease activity were demonstrated, providing further evidence that these subjects have a zinc-
deficient status [31]. After zinc therapy, the plasma zinc content
increased and the activity of the plasma ribonuclease decreased.
Also, in a child with acrodermatitis enteropathica, Hambidge et al.
[14] found an elevated ribonuclease activity in serum, besides a
reduced vitamin A content; both the vitamin A content in the serum
and the ribonuclease activity returned to normal after zinc therapy
was begun. In volunteers in whom a mild zinc deficiency was produced
experimentally, the activity of the plasma ribonuclease was nearly
twice as high during the phase of zinc restriction as during the

zinc-repletion phase [12]. These results are similar to those obtained with zinc-deficient rats [49].

Determination of the activity of a zinc-dependent enzyme, such as the ribonuclease in plasma, could therefore be an additional parameter useful for the diagnosis and assessment of the zinc status in human beings. Opposed to this conclusion stands a recent report by Chesters and Will [50], who state that ribonuclease activity, total ribonuclease concentration, and free ribonuclease inhibitor concentration in liver, kidneys, and testes of zinc-deficient rats were not altered compared to those of pair-fed control animals. They assume that the effect of a dietary zinc deficiency on the ribonuclease activities of tissues is a general response to the altered growth rate rather than a direct effect of zinc on the enzyme system.

2.1.7. Thymidine Kinase

More recent studies showed that zinc is also required for the activity of thymidine kinase, an enzyme essential for DNA synthesis and hence also for cell division. Thymidine kinase could therefore be the enzyme that is responsible for the symptoms of early zinc depletion. Lieberman et al. [51] had indicated that zinc-deficient mammalian cells had a lower thymidine kinase activity after in vitro incubation than did control cells. In zinc-deficient rats the activity of thymidine kinase in fast-regenerating connective tissue was, after 6 days of zinc-deficient nutrition, significantly reduced below the values for pair-fed or ad libitum control animals [52]. Similar results were obtained after 13 days, and after 17 days of zinc depletion the thymidine kinase was no longer detectable. The incorporation of [^{14}C]thymidine into the DNA of connective tissue was also reduced after 6 days on the zinc-deficient diet. In hepatectomized rats, which were given postoperatively a zinc-deficient diet, Duncan and Dreosti [53] found decreased activity of thymidine kinase, compared with that of control animals, before a reduction in DNA and protein synthesis occurred. Whereas the synthesis of protein was

reduced after 48 h, and that of DNA after 18 h, of postoperative
zinc deficiency, the thymidine kinase activity was diminished after
only 10 h. Similarly, there was a markedly reduced activity of
thymidine kinase in 12-day-old embryos of zinc-deficient rats compared with those of pair-fed control animals [54]. Reduced activity
of thymidine kinase became evident even when pregnant females were
given the zinc-deficient diet from days 9 to 12 of pregnancy only.

These results show that either the thymidine activity or the
biochemical processes necessary for the activity of thymidine kinase
are extremely sensitive to zinc depletion. In the metabolic sequence
the activity of this enzyme stands before the polymerization of DNA
and before cell division. Reduced activity or delayed response of
the activity of this enzyme is one of the earliest and most sensitive
metabolic alterations and could ultimately prove to be responsible
for the rapid decline in growth rate, the anorexia, the delayed
wound healing, and so on, in zinc-deficient subjects [23,55].

2.2. Reduced Enzyme Activities and Zinc Deficiency Symptoms

The results presented clearly reveal that among the zinc metalloenzymes only specific ones change their activity, and then only in
certain tissues. Therefore, the probability of detecting biochemical
changes is highest in the tissues that respond rather sensitively to
a lack of available zinc (e.g., serum, bones, pancreas, and intestinal
mucosa). It is not to be expected that zinc-dependent enzymes are
affected to the same extent in all tissues of a zinc-deficient animal.
Differences in the sensitivity of enzymes evidently result from differences in both the zinc-ligand affinities of the various zinc
metalloenzymes and their turnover rates in the cells of the affected
tissues [36]. Thus the zinc metalloenzymes that bind zinc with a
very high affinity should still be fully active even in extreme stages
of zinc deficiency.

Swenerton [56-58] could not find reduced activity of lactate,
glutamate, or alcohol dehydrogenase in the liver of zinc-depleted

rats, nor were there reduced activities of malate and lactate
dehydrogenases in the testes showing histological lesions. Therefore, they did not agree with the hypothesis that reduced enzyme
activities are responsible for the severe physiological and morphological changes observed in zinc-deficient animals. The complete
lack of responsiveness of the activity of these enzymes, even when
symptoms of severe zinc deficiency are apparent, may again be
explained on the basis that the affinity of zinc for these enzymes
is high, and consequently, that the turnover rate in these tissues
remains unaltered.

In addition, the possibility cannot be excluded that a specific
enzyme requiring zinc for its activity is replaced by another enzyme
that does not contain zinc but nevertheless catalyzes the same reaction. Furthermore, numerous metalloenzymes are able to remain functionally intact despite a deficiency of a particular metal, because
some other metal may substitute [59].

Since many enzymes need zinc for their physiological functions,
the manifestation of severe zinc-deficiency symptoms may be associated
with reduced activities of a number of zinc-containing enzymes. Since
tissues bind zinc with different affinities, dietary zinc depletion
may rapidly lead to a deficiency in labile zinc, especially in certain
organs, and to a corresponding loss in the activities of specific zinc
metalloenzymes. Adequate zinc supplementation rapidly overcomes this
deficiency and raises the activity of these zinc metalloenzymes to
normal levels. The extent to which a metalloenzyme loses its activity
also depends on the functional role of zinc in maintaining the enzyme
structure. With some zinc-dependent enzymes (e.g., alkaline phosphatase) zinc deficiency may induce structural changes that increase the
chance for degradation. The consequence is an increased turnover
rate and a lower activity of the enzyme in the tissues [60].

According to Prasad et al. [61,62], many metabolic processes
are regulated by zinc metalloenzymes, which in turn depend on the
tissue levels of zinc available for the control of their synthesis
and activity. None of the currently known enzymes is sufficiently
sensitive or important in metabolism to be solely responsible for

the very early and basic effects of zinc depletion on growth and
appetite. Zinc metalloenzymes are, however, fundamentally involved
in many intracellular biochemical mechanisms, as they, in particular,
regulate the various stages of protein synthesis and the early nucleic
acid synthesis before cell division [63]. In newer studies [5,64]
particular attention is called to the important physiological role
of zinc for the structure and function of biomembranes. There is
evidence that the defects in membrane structure and function observed
during zinc deficiency may ultimately be responsible for many of the
known deficiency symptoms.

Two enzymes that affect protein digestion are the pancreas
carboxypeptidases A and B. The activities of these two enzymes are
reduced in zinc deficiency; pancreas carboxypeptidase A lost about
one-fourth of its activity within 2 days [65]. However, the digestibility of the dietary dry matter and crude protein was reduced just
slightly (1.5 and 3%, respectively) in young zinc-deficient rats
compared with pair-fed control animals [66]. On the other hand, a
pronounced decrease in feed efficiency and growth rate is generally
observed in zinc deficiency. Thus Kirchgessner and Roth [65] and
Weigand and Kirchgessner [67] recorded the food expenditure for body
weight gain to be three to seven times higher in zinc-deficient rats
than in control animals. Consequently, the reasons for these drastic
effects of zinc deficiency must be sought not so much in the intestinal tract as in the biological processes of the cells in the body
tissues.

In addition to carboxypeptidases A and B, alkaline phosphatase
may be related to the many disorders caused by zinc deficiency. The
zinc-containing metalloenzyme is important for a number of processes
in the biochemistry of normal bone formation [68-70]. In zinc-
deficient rats the alkaline phosphatase of the femur bone is reduced
by two-thirds compared with pair-fed control animals [71], so that,
in turn, the alkaline phosphatase in serum is reduced, since bones
are the major source of serum alkaline phosphatase [70,72]. Westmoreland [73,74] reported on the possible consequences of reduced

alkaline phosphatase for the growth and calcification of bones and for the leg abnormalities observed in zinc-deficient chicks. Zinc deficiency causes hair defects and a reduction of the alkaline phosphatase in rat skin because of a suppressed hair cycle and of a simultaneous decrease in the hair follicles, which are rich in alkaline phosphatase [72]. Moreover, a reduced activity of the alkaline phosphatase found in high concentrations in the membranes of the taste buds [75,76] could possibly be responsible for the blunting of the sense of taste in zinc deficiency [77]. These studies demonstrate that changes in alkaline phosphatase could indeed be partly responsible for many of the defects observed in zinc deficiency. Thus significant growth promotion could be obtained by injecting zinc-deficient rats with alkaline phosphatase [78]; however, the impaired growth could be only partly reversed.

To understand better the biochemical and physiological aspects of zinc, additional research is needed to elucidate the fundamental mechanisms ultimately responsible for the clinical symptoms of zinc deficiency in human beings and animals.

2.3. Possibilities on the Diagnosis of Zinc Deficiency

The animal body possesses a high capacity for regulating the metabolism of trace elements. This homeostatic adaptation makes it possible to maintain the concentration of zinc in tissues at a relatively constant level, because both absorption and endogenous excretion homeostatically change over a wide range of different zinc intake. For example, the zinc concentration in the bone pool shows little dependence on the dietary zinc content in the case of normal supply [79]. Long-lasting severe undernutrition with zinc, however, eventually brings about a pronounced reduction of the zinc concentration in bones. Such extreme and distinct zinc deficiency states as could be induced in experimental studies are, however, unlikely to occur under practical conditions. More often, the latent forms of zinc deficiency that are difficult to diagnose because they do not manifest themselves in clinical deficiency symptoms, such as growth cessation and skin lesions, are encountered in the field. They, however, also impair health, performance and reproduction, and the bodily development of humans and animals. Therefore, some sensitive parameters are needed to diagnose these deficiency states. As newer studies show, these states of marginal deficiency are more widespread than thus far believed.

2.3.1. Alkaline Phosphatase Response

For diagnostic purposes, zinc undernutrition has frequently been assessed by analyzing the zinc content in plasma or serum, erythrocytes, hair, urine, and saliva. Hair analysis as an indicator of the zinc status would certainly offer the major advantage of easily obtaining samples from living animals. But in studies with lactating dairy cows [80,81] and young rats [82], the hair test was of very questionable and limited diagnostic value in the case of zinc. Lately, the serum or plasma zinc content has been proposed repeatedly as an indicator of zinc status. Under practical conditions these values greatly vary and are influenced by many other factors. The greatest disadvantage of using the serum zinc content for assessing the supply status, is, however, the fact that serum zinc rises steadily with increasing dietary zinc supply (Fig. 3) and that it does not assume a plateau when the zinc supply is adequate, in contrast to growth, which does not differ at levels between 12 and 100 ppm zinc in the diet [10].

In model studies, therefore, preference was given to the activity of the serum alkaline phosphatase (Table 3), a zinc metalloenzyme that quickly responds to dietary zinc depletion with a marked loss of activity. Under practical conditions, however, the activity of the alkaline phosphatase, similarly to the serum zinc content, not only varies greatly but is also affected by many other factors, so that it is not possible to state a normal activity that would indicate adequacy of supply. This great problem of designating norm

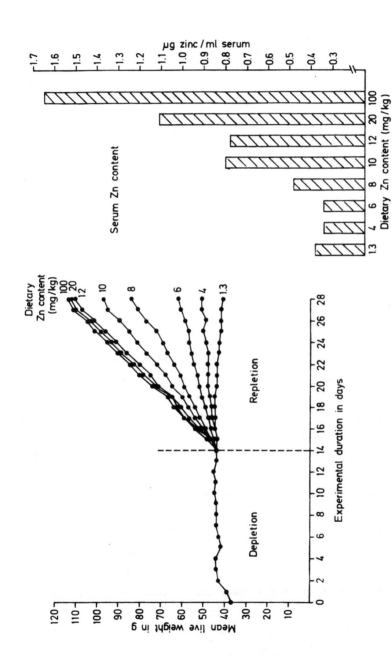

FIG. 3.   Growth response and serum zinc content of rats following zinc supplementation at graded levels.   (Compiled from Ref. 10 and reproduced from Ref. 211.)

TABLE 3

Activity of the Alkaline Phosphatase in Serum of Rats Before
and 3 Days After Zinc Injection (0.8 mg Zinc per Animal)

Dietary zinc content (ppm)	Alkaline phosphatase (mU/ml)		Difference (ΔmU/ml)	Relative activity increase (%)
	Before zinc injection	After zinc injection		
1.3	23 ± 8	152 ± 36	129	560
4	39 ± 10	162 ± 35	123	315
6	88 ± 41	195 ± 61	107	122
8	116 ± 19	185 ± 72	69	59
10	139 ± 29	193 ± 33	54	39
12	183 ± 44	245 ± 62	62	34
20	175 ± 39	249 ± 66	74	42
100	239 ± 27	246 ± 50	7	3

Source: Ref. 10.

values was to be avoided by measuring the activity before and 3 days
after zinc injection. Rats were brought to definite supply states
by giving them diets with different dietary zinc contents to investigate whether suboptimum zinc supply can be diagnosed by this response
technique. There was a close correspondence between the enzyme's
activity and the zinc supply in these rats. The range between minimum and optimum supply covered the factor of 10. Even in the range
of extreme deficiency from 1.3 to 6 ppm zinc in the diet, the alkaline
phosphatase showed a corresponding grading of its activity and it
increased manifold following zinc injection. Similar to the response
of the serum zinc, the alkaline phosphatase did not yet reach an
optimal activity when the dietary zinc content ranged between 12 and
20 ppm. In the case of 20 ppm zinc, the activity still rose by about
40% following zinc injection. The optimal level of the activity of the
alkaline phosphatase in serum probably amounts to about 250 milliunits
per milliliter (mU/ml) in these studies. Accordingly, zinc injection

did not bring about increased activity in animals given a diet with
100 ppm zinc; by contrast, the serum zinc content does not assume
this plateau.

Although the activity of the alkaline phosphatase decreases
with advancing age, especially toward the end of the growing phase,
a reliable evaluation of the zinc supply status was also possible
for rats at three different ages [83]. The activity of the alkaline
phosphatase increased in rats initially weighing 50 g as well as in
those weighing 100 g in response to zinc injection when the diet was
supplied between 1.3 and 9 ppm zinc. In the case of the animals
given the adequate supply of 60 ppm zinc, the activity did not
increase in either age group, while the serum zinc content continued
to rise. Even in the group with 300-g heavy rats, the response was
basically the same. Zinc injection into those animals given diets
with 1.3, 2, 3, and 5 ppm zinc resulted in a higher activity of the
alkaline phosphatase 3 days later, whereas in the animals with 60 ppm
zinc in their diet, the activity was unaffected. Ergo, if zinc injection causes an increase in activity, a more or less pronounced zinc
deficiency must have been present depending on the extent of the
response, since additional zinc does not bring about increased
activity in the case of optimum supply.

2.3.2. ^{65}Zn Uptake

Not even the in vitro ^{65}Zn uptake by erythrocytes [84,85] showed any
dependence on the zinc status in studies with rats [86] given diets
with different zinc contents (Table 4). When these values are cor-
rected to a uniform hematocrit of 40%, because the hematocrit can
in part be greatly increased during zinc deficiency, it is evident
that there is an increased ^{65}Zn uptake in the case of the deficient
to suboptimal supply ranging from 6 to 12 ppm dietary zinc compared
to animals given the higher supplies of 20 or 100 ppm dietary zinc.
The low ^{65}Zn uptake in the extreme zinc deficiency range between
1.3 and 4 ppm may be attributed to the damage that has already

TABLE 4

In Vitro ^{65}Zn Uptake by Erythrocytes of Rats
with Varied Dietary Zinc Supply

Dietary zinc content (ppm)	^{65}Zn uptake by erythrocytes (%)		Hematocrit
	Absolute	Corrected to 40% hematocrit	
1.3	26.2 ± 9.3	23.5 ± 8.4	44.8 ± 4.9
4	23.7 ± 4.4	23.5 ± 4.0	40.1 ± 2.0
6	32.9 ± 4.8	35.2 ± 3.7	36.1 ± 3.0
8	25.7 ± 3.9	36.3 ± 7.2	28.4 ± 1.2
10	21.7 ± 2.5	30.7 ± 3.3	28.3 ± 1.9
12	23.5 ± 2.9	34.9 ± 5.0	27.1 ± 3.3
20	18.0 ± 2.6	24.8 ± 2.1	28.9 ± 1.9
100	17.4 ± 1.7	22.0 ± 2.9	31.3 ± 3.1

Source: Compiled from Ref. 86 and reproduced from Ref. 211.

occurred to the erythrocyte membrane. Obviously, this in vitro
radioisotope method is of no major diagnostic value either.

2.3.3. Zinc-Binding Capacity

As a rule, however, it is not satisfactory to diagnose such a complex
criterion as the zinc supply status by just one parameter. Therefore,
another method based on the percent zinc-binding capacity of serum
has been proposed [87]. As is shown in Table 5, the percent zinc-
binding capacity of rats was nearly 90 in the range of extreme zinc
deficiency with diets containing between 1.3 and 4 ppm zinc and it
decreased to somewhat less than 60 as the dietary zinc content increased. Hence the percent zinc-binding capacity of serum exhibited
an inversely proportional response to the dietary zinc content.
Within 3 days after zinc injection, the zinc-binding capacity had
significantly decreased to values near 70% in all the animals of the
groups given the diets with 1.3-12 ppm zinc. In the rats on the

TABLE 5

Percent Zinc-Binding Capacity of Serum from Rats Before and
3 Days After Zinc Injection (0.8 mg Zinc per Animal)

Dietary zinc content (ppm)	Zinc-binding capacity (%)	
	Before zinc injection	After zinc injection
1.3	87.1 ± 2.7	70.4 ± 2.0
4	89.1 ± 3.5	76.6 ± 2.3
6	84.3 ± 2.1	71.4 ± 3.8
8	83.8 ± 1.1	74.0 ± 3.5
10	74.2 ± 4.8	67.4 ± 3.1
12	74.8 ± 3.1	59.8 ± 2.4
20	69.2 ± 3.7	70.0 ± 3.3
100	58.9 ± 5.3	61.2 ± 3.8

Source: Compiled from Ref. 87 and reproduced from Ref. 211.

diets with 20 and 100 ppm zinc, the percent zinc-binding capacity of
serum was not altered by zinc injection. Because of the constancy
of the percent zinc-binding capacity of serum, as it became evident
from the studies with different ages, it is not necessary to inject
zinc into the test animals. Thus the zinc supply status of animals
or of humans can be assessed by an in vitro test with serum. This
means that the percent zinc-binding capacity should reach a value
between 60 and 70% when zinc supply is optimal. Basically, however,
it must yet be investigated to what extent other factors that, for
example, influence serum zinc concentration also might affect the
percent zinc-binding capacity.

3. EFFECTS OF ZINC ON HORMONES

Relationships between zinc and hormones not only are known for
insulin but are also considered for other hormones, such as glucagon,

growth hormone, and sex hormones [3]. The importance of zinc status
on the growth and sexual development of males is well documented for
various species, including human beings [2,88]. Trace elements can
affect hormones at various action sites, such as their secretion,
activity, and tissue binding sites. Conversely, hormones can also
affect the metabolism of the trace elements at various sites, for
example, their excretion and transport sites.

3.1. Insulin

The pancreas is one of the most sensitive tissues of zinc metabolism,
and its β cells of the islets of Langerhans have a high capacity for
accumulating, turning over, and storing the retained zinc. This
tissue responds to dietary zinc deprivation with a quick and severe
zinc loss [89]. Since Scott [90] discovered that crystalline insulin
contains appreciable amounts of zinc, numerous studies have indicated
a close functional and morphological relationship between zinc and
insulin. Insulin is stored in the pancreatic β cells in the form of
secretory granules containing the zinc insulin in a crystalline or
paracrystalline form [91,92]. Changes in the zinc status of the body
could therefore affect the synthesis, storage, secretion, and hormonal
potency of insulin. Although the etiology of disorders of this endocrine system have been investigated extensively, only a few studies
are available regarding the influence of dietary factors on insulin
metabolism, such as, for example, the influence of the dietary zinc
content. On the basis of these experimental findings, the possible
effects of zinc supply on insulin metabolism will be presented and
discussed [4,93].

3.1.1. Glucose Tolerance

Among the best known functions of insulin is its blood glucose-
lowering effect. Accordingly, Hove et al. [94] had already conducted
studies on glucose tolerance in zinc-deprived rats as early as 1937.

They found just minor differences in glucose tolerance curves after
oral glucose dosing between zinc-deficient and ad libitum-fed control
rats, while Hendricks and Mahoney [95] found no difference in the
ability to metabolize orally administered glucose between zinc-
deficient and zinc-supplied rats.

Using intraperitoneal glucose injection, Quarterman et al.
[96], however, found a reduced glucose tolerance of zinc-deficient
rats compared with pair-fed control animals after an extended experimental period and an overnight fast of the animals. This has been
confirmed by studies on Chinese hamsters of Boquist and Lernmark
[97], and studies on rats of Hendricks and Mahoney [95], Huber and
Gershoff [98], Roth et al. [99], and Roth and Kirchgessner [89,93].
Figure 4 shows representative glucose tolerance curves for zinc-
deficient rats in comparison with pair-fed and ad libitum-fed control
animals. In these studies, the rats were injected with 80 mg of
glucose per 100 g of body weight into the femoral muscles after they
had been zinc-depleted on a semisynthetic low-zinc diet (2 ppm of
zinc) for a period of 34 days followed by a 12-h fast.

Although the glucose concentrations were initially the same,
the zinc-depleted rats showed a significantly reduced glucose tolerance compared to the control rats. With the aid of pair-fed animals
it could be demonstrated that the lowered glucose tolerance of zinc-
deficient animals is not the consequence of the reduced food intake
or the associated weakening of the animals. The pair-fed animals
exhibited an even greater glucose tolerance than did the ad libitum-
fed control animals.

In human beings with dwarfism and symptoms of zinc deficiency,
abnormal glucose tolerance curves were also observed after oral
glucose [88,100]. In contrast to these findings, the studies of
Macapinlac et al. [101] and Quarterman and Florence [102] adminis-
tering glucose intraperitoneally and of Brown et al. [103] adminis-
tering glucose orally could not detect any effect of zinc deficiency
on glucose tolerance. Differences in the results between oral glucose dosing and intraperitoneal or intravenous glucose injection

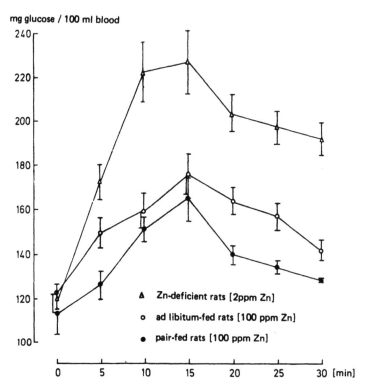

FIG. 4. Glucose tolerance curves of zinc-deficient rats compared to
ad libitum- and pair-fed control rats. Vertical bars represent the
standard errors of the mean of six animals. (From Ref. 1.)

might owe to a greater stimulation of insulin secretion in response
to oral glucose [97,104].

Quarterman and Florence [102] suggested that the lowered glucose tolerance of zinc-deficient rats is the consequence of the
different pattern in food intake, because zinc-deficient animals
consume their food intermittently over the day in contrast to pair-
fed or pair-weight control animals that eat their daily ration within
a relatively short time. They postulated that the level of food
intake during the day before the glucose tolerance test is of decisive importance. This explanation must, however, be regarded as
unlikely on the basis of newer studies [93,99] in which food intake

was kept the same. The difference seems rather to owe to experimental procedures. If, for example, a high glucose dose is injected,
the homeostatic regulation maintaining the blood glucose concentration at a constant level is overloaded in both zinc-depleted and
control animals, while after the injection of too low a glucose dose,
both groups can readily cope with this load and hence no difference
becomes apparent. For a normal glucose tolerance test, the chromium
status is also of importance [105]. Together with a few studies
involving use of the glucose tolerance test during zinc deficiency,
dietary chromium levels have also been assessed [106]. Without
supplemental chromium, the diets contained 2.2 ppm chromium, a level
that should adequately assure an optimum supply [106-109].

3.1.2. Insulin Secretion

To clarify the question of whether the reduced glucose tolerance of
zinc-deficient rats owes to a diminished insulin secretion by the β
cells of the pancreatic tissue, the serum insulin concentration has
been determined [89]. Deficient animals regularly display reduced
serum insulin concentrations compared with ad libitum-fed controls,
but not compared with pair-fed controls in which serum insulin
levels were reduced in only one of three experiments. These insulin
values rose within 15 min in response to glucose stimulation. However, no significant differences were found between the zinc-depleted
groups showing reduced glucose tolerance and their zinc-supplemented
control groups. Since zinc deficiency always leads to a depression
of voluntary food intake in growing animals, and insulin levels are
known to fall in response to fasting or lower food intake, it was
difficult to decide whether the reduced serum insulin content came
about primarily because of zinc deficiency or because of lowered
food consumption. Injections of sulfonylurea (Rastinon) in such
doses that there the blood sugar level fell equally in both zinc-
deficient and control animals stimulated a much greater release of
endogenous insulin in zinc-deficient rats than in their controls
[89]. It was therefore postulated that reduced blood glucose

TABLE 6

Serum Insulin Concentrations of Zinc-Deficient and Pair-Fed
Control Rats After Intraperitoneal Glucose Injection[a]

| Time (min) | Insulin/ml serum (µU) | |
	Depleted animals	Pair-fed controls
0	24.0 ± 11.4	15.0 ± 4.2
5	16.9 ± 6.8	44.5 ± 28.8
10	24.9 ± 17.3	14.3 ± 4.5
15	20.2 ± 12.0	40.7 ± 23.9
20	19.9 ± 7.7	19.8 ± 14.7
25	16.6 ± 5.9	17.0 ± 10.1
30	20.1 ± 5.0	19.3 ± 5.8

[a]Concentrations were 200 mg per 100 g metabolic body.
Source: Ref. 93.

homeostasis in zinc-deficient rats is evidently not the consequence
of a lower insulin secretion by the β cells, but rather of a reduced
physiological potency of insulin at its peripheral sites of action.
In studies with mature dairy cows, in which zinc depletion did not
bring about a depression of food intake despite the manifestation of
distinct zinc deficiency symptoms, it was found that the serum insulin
level was markedly reduced and could be raised again by sufficiently
high zinc supplementation [110]. Also, in recent studies with young
rats [93], when utmost care was taken to equalize food consumption,
the fasting times and frequency of feeding of the pair-fed animals,
especially toward the end of the experiment, allowed demonstration
of a lower serum insulin content for zinc-deficient versus pair-fed
animals after glucose stimulation (Table 6).

In the pair-fed control animals, the serum insulin levels
showed a typical biphasic response after intraperitoneal (ip) glucose
injection (see Table 6). This characteristic insulin release pattern
is usually induced by glucose stimulation alone, as has been often
reported [111-113]. Twenty minutes after glucose injection, the

serum insulin concentration of the pair-fed rats had almost returned
to the initial level. In the case of the zinc-deficient animals,
however, there was no additional insulin secretion in response to
glucose injection. This difference in insulin response to glucose
stimulation observed in this study clearly indicates that the reduced
glucose tolerance of zinc-deprived animals is due to reduced insulin
secretion. Regarding the typical biphasic insulin secretion of the
pair-fed controls after glucose stimulation, the first rapid phase
corresponds to the release of insulin from a readily available
reserve of β granules and can be induced by a number of stimulants,
such as glucose, amino acids, sulfonylurea, and glucagon [111]. The
second, delayed, secretory phase also is due mainly to release of
stored, previously produced insulin [114]. This phase is induced
only by glucose. The reduced concentrations of circulating insulin
in this study with zinc-deficient rats can only be explained by a
diminished synthesis, storage, or release of insulin by the pancreas
or by an augmented degradation of circulating insulin in the state
of zinc deficiency. According to Reinhold and Kfoury [60], zinc
may help to stabilize its protein structure. In a similar manner,
removal of zinc from carbonic anhydrase, a zinc metalloenzyme, lowers
the stability of this enzyme and hence enhances its denaturation.
The studies of Quarterman et al. [96] demonstrated that zinc-deficient
rats given insulin are much more resistant to hypoglycemic coma than
are zinc-supplied control rats. This finding supports the hypothesis
of Hendricks and Mahoney [95] that the rate of insulin degradation is
increased in zinc-deprived rats. It is, however, equally possible
that an increased insulin resistance of the peripheral tissues is
encountered in zinc deficiency, similar to the situation encountered
in some cases of diabetes mellitus [115]. Recent studies linking
insulin degradation to receptor binding [116] would suggest, however,
that any effect of zinc deficiency on insulin degradation might be
mediated through altered receptor levels or affinity.

Engelbart and Kief [117] found that acute stimulation of
insulin secretion also lowered the zinc content of the β cells of
rat pancreas. Since it may be assumed that zinc participates in

the synthesis and storage of insulin in the β cells, it could be
that a smaller amount of insulin is stored during zinc deficiency.
Further studies should be devoted particularly to those aspects of
the formation, potency, and degradation of active insulin that will
help clarify the relationship between zinc deficiency and insulin.

The pancreas of rats suffers a much greater decrease in zinc
content than most other tissues when these animals become zinc-
depleted [89,118]. In the β cells, zinc participates in the polymerization process of insulin [119] and forms an insoluble zinc-
insulin complex [120-122] that is then stored in the secretory
granules. Electron microscopic studies on insulin release from the
β cells of the pancreas [123] showed that the zinc ions identify the
place and constitute the visible form of insulin within the β cells.
The fact that changes in the zinc and insulin content of islet tissues are identical provides evidence for a close functional relationship [124]. During glucose loading, the number of granules in the
β cells decreases proportionately to the loss of insulin. Under
these circumstances, the disappearance of insulin from the β cells
is assumed to be accompanied by a loss of zinc [117]. Accordingly,
the administration of hypoglycemic sulfonamides led to a marked loss
of zinc from the β cells, while the zinc content of the α cells
remained unaltered [124]. If insulin requires zinc or chromium to
maintain its physiological activity, changes or shifts in the zinc
and chromium concentrations of the tissues are likely to occur
together with the secretion of substantial amounts of insulin
following glucose stimulation.

In the relevant studies [106], the serum zinc concentration
of the zinc-depleted animals was reduced by about half of that of
the pair-fed control animals (see Table 7). In the zinc-depleted
animals, there was no major change in the serum zinc content or in
the insulin levels within 30 min following glucose injection. In
the case of the pair-fed control animals, however, the serum zinc
level markedly increased above the initial value at the start of
the experiment (P < 0.001) within 10-15 min after glucose injection.

TABLE 7

Zinc Concentration in Serum of Zinc-Deficient and Pair-Fed
Control Rats After Intraperitoneal Glucose Injection[a]

Time (min)	μg zinc/ml serum	
	Zinc-depleted animals	Pair-fed controls
0	0.77 ± 0.17	1.49 ± 0.17
5	0.70 ± 0.11	1.56 ± 0.21
10	0.73 ± 0.20	1.83 ± 0.08
15	0.86 ± 0.06	1.90 ± 0.13
20	0.81 ± 0.08	1.42 ± 0.29
25	0.68 ± 0.03	1.61 ± 0.14
30	0.80 ± 0.05	1.61 ± 0.17

[a]Concentrations were 200 mg per 100 g metabolic body.
Source: Ref. 106.

It was also exactly during this time span of the high zinc concentration that the highest blood glucose levels were recorded. After
20 min the serum zinc content had again returned to normal. This
suggests that the secretion of insulin leads to elevated serum zinc
contents. The extent to which insulin would require zinc for maintaining its physiological potency is difficult to evaluate. Even
though the primary structure of insulin is known, the relationship
of its ternary and quarternary structure to zinc and to the physiological activity of this hormone is still not clear. The difficulty
in demonstrating such a relationship lies in the complexity of the
systems required to study this problem [146], especially since
insulin also plays a role not only in carbohydrate metabolism but
also in fat and protein metabolism [125,126].

    The concentration of chromium in the serum of the pair-fed
control rats showed a similar response curve following glucose stimulation as did the zinc level. It had reached its maximum within 10
min (see Table 8). In contrast to zinc, chromium concentration also

TABLE 8

Chromium Concentration in Serum of Zinc-Deficient and Pair-Fed
Control Rats After Intraperitoneal Glucose Injection[a]

Time (min)	ng chromium/ml serum	
	Zinc-depleted animals	Pair-fed controls
0	0.77 ± 0.23	1.03 ± 0.03
5	0.95 ± 0.12	1.18 ± 0.10
10	0.88 ± 0.14	1.28 ± 0.09
15	0.79 ± 0.14	1.14 ± 0.09
20	1.00 ± 0.08	1.10 ± 0.14
25	1.00 ± 0.06	1.08 ± 0.09
30	1.16 ± 0.07	1.04 ± 0.05

[a]Injection glucose concentrations were 200 mg per 100 g metabolic
body.
Source: Ref. 106.

significantly increased in the serum of deficient animals, especially
toward the end of the glucose tolerance test. Overall, the serum
chromium contents were somewhat lower in the zinc-deficient rats than
in the control animals, although their diets both supplied chromium
at the same level of 2.2 ppm.

The biochemical significance of chromium lies in its influence
on glucose tolerance and has recently been studied extensively by
Mertz and co-workers [105,127-130] and Hambidge [131,132]. Chromium
is assumed to function as an essential cofactor in the reaction of
insulin with its receptors in cell membranes. Thus insulin synthesis
and metabolism would not be impaired by chromium deficiency. In rats
and healthy young human subjects, sugar loading results in an acute
rise of the chromium concentration in serum up to levels that are two
to seven times higher than its fasting concentrations [127,131,133,
134]. The specific pool from which this chromium is derived is not
yet known for certain; it could be the liver, which had greatly
elevated chromium content in the case of the aforementioned zinc-

deficient rats. The failure of the chromium level to increase after glucose loading indicates exhausted chromium reserves and can serve as a criterion to diagnose the chromium status [129]. Normally, the chromium content in serum increases in parallel to the rise in immuno-reactive insulin. This was not the case in zinc-deficient rats [106]. Similarly, in 7 of 12 pregnant diabetic women with reduced glucose tolerance, the serum chromium contents were not found to rise in response to oral glucose loading, although there was an increase in immunoreactive insulin [131].

3.1.3. Proinsulin

The biosynthesis of insulin is known to proceed via proinsulin, the higher-molecular-weight, one-chain peptide, as shown originally by Steiner et al. [135]. It is converted to insulin by proteolytic cleavage and release of a sequence of about 30 amino acids, the connecting peptide (C peptide) which is inserted between the COOH terminus of the β chain and the NH_2 terminus of the A chain. Cleavage of the C peptide takes place during the transport of proinsulin from the rough endoplasmic reticulum via the Golgi apparatus to the secre-tory granules within the β cells of the islets of Langerhans. The pancreatic enzymes identified to be able to do so are trypsin and carboxypeptidase B [136]. However, the carboxypeptidase B of the pancreas is a zinc metalloenzyme that loses about 50% of its activity in zinc-deficient rats [35]. It is thus possible that in zinc-deficient rats, the capacity of the pancreatic islets to convert proinsulin to insulin may be diminished or that a predominant secre-tion of proinsulin occurs directly from the β cells by a route which evades the granules where the conversion to insulin normally takes place. By contrast, the trypsinlike activity in the duodenal fluid of rats is almost doubled during zinc deficiency [125]. The fact that proinsulin possesses just a slight biological activity [137] has led to the speculation that the carbohydrate intolerance is possibly the consequence of an augmented proinsulin secretion. However, the proinsulin contents in serum (see Table 9) did not

TABLE 9

Serum Proinsulin Contents of Zinc-Deficient and Pair-Fed
Control Rats After Intraperitoneal Glucose Injection[a]

Time (min)	ng proinsulin/ml serum	
	Zinc-depleted animals	Pair-fed controls
0	2.4 ± 1.4	1.1 ± 0.5
5	1.9 ± 0.7	2.3 ± 0.7
10	2.5 ± 0.8	2.5 ± 0.5
15	2.1 ± 0.6	2.5 ± 0.9
20	3.5 ± 1.3	3.5 ± 2.0
25	2.4 ± 1.2	2.0 ± 1.1
30	2.6 ± 1.5	2.8 ± 1.8

[a]Glucose injections were 200 mg per 100 g metabolic body.
Source: Ref. 93.

show any difference in response to zinc deficiency during the course of the glucose tolerance test with depleted and control rats [93]. In both groups there was a slight increase after 20 min. Kobayashi [138] found the same values and a similar course for proinsulin in pigs on a normal diet when subjected to a glucose tolerance test. Hence the conversion of proinsulin to insulin is largely unaffected by dietary zinc content and by glucose stimulation. The latter finding has also been reported by Brosky and Heuck [139], Brunner et al. [140], and Schultz et al. [141]. Similarly, Howell et al. [142] could not find a negative effect of incubating a culture of isolated islets of Langerhans from mice with a severely zinc-deficient medium on the biosynthesis of proinsulin or its conversion to insulin, nor on the capacity of the cells to store newly formed insulin inside the granules.

Islet cells contain much zinc, which is of importance for the storage of insulin. As soon as insulin is liberated from proinsulin, it tends to crystallize with zinc [136]. Proinsulin also binds zinc.

Although porcine insulin binds only 1 mol Zn^{2+} per mol and, as a
consequence, completely precipitates from solutions, porcine proinsulin binds more than 5 mol Zn^{2+} per mol by forming soluble
polymers in the presence of zinc [143]. This fivefold difference
in zinc binding between insulin and proinsulin is attributed to the
γ-carboxyl groups of the glutamic acid side chains in the C peptide
of proinsulin. This indicates that proinsulin might play a role in
zinc accumulation by the islet cells. Besides the aforementioned
influence on the conversion of proinsulin, this metal cation binds
the newly built insulin in an osmotically inactive and biochemically
stable crystalline form. According to Engelbart and Kief [117], the
zinc detectable in the granules of the β cells is likely to correspond to the storage form of insulin and could be responsible for
poor in vivo solubility of insulin because of the reduced solubility
of insulin-zinc complexes.

Hence islet zinc could play a role in the mechanism of insulin
secretion, in that it must be dissolved from the zinc-insulin complexes released during secretion. In this process metabolites of
the stimulated β cells, such as citrate, oxaloacetate, or organic
phosphorus compounds, which are stronger complexing agents than
insulin, could also play a role [144]. In the case of reduced zinc
concentration in the pancreas of zinc-deficient rats, the amount of
insulin stored and secreted by these animals might be reduced [89]
if zinc is indeed necessary for the storage of insulin in the β cells
[144].

Insulin crystallizes from dilute NaCl solutions to rhombohedra
containing two atoms of Zn^{2+} per hexamer. When the NaCl concentration is raised above 6%, a rhombohedral, slower-acting 4Zn insulin
crystallizes [145]. Although the conditions for this reversible
conversion are very different from physiological, it is still remarkable that the 4Zn insulin acts more slowly than the 2Zn form. An
analogous structural conversion could favor its solubility, and
hence the release of insulin inside the tissue, because zinc-free
insulin seems to have a structure that is unfolded to a greater

extent than zinc insulin [146]. The role of zinc in the regulation
of granule formation has not been studied extensively. Investigations show that most islet zinc is present in the secretory granules
and is released proportionately to insulin during the discharge of
the secretory granules. However, the mechanism for the accumulation
of zinc and its exact biochemical function in the granules are still
unknown. It must be pointed out that it is the structure of the
crystalline hexameric insulin that is known and not the structure of
the monomeric insulin present in solution. The studies by Arquilla
et al. [147] showed that the removal of zinc induces significant
changes in the ternary structure of insulin and reduces the immunological activity of insulin by altering the antigenic determinants
of insulin. It is likely that the zinc-insulin complex is not essential for the hormonal activity [148]. However, as noted earlier,
the possibility that Zn^{2+} alters the action and conversion rate of
insulin cannot be excluded.

3.1.4. Insulinlike Activity

Serum from rats, pigs, and humans contains insulinlike substances
that possess a high specific biological activity [149-152] and differ
chemically from insulin. The exact site of biosynthesis and storage
of these substances, which are generally referred to as nonsuppres-
sible insulinlike activity in serum because their biological activity
cannot be neutralized by insulin antibodies, is yet unknown. According to Liske and Reber [153], the exocrine portion of the pancreas
and the salivary glands are presumably the major sites for the synthesis of these insulinlike substances, not the liver as has been
assumed previously. Since the radioimmunological insulin determination does not measure the actual biological potency of insulin, the
total insulinlike activity (TILA) in serum was determined in relation
to dietary zinc supply by using isolated cells of epididymal adipose
tissue [93]. A significantly higher TILA was found in zinc-deficient
animals than in pair-fed controls between 10 and 15 min after glucose
stimulation (see Table 10).

TABLE 10

Total Insulinlike Activity in Serum of Zinc-Deficient and Pair-Fed
Control Rats After Intraperitoneal Glucose Injection[a]

Time (min)	Depleted animals (cpm)	Pair-fed controls (cpm)
0	363 ± 116	443 ± 153
5	247 ± 84	292 ± 89
10	416 ± 100	253 ± 121
15	314 ± 95	216 ± 47
20	214 ± 74	223 ± 61
25	311 ± 126	228 ± 71
30	261 ± 59	319 ± 151

[a]Glucose injections were 200 mg per 100 g metabolic body.
Source: Ref. 93.

It was at the time of the highest glucose concentration that
the TILA values were significantly higher for the depleted animals
than for the pair-fed controls. Evidently, the bodies of zinc-
deficient animals may attempt to compensate partially for the lower
insulin levels by secreting and activating other insulinlike substances during a glucose load. Hence zinc deficiency does not seem
to affect adversely the synthesis and secretion of these insulinlike
substances, although they do not seem to be able to overcome the
glucose intolerance of zinc deficiency. This finding is also supported by observations that glucose loading brings about a decrease
in protein concentration in the pancreatic tissue to the same extent
in the case of both the depleted animals and the pair-fed control
animals [93]. The total protein content in the pancreas of the zinc-
deficient rats was, on the average, 34% lower than in that of pair-
fed control animals that had 43% higher live weights. As a result
of glucose stimulation there was a decrease in the protein content
within 30 min of the experimental period; however, this decrease
was significant only for the pair-fed animals. By contrast, protein

concentration of the pancreas significantly decreased by 30% in the depleted animals and by 37% in the pair-fed control animals during the glucose tolerance test.

The mechanism of insulin secretion also depends on the method of glucose stimulation. Zinc deficiency seems to reduce just the intravenous or intraperitoneal glucose tolerance, while in the case of oral glucose stimulation, a zinc dependence was not observed with rats [95,103]. Although after intravenous or intraperitoneal glucose loading, it is exclusively the rise of the blood sugar level that brings about insulin secretion, intestinal factors play a physiological role in stimulating the secretion of endogenous insulin after oral dosing [154]. The duodenal mucosa is able to produce a number of hormones, among which gastrin, secretin, pancreozymin, and an islet factor with glucagonlike immunoreactivity are additional factors, apart from the rise in blood sugar to stimulate the release of insulin and insulinlike substances from the pancreas and salivary glands [104].

3.1.5. Diabetes mellitus

Since zinc and insulin are bound as a complex in the granules of the islet cells, it was obvious that changes in zinc metabolism might be found in diabetes mellitus. Pidduck et al. [155] found a significant hyperzincuria in diabetics, in whom the amount of zinc excreted was parallel with the severity of diabetes. Although the studies of Kumar and Rao [156] and Mateo et al. [157] reported a reduction of the plasma zinc contents of diabetics with hyperzincuria, Davies et al. [158], Rosner and Gorfien [159], and Chooi et al. [160] found no significant differences between diabetics and normal healthy individuals. Also, in the case of diabetic children, a very low hair zinc level became normal again after insulin therapy [161]. In an early study, Scott and Fisher [162] reported that the insulin content in the pancreas of diabetics was reduced to one-fourth, and the zinc content to one-half that of healthy subjects. In 13 men and 7 women 62-81 years old showing diminished glucose tolerance,

Heinitz [163] could bring about a significant reduction in the
capillary blood sugar values, which were elevated above marginal
levels before treatment, by oral administration of 100 mg of zinc
aspartate ($\hat{=}$ 19 mg Zn^{2+}) three times daily for 6 weeks. Also, in
the case of chromium deficiency resulting from long-term parenteral
nutrition, the markedly impaired glucose tolerance could be reverted
to normal by the substitution of organic chromium compounds [164].

3.2. Adrenocorticotropin and Growth Hormones

Homan et al. [165] demonstrated that the addition of zinc salts
increases and prolongs the physiological potency of corticotropin
preparations. In in vitro studies with human cell cultures, adrenal
steroid hormones with glucocorticoid activity increased the uptake
of Zn^{2+} [166]. In studies by Flynn et al. [167], the in vitro effect
of zinc on the stimulation of corticosterone synthesis by adrenocorticotropin hormone (ACTH) was tested with isolated adrenal glands.
It was found that ACTH does not stimulate corticosterone synthesis
in the presence of a zinc-chelating agent in the incubation medium.
However, if zinc is added in excess to the chelating agent, the
activity of ACTH is restored. This observation shows that ACTH is
functionally dependent on zinc. If the ACTH activity in the intact
animal depends on the extracellular zinc, reduced corticosterone
synthesis should be evident in zinc-deficient rats.

In zinc-deficient patients in Iran and Egypt, accurate analysis
of their endocrine functions of the anterior hypophysis revealed that
these were in part suboptimal [88]. Growth failure, including dwarfism, and hypogonadism were the most striking features. More than
half of a group of human zinc-deficient dwarfs under investigation
showed a reduced hypophyseal ACTH reserve and responded to ACTH
injection with an abnormal delay of the renal output of 17-hydroxy-
steroids. After prolonged treatment with zinc, these patients
responded to ACTH injection with a normal excretion pattern of renal
steroids. However, this could also be an indirect effect of the

zinc application, inasmuch as the zinc therapy resulted in a general improvement of the patients' health.

Henkin et al. [168] and Lifschitz and Henkin [169] also reported on changes in the zinc metabolism of several patients with abnormalities in the adrenocorticotropin metabolism. Patients with an isolated deficiency of growth hormone (GH), having plasma GH levels below the analytical limit of detection, showed significantly higher serum zinc concentrations and lower renal zinc excretions than did control persons [170]. After treatment with exogenous GH the serum zinc concentration fell again, while the renal zinc excretion rose. Conversely, an increased concentration of circulating GH, as in untreated acromegaly, leads to a reduction of the serum zinc and an elevation of the renal zinc excretion [170]. Again, treatment of these patients by surgical hypophysectomy or with x-rays brings about a decrease in the circulating GH level, an increase in the serum zinc concentration, and a reduction in the renal zinc excretion. In adrenalectomized or hypophysectomized cats [171] and rats [62], the serum zinc concentration also increased, while the renal zinc excretion decreased. These results demonstrate that the GH level in plasma is inversely proportional to the serum zinc content and directly proportional to the renal zinc excretion.

These changes are presumably the consequence of a direct and/or indirect influence of GH on the binding of zinc to macromolecular and micromolecular ligands in the blood, so that renal excretion of the zinc bound by micromolecular ligands is affected [172-174]. By contrast, the circulating zinc ions are bound primarily to the histidine residues of albumin, the major macromolecular ligand. When there is an increase in the concentration of circulating histidine, the major micromolecular ligand, the zinc shifts from albumin to the amino acid histidine, which, because of its size, permeates the renal glomerular membrane, whereas albumin normally cannot do so. This view is supported by clinical observations at elevated contents of circulating GH that brought about a higher cellular turnover rate and an increased content of serum and urine amino acids, including

histidine [170]. Consequently, an intact pituitary-adrenal cortex
system is required to maintain normal circulating zinc levels and to
mobilize body zinc depots [175]. In zinc-deficient rats, however,
Reeves et al. [176] found that the serum corticosterone concentration
depended neither on the dietary zinc content nor on the zinc status
of the animals. Consequently, there was no response if the previously
zinc-depleted animals were repleted. Furthermore, zinc depletion also
had no effect on the ACTH-induced serum corticosterone levels. It
must therefore be assumed that short-term zinc supplementation of
depleted animals does not influence the serum corticosterone level.
Nor did the administration of ACTH have any influence on the serum
zinc concentration, whether or not the animals had been given an
adequate zinc supply.

Treating zinc-deficient rats with bovine GH did not improve
weight gains [101]. Similarly, Ku [177] reported that GH given to
zinc-deficient pigs did not improve growth and food intake, nor did
it influence their serum zinc levels, serum alkaline phosphatase
activity, or parakeratotic lesions. The administration of bovine GH
to zinc-deficient, nonhypophysectomized rats in the studies by Prasad
et al. [62] also failed to enhance growth, whereas growth rates
greatly increased after zinc supplementation. The growth rates of
hypophysectomized rats, however, responded to both hormone and zinc
supplementation, regardless of zinc status. Here the effects of the
hormone and the zinc were additive but independent of each other.

Since growth depression is known as an early and characteristic
deficiency symptom when young animals are subjected to alimentary
zinc depletion, the concentration of growth hormone in serum was
determined radioimmunologically in two experimental series (Table 11).
The growth hormone concentration of the zinc-deficient rats was
reduced greatly compared to the ad libitum-fed control animals. The
same reduction, however, was also noted in the case of the pair-fed
control animals. Accordingly, the reduction of the growth hormone
concentration must be regarded as the result of the diminished feed
intake and not of zinc deficiency per se.

TABLE 11

Growth Hormone (GH) Concentration in Serum of Zinc-Deficient,
Pair-Fed, and Ad Libitum-Fed Control Rats (ng GH/ml serum)[a]

Depleted animals (1.3 mg zinc/kg DM)	Pair-fed controls (100 mg zinc/kg DM)	Ad libitum-fed controls (100 mg zinc/kg DM)
32.7 ± 18.5	30.0 ± 6.7	93.5 ± 82.5
59.8 ± 16.1	55.3 ± 9.5	93.5 ± 47.7

[a]DM, dry matter.
Source: Ref. 178.

3.3. Sex Hormones

It has been shown by Bischoff [179,180] and Maxwell [181] that the
activities of the hypophyseal follicle-stimulating hormone (FSH) and
luteinizing hormone (LH) could be increased when zinc salts were
added to extracts of the anterior pituitary before they were injected
into sexually immature or hypophysectomized rats. Injections of
gonadotropin and testosterone stimulated the growth of all the acces-
sory sex organs under zinc deficiency but did not prevent tubular
atrophy of the testes, which is considered to be a typical zinc
deficiency symptom [182]. On the other hand, few if any, changes
were evident in the serum or urinary zinc after estrogen administration to rats, while the serum zinc concentration decreased after
progesterone administration [183]. Briggs et al. [184] reported a
significant decrease in the serum zinc in women after estrogen
therapy. Many other authors [184-189] reported a similar response
of the plasma or serum zinc levels to the use of oral contraceptives.

Apgar [190] was able to maintain pregnancy to term in about
50% of zinc-deficient rats by the administration of progesterone and
estrone. According to studies by Pories et al. [191], progesterone
is regarded as the factor that may be responsible for the mobilization of zinc from tissues. They found an increase in serum zinc
content during pregnancy in women and a decrease after parturition;

this may be attributed to the progesterone activity [192]. Also,
the effects of estrogen during pregnancy must be considered as an
additional factor influencing serum zinc contents. The changes in
the zinc contents in the serum and organs during the course of pregnancy and lactation are different, however, depending on the zinc
nutrition [193,194].

Gombe et al. [195] observed that the LH content in pooled
pituitaries of zinc-deficient female rats was not different from
that of pair-fed and ad libitum-fed control animals. The levels of
LH and progesterone, however, were reduced in the plasma of the zinc-
deficient and also the restricted-fed animals compared to that of
the ad libitum-fed controls. The lower plasma LH levels of zinc-
deficient rats and their restricted-fed mates do not seem to be due
to a lack of the LH-releasing factor, since its level was comparable
in all three groups. Lactating dairy cows, in which experimental
zinc depletion resulted in the appearance of distinct deficiency
symptoms without their feed consumption being depressed, showed
unaltered basal levels of LH and FSH in their serum, compared to
repleted animals [196]. Also, in human blood donors selected according to low and high serum zinc contents, the serum testosterone correlated with the serum zinc content only between the ages of 36 and
60, while there was no relationship to the FSH and LH [197]. This
is evidence that, contrary to previous beliefs, a slight zinc deficiency does not reduce the hypophyseal gonadotropins, but rather
effects the testicular contents. Lei et al. [198] found in zinc-
deficient rats an increased response of the serum LH and FSH to an
introvenous injection of LH-releasing factor, whereas the serum
testosterone response was reduced in comparison with that of restrictively fed control rats. This also indicates that the role of zinc
in the male reproductive system concerns mainly the testicular site.

In patients with sickle-cell anemia who suffer, in part, from
zinc deficiency [22,24,31], the average basal levels of serum testosterone, dehydrotestosterone, and androtestosterone were significantly lower than in normal persons [199]. After stimulation with

gonadotropin-releasing hormone (Gn-RH), the rise in the serum LH
and FSH was higher in the patients with sickle-cell anemia, whereas
the testosterone response to Gn-RH was sluggish [200]. The basal
serum LH and FSH values were also higher than in the control patients.
The androgen deficiency, always a characteristic symptom in patients
with sickle-cell anemia, appears to be primarily the consequence of
testicular rather than hypophyseal atrophy. Alternatively, it could
be assumed that the androgen catabolism within the testes is increased
during zinc deficiency. The basic abnormalities responsible for the
defective testicular function have not yet been fully clarified.

Acrodermatitis enteropathica, an autosomal recessive disease,
is characterized by abnormal metabolism of the essential fatty acids
[201-203] and by disorders in zinc absorption [13,15,204]. Evans
et al. [205] found a low-molecular-weight zinc-binding ligand in
human milk to be essential for the maintenance of normal zinc absorption. This ligand was identified by Song and Adham [206] as prostaglandin. In cow's milk this zinc-prostaglandin complex could not be
detected, a finding that may explain the positive effect of human
breast milk in the treatment of patients with acrodermatitis. The
reduced arachidonic acid levels found in the serum of patients with
acrodermatitis enteropathica [201-203] indicate a disturbed prostaglandin synthesis, because arachidonic acid serves as the precursor
of prostaglandins [18,207]. As shown in studies on rats, the essential role of prostaglandin E_2 in zinc absorption lies not only in
its chelation of zinc but, more important, in an increase in the
zinc transport across the intestinal mucosa [208,209]. These results
show that the symptoms of acrodermatitis enteropathica result from
inability to synthesize prostaglandin, which stimulates intestinal
zinc absorption. Pathological symptoms similar to those seen in
zinc deficiency were evident in female rats given aspirin at the end
of pregnancy [210]. Aspirin inhibits prostaglandin biosynthesis in
various tissues and hence results in impaired zinc absorption in rats
[207]. Because of the similarity of the defects that can be brought
about by a prostaglandin inhibitor such as aspirin and by zinc

deficiency in pregnant rats, one could speculate that zinc participates in the biosynthesis or function of prostaglandin [210]. However, additional studies are needed to show whether or not zinc and
the prostaglandins are related by a common metabolic pathway that,
when disturbed, leads to the similarity in pathology.

In summary, it may be said, in reference to zinc and hormones,
that there is a great need for further research on the role of zinc
in hormone metabolism, especially in respect to its functions in the
synthesis and secretion of various hormones. Specifically, additional studies must clarify the involvement of zinc in the biopotency
of these hormones.

REFERENCES

1. M. Kirchgessner, H.-P. Roth, and E. Weigand, in *Trace Elements
    in Human Health and Disease*, Vol. 1 (A. S. Prasad, ed.), Academic
    Press, New York, 1976, p. 189ff.

2. E. J. Underwood, in *Trace Elements in Human and Animal Nutrition*,
    4th ed. (E. J. Underwood, ed.), Academic Press, New York, 1977,
    p. 196ff.

3. M. Kirchgessner and H.-P. Roth, in *Zinc in the Environment*,
    Part II: *Health Effects* (J. O. Nriagu, ed.), Wiley, New York,
    1980, p. 71ff.

4. H.-P. Roth and M. Kirchgessner, *Biol. Trace Elem. Res.*, *3*, 13
    (1981).

5. B. L. O'Dell, in *Trace Element Metabolism in Man and Animals*,
    Vol. 4 (J. McHowell, ed.), 4th Int. Symp., Perth, Western
    Australia, 1981, p. 319ff.

6. D. J. Plocke, C. Levinthal, and B. L. Vallee, *Biochemistry*, *1*,
    373 (1962).

7. R. T. Simpson and B. L. Vallee, *Biochemistry*, *7*, 4343 (1968).

8. H.-P. Roth and M. Kirchgessner, *Z. Tierphysiol. Tierernaehr.
    Futtermittelkd.*, *32*, 289 (1974).

9. M. Kirchgessner, W. A. Schwarz, and H.-P. Roth, *Z. Tierphysiol.
    Tierernaehr. Futtermittelkd.*, *35*, 191 (1975).

10. H.-P. Roth and M. Kirchgessner, *Res. Exp. Med.*, *174*, 283 (1979).

11. R. I. Henkin, P. J. Schechter, W. T. Friedewald, D. L. Demets,
    and M. Raff, *Am. J. Med. Sci.*, *272*, 285 (1976).

12. A. S. Prasad, P. Rabbani, A. Abbasi, E. Bowersox, and M. R. S. Fox, in *Trace Element Metabolism in Man and Animals,* Vol. 3 (M. Kirchgessner, ed.), Arbeitskreis Tierernährungsforschung Weihenstephan, Freising-Weihenstephan, West Germany, 1978, p. 280ff.

13. J. Lombeck, H. G. Schnippering, K. Kasparek, F. Ritzl, H. Kästner, L. E. Feinendegen, and H. J. Bremer, *Z. Kinderheilkd., 120,* 181 (1975).

14. K. M. Hambidge, P. A. Walravens, and K. H. Neldner, in *Zinc Metabolism: Current Aspects in Health and Disease* (G. J. Brewer, and A. S. Prasad, eds.), Alan R. Liss, New York, 1977, p. 329ff.

15. K. H. Neldner and K. M. Hambidge, *N. Engl. J. Med., 292,* 879 (1975).

16. H.-P. Roth and M. Kirchgessner, *World Rev. Nutr. Diet., 34,* 144 (1980).

17. W. A. Van Vloten and L. P. Bos, *Dermatologica, 156,* 175 (1978).

18. K. Weismann and T. Flagstad, *Acta Dermato-Venerol., 56,* 151 (1976).

19. J. Kroneman, G. J. W. Mey, and A. Helder, *Zentralbl. Veterinaermed. Reih. A, 22,* 201 (1975).

20. M. Stöber, *Dtsch. Tieraerztl. Wochenschr., 78,* 257 (1971).

21. P. A. Simkin, *Lancet, 2,* 539 (1976).

22. A. S. Prasad, E. B. Schoomaker, J. Ortega, G. J. Brewer, D. Oberleas, and F. J. Oelshlegel, Jr., Abstr. First Natl. Symp. Sickle Cell Dis., 1974, p. 33ff.

23. A. S. Prasad, in *Trace Elements in Human Health and Disease,* Vol. 1 (A. S. Prasad, ed.), Academic Press, New York, 1976, p. 1ff.

24. G. J. Brewer, A. S. Prasad, F. J. Oelshlegel, Jr., E. B. Schoomaker, J. Ortega, and D. Oberleas, in *Trace Elements in Human Health and Disease,* Vol. 1 (A. S. Prasad, ed.), Academic Press, New York, 1976, p. 283ff.

25. (a) A. S. Prasad, J. A. Halsted, and M. Nadimi, *Am. J. Med., 31,* 532 (1961). (b) A. S. Prasad, A. Miale, Jr., Z. Farid, A. R. Schulert, and H. H. Sandstead, *J. Lab. Clin. Med., 61,* 537 (1963).

26. D. Keilin and T. Mann, *Biochem. J., 34,* 1163 (1940).

27. D. Keilin and T. Mann, *Nature (Lond.), 145,* 304 (1940).

28. H.-P. Roth and M. Kirchgessner, *Z. Tierphysiol. Tierernaehr. Futtermittelkd., 32,* 296 (1974).

29. M. Kirchgessner, A. E. Stadler, and H.-P. Roth, *Bioinorg. Chem., 5,* 33 (1975).

30. M. Iqbal, *Enzyme*, *12*, 33 (1971).

31. A. S. Prasad, E. B. Schoomaker, J. Ortega, G. J. Brewer, D. Oberleas, and F. J. Oelshlegel, Jr., *Clin. Chem.*, *21*, 582 (1975).

32. J. M. Hsu, J. K. Anilane, and D. E. Scanlan, *Science*, *153*, 882 (1966).

33. C. F. Mills, J. Quarterman, R. B. Williams, A. C. Dalgarno, and B. Panic, *Biochem. J.*, *102*, 712 (1967).

34. A. S. Prasad and D. Oberleas, *J. Appl. Physiol.*, *31*, 842 (1971).

35. H.-P. Roth and M. Kirchgessner, *Z. Tierphysiol. Tierernaehr. Futtermittelkd.*, *33*, 62 (1974).

36. A. S. Prasad, D. Oberleas, E. R. Miller, and R. W. Luecke, *J. Lab. Clin. Med.*, *77*, 144 (1971).

37. W. E. C. Wacker, *Biochemistry*, *1*, 859 (1962).

38. M. C. Scrutton, C. W. Wu, and D. A. Goldthwait, *Proc. Natl. Acad. Sci. USA*, *68*, 2497 (1971).

39. J. P. Slater, A. S. Mildvan, and L. A. Loeb, *Biochem. Biophys. Res. Commun.*, *44*, 37 (1971).

40. C. F. Springgate, A. S. Mildvan, and L. A. Loeb, *Fed. Proc., Fed. Am. Soc. Exp. Biol.*, *32*, 451 (1973).

41. C. F. Springgate, A. S. Mildvan, and R. Abramson, *J. Biol. Chem.*, *248*, 5987 (1973).

42. D. S. Auld, H. Kawaguchi, D. M. Livingston, and B. L. Vallee, *Biochem. Biophys. Res. Commun.*, *57*, 967 (1974).

43. D. S. Auld, H. Kawaguchi, D. M. Livingston, and B. L. Vallee, *Proc. Natl. Acad. Sci. USA*, *71*, 2091 (1974).

44. I. Lieberman and P. Ove, *J. Biol. Chem.*, *237*, 1634 (1962).

45. M. W. Terhune and H. H. Sandstead, *Science*, *177*, 68 (1972).

46. H. H. Sandstead, D. D. Gillespie, and R. N. Brady, *Pediatr. Res.*, *6*, 119 (1972).

47. F. Fernandez-Madrid, A. S. Prasad, and D. Oberleas, *J. Lab. Clin. Med.*, *82*, 951 (1973).

48. M. Somers and E. J. Underwood, *Aust. J. Biol. Sci.*, *22*, 1277 (1969).

49. A. S. Prasad and D. Oberleas, *J. Lab. Clin. Med.*, *82*, 461 (1973).

50. J. K. Chesters and M. Will, *Br. J. Nutr.*, *39*, 375 (1978).

51. I. Lieberman, R. Abrams, N. Hunt, and P. Ove, *J. Biol. Chem.*, *238*, 3955 (1963).

52. A. S. Prasad and D. Oberleas, *J. Lab. Clin. Med.*, *83*, 634 (1974).

53. J. R. Duncan and I. E. Dreosti, *J. Comp. Pathol.*, *86*, 81 (1976).

54. I. E. Dreosti and L. S. Hurley, *Proc. Soc. Exp. Biol. Med.*, *150*, 161 (1975).

55. D. Oberleas and A. S. Prasad, in *Trace Element Metabolism in Animals*, Vol. 2 (W. G. Hoekstra, J. W. Suttie, H. E. Ganther, and W. Mertz, eds.), University Park Press, Baltimore, 1974, p. 730ff.

56. H. Swenerton, *Diss. Abstr. Int. B*, *31*, 5443 (1971).

57. H. Swenerton, R. Shrader, and L. S. Hurley, *Proc. Soc. Exp. Biol. Med.*, *141*, 283 (1972).

58. H. Swenerton and L. S. Hurley, *J. Nutr.*, *95*, 8 (1968).

59. J. F. Riordan, *Med. Clin. N. Am.*, *60*, 661 (1976).

60. J. G. Reinhold and G. A. Kfoury, *Am. J. Clin. Nutr.*, *22*, 1250 (1969).

61. A. S. Prasad, D. Oberleas, P. Wolf, J. P. Horwitz, E. R. Miller, and R. W. Luecke, *Am. J. Clin. Nutr.*, *22*, 628 (1969).

62. A. S. Prasad, D. Oberleas, P. Wolf, and J. P. Horwitz, *J. Lab. Clin. Med.*, *73*, 486 (1969).

63. G. S. Feil and R. R. Burns, *Proc. R. Soc. Med.*, *69*, 474 (1976).

64. W. J. Bettger and B. L. O'Dell, *Life Sci.*, *28*, 1425 (1981).

65. M. Kirchgessner and H.-P. Roth, *Zentralbl. Veterinaermed. Reih. A*, *22*, 14 (1975).

66. J. Pallauf and M. Kirchgessner, *Arch. Tierernaehr.*, *26*, 457 (1976).

67. E. Weigand and M. Kirchgessner, *Z. Tierphysiol. Tierernaehr. Futtermittelkd.*, *39*, 16 (1977).

68. R. Ellul-Micallef, A. Galdes, and F. F. Feneck, *Postgrad. Med. J.*, *52*, 148 (1976).

69. G. H. Bourne, in *The Biochemistry and Physiology of Bone* (G. H. Bourne, ed.), Academic Press, New York, 1972, p. 177ff.

70. D. G. Thawley and R. A. Willoughby, *Can. J. Comp. Med.*, *41*, 84 (1977).

71. H.-P. Roth and M. Kirchgessner, *Z. Tierphysiol. Tierernaehr. Futtermittelkd.*, *33*, 57 (1974).

72. C. W. Lin, *Diss. Abstr. Int. B*, *31*, 515 (1970).

73. N. P. Westmoreland, *Diss. Abstr. Int. B*, *30*, 63 (1969).

74. N. P. Westmoreland, *Fed. Proc.*, *30*, 1001 (1971).

75. B. Trefz, *J. Dent. Res.*, *51*, Suppl. to No. 5, 1203 (1972).

76. C. K. Lum and R. I. Henkin, *Biochim. Biophys. Acta*, *421*, 362 (1976).

77. F. A. Catalanotto and R. Nanda, *J. Oral Pathol.*, *6*, 211 (1977).

78. H.-P. Roth and M. Kirchgessner, *Zentralbl. Veterinaermed. Reih. A*, *23*, 578 (1976).

79. M. Kirchgessner, E. Weigand, A. Schnegg, E. Grassmann, F. J. Schwarz, and H.-P. Roth, in *Ernährungslehre und Diätetik*, Vol. I/2 (H.-D. Cremer, D. Hötzel, and J. Kühnau, eds.), Georg Thieme, Stuttgart, 1980, p. 275ff.

80. M. Kirchgessner and W. Schwarz, *Zentralbl. Veterinaermed. Reih. A*, *22*, 572 (1975).

81. W. Schwarz and M. Kirchgessner, *Dtsch. Tieraerztl. Wochenschr.*, *82*, 141 (1975).

82. J. Pallauf and M. Kirchgessner, *Zentralbl. Veterinaermed. Reih. A*, *20*, 100 (1973).

83. H.-P. Roth and M. Kirchgessner, *Zentralbl. Veterinaermed. Reih. A*, *27*, 290 (1980).

84. J. K. Chesters and M. Will, in *Trace Element Metabolism in Man and Animals*, Vol. 3 (M. Kirchgessner, ed.), Arbeitskreis Tierernährungsforschung Weihenstephan, Freising-Weihenstephan, West Germany, 1978, p. 211ff.

85. J. K. Chesters and M. Will, *Br. J. Nutr.*, *39*, 297 (1978).

86. H.-P. Roth and M. Kirchgessner, *Z. Tierphysiol. Tierernaehr. Futtermittelkd.*, *42*, 95 (1979).

87. H.-P. Roth and M. Kirchgessner, *Res. Exp. Med.*, *177*, 213 (1980).

88. H. H. Sandstead, A. S. Prasad, A. R. Schulert, Z. Farid, A. Miale, Jr., S. Bassilly, and W. J. Darby, *Am. J. Clin. Nutr.*, *20*, 422 (1967).

89. H.-P. Roth and M. Kirchgessner, *Int. Z. Vitamin- Ernaehrungsforsch.*, *45*, 201 (1975).

90. D. A. Scott, *Biochem. J.*, *28*, 1592 (1934).

91. M. H. Greider, S. L. Howell, and P. E. Lacy, *J. Cell Biol.*, *41*, 162 (1969).

92. S. L. Howell, D. A. Young, and P. E. Lacy, *J. Cell Biol.*, *41*, 167 (1969).

93. H.-P. Roth and M. Kirchgessner, *Z. Tierphysiol. Tierernaehr. Futtermittelkd.*, *42*, 287 (1979).

94. E. Hove, C. A. Elvehjem, and E. B. Hart, *Am. J. Physiol.*, *119*, 768 (1937).

95. D. G. Hendricks and A. W. Mahoney, *J. Nutr.*, *102*, 1079 (1972).

96. J. Quarterman, C. F. Mills, and W. R. Humphries, *Biochem. Biophys. Res. Commun.*, *25*, 354 (1966).

97. L. Boquist and A. Lernmark, *Acta Pathol. Microbiol. Scand.*, *76*, 215 (1969).

98. A. M. Huber and S. N. Gershoff, *J. Nutr.*, *103*, 1739 (1973).

99. H.-P. Roth, U. Schneider, and M. Kirchgessner, *Arch. Tierernaehr.*, *25*, 545 (1975).

100. H. H. Sandstead, A. S. Prasad, Z. Farid, A. Schulert, A. Miale, Jr., S. Bassilly, and W. J. Darby, in *Zinc Metabolism* (A. S. Prasad, ed.), Charles C Thomas, Springfield, Ill., 1966, p. 304ff.

101. M. P. Macapinlac, W. N. Pearson, and W. J. Darby, in *Zinc Metabolism* (A. S. Prasad, ed.), Charles C Thomas, Springfield, Ill., 1966, p. 142ff.

102. J. Quarterman and E. Florence, *Br. J. Nutr.*, *28*, 75 (1972).

103. E. D. Brown, J. C. Penhos, L. Recant, and J. C. Smith, *Proc. Soc. Exp. Biol. Med.*, *150*, 557 (1975).

104. J. Fasel, M. D. H. Hadjikhani, and J. P. Felber, *Gastroenterology*, *59*, 109 (1970).

105. W. Mertz, *Physiol. Rev.*, *49*, 164 (1969).

106. H.-P. Roth and M. Kirchgessner, *Z. Tierphysiol. Tierernaehr. Futtermittelkd.*, *42*, 277 (1979).

107. H. Djahanschiri and H. Brune, *Z. Tierphysiol. Tierernaehr. Futtermittelkd.*, *35*, 40 (1975).

108. H. Djahanschiri and H. Brune, *Z. Tierphysiol. Tierernaehr. Futtermittelkd.*, *35*, 201 (1975).

109. A. M. Preston, R. P. Dowdy, M. A. Preston, and J. N. Freeman, *J. Nutr.*, *106*, 1391 (1976).

110. M. Kirchgessner, H.-P. Roth, and W. A. Schwarz, *Z. Tierphysiol. Tierernaehr. Futtermittelkd.*, *36*, 175 (1976).

111. K. D. Hepp, *Umschau*, *72*, 513 (1972).

112. E. Cerasi, S. Efendic, and R. Luft, *Lancet*, *1*, 794 (1973).

113. E. Cerasi, *Q. Rev. Biophys.*, *8*, 1 (1975).

114. H. Sando and G. M. Grodsky, *Diabetes*, *22*, 354 (1974).

115. N. N., *Nutr. Rev.*, *34*, 332 (1976).

116. S. Terris, C. Hoffmann, and D. F. Steiner, *Can. J. Biochem.*, *57*, 459 (1979).

117. K. Engelbart and H. Kief, *Virchows Arch. B*, *4*, 294 (1970).

118. R. B. Williams and C. F. Mills, *Br. J. Nutr.*, *24*, 989 (1970).

119. M. Chvapil, *Life Sci.*, *13*, 1041 (1973).

120. B. L. Vallee, *Physiol. Rev.*, *39*, 443 (1959).

121. H. Maske, *Z. Naturforsch.*, *8 b*, 96 (1953).

.2. H. Maske, *Diabetes*, *6*, 335 (1957).

123. T. Yoshinaga and S. Ogawa, *Acta Histochem.*, *53*, 161 (1975).

124. Ya. A. Lazaris, L. G. Kuts, and Z. E. Bavel'skii, *Exp. Biol. Med.*, *77*, 647 (1974).

125. H.-P. Roth and M. Kirchgessner, *Int. Z. Vitamin- Ernaehrungsforsch.*, *47*, 277 (1977).

126. Y. Kanter, *Int. J. Biochem.*, *7*, 253 (1976).

127. W. Mertz and E. E. Roginski, in *Newer Trace Elements in Nutrition* (W. Mertz and W. E. Cornatzer, eds.), Marcel Dekker, New York, 1971, p. 123ff.

128. W. Mertz, E. W. Toepfer, E. E. Roginski, and M. M. Polansky, *Fed. Proc.*, *33*, 2275 (1974).

129. W. Mertz, in *Spurenelemente in der Entwicklung von Mensch und Tier* (K. Betke and F. Bidlingmaier, eds.), Urban & Schwarzenberg, Munich, 1975, p. 189ff.

130. W. Mertz, in *Infusionstherapie, 4*, 181 (1977).

131. K. M. Hambidge, in *Newer Trace Elements in Nutrition* (W. Mertz and W. E. Cornatzer, eds.), Marcel Dekker, New York, 1971, p. 169ff.

132. K. M. Hambidge, *Am. J. Clin. Nutr.*, *27*, 505 (1974).

133. D. Behne and F. Diehl, in *Nuclear Activation Techniques in the Life Sciences*, International Atomic Energy Agency, Vienna, 1972, p. 407.

134. W. H. Glinsmann, F. J. Feldman, and W. Mertz, *Science*, *152*, 1243 (1966).

135. D. F. Steiner, J. L. Clark, C. Nolan, A. H. Rubinstein, E. Margoliash, B. Aten, and P. E. Oyer, *Recent Prog. Horm. Res.*, *25*, 207 (1969).

136. D. F. Steiner, W. Kemmler, H. S. Tager, and J. D. Peterson, *Fed. Proc.*, *33*, 2105 (1974).

137. A. E. Kitabchi, W. C. Duckworth, and B. Benson, *Diabetes, 21*, 935 (1972).

138. K. Kobayashi, *Endocrinol. Jpn.*, *22*, 489 (1975).

139. G. M. Brosky and C. C. Heuck, *Endokrinologie, 66*, 46 (1975).

140. J. Brunner, H.-P. Anders, and E. Gerhards, *Arzneim.-Forsch.*, *25*, 1429 (1975).

141. B. Schultz, D. Michaelis, M. Ziegler, W. Teichmann, W. Nowak, G. Albrecht, and H. Bibergeil, *Endokrinologie, 68*, 309 (1976).

142. S. L. Howell, M. Tyhurst, H. Duvefelt, A. Andersson, and C. Hellerström, *Cell Tissue Res.*, *188*, 107 (1978).

143. T. L. Coombs, P. T. Grant, and B. H. Frank, *Biochem. J.*, *125*, 62 (1971).

144. H. Maske, in *Diabetes* (R. H. Williams, ed.), Hoeber, New York, 1960, p. 46.

145. G. Bentley, E. Dodson, G. Dodson, D. Hodgkin, and D. Mercola, *Nature (Lond.), 261*, 166 (1976).

146. L. Weil, T. S. Seibles, and T. T. Herskovits, *Arch. Biochem. Biophys., 111*, 308 (1965).

147. E. R. Arquilla, P. Thiene, T. Brugman, W. Ruess, and S. Sugiyama, *Biochem. J., 175*, 289 (1978).

148. J. Goldman and F. H. Carpenter, *Biochemistry, 13*, 4566 (1974).

149. P. L. Poffenbarger, A. Espinosa De Los Monteros Mena, and J. Steinke, *Metabolism, 19*, 509 (1970).

150. P. L. Poffenbarger, *J. Clin. Invest., 56*, 1455 (1975).

151. P. L. Poffenbarger and M. J. Prince, *Growth, 40*, 83 (1976).

152. P. L. Poffenbarger and M. A. Haberal, *Surgery, 80*, 608 (1976).

153. R. Liske and K. Reber, *Horm. Res., 7*, 214 (1976).

154. K. F. Weinges, *Aerztl. Praxis, 20*, 1151 (1968).

155. H. G. Pidduck, P. J. J. Wren, and D. A. Price Evans, *Diabetes, 19*, 240 (1970).

156. S. Kumar and K. S. J. Rao, *Nutr. Metab., 17*, 231 (1974).

157. M. C. M. Mateo, J. B. Bustamante, J. F. B. De Quiros, and O. O. Manchado, *Biomedicine, 23*, 134 (1975).

158. J. J. T. Davies, M. Musa, and T. L. Dormandy, *J. Clin. Pathol., 21*, 359 (1968).

159. F. Rosner and P. C. Gorfien, *J. Lab. Clin. Med., 72*, 213 (1968).

160. M. K. Chooi, J. K. Todd, and N. D. Boyd, *Nutr. Metabol., 20*, 135 (1976).

161. K. Baerlocher and W. Weissert, *Helv. Paediatr. Acta, 31*, 99 (1976).

162. D. A. Scott and A. M. Fisher, *J. Clin. Invest., 17*, 725 (1938).

163. M. Heinitz, *Med. Welt, 28*, 634 (1977).

164. K. N. Jeejeebhoy, R. Chu, E. B. Marliss, G. R. Greenberg, and A. Bruce-Robertson, *Clin. Res., 23*, 636 A (1975).

165. J. D. H. Homan, G. A. Overbeek, J. P. J. Neutelings, L. J. Booiy, and J. Van Der Vies, *Lancet, 2*, 541 (1954).

166. R. P. Cox and A. Ruckenstein, *J. Cell. Physiol., 77*, 71 (1971).

167. A. Flynn, W. H. Strain, and W. J. Pories, *Biochem. Biophys. Res. Commun., 46*, 1113 (1972).

168. R. I. Henkin, S. Meret, and J. B. Jacobs, *J. Clin. Invest., 48*, 38 a (1969).

169. M. D. Lifschitz and R. I. Henkin, *J. Appl. Physiol.*, *31*, 88 (1971).

170. R. I. Henkin, in *Trace Element Metabolism in Animals*, Vol. 2 (W. G. Hoekstra, J. W. Suttie, H. E. Ganther, and W. Mertz, eds.), University Park Press, Baltimore, 1974, p. 652ff.

171. R. I. Henkin, in *Trace Element Metabolism in Animals*, Vol. 2 (W. G. Hoekstra, J. W. Suttie, H. E. Ganther, and W. Mertz, eds.), University Park Press, Baltimore, 1974, p. 647ff.

172. E. L. Giroux and R. I. Henkin, *Biochim. Biophys. Acta*, *273*, 64 (1972).

173. R. I. Henkin, in *Protein-Metal Interactions* (M. Friedman, ed.), Plenum Press, New York, 1974, p. 299ff.

174. R. I. Henkin, *Med. Clin. N. Am.*, *60*, 779 (1976).

175. A. Flynn, W. J. Pories, W. H. Strain, and O. A. Hill, Jr., *Lancet*, *2*, 235 (1972).

176. P. G. Reeves, S. G. Frissell, and B. L. O'Dell, *Proc. Soc. Exp. Biol. Med.*, *156*, 500 (1977).

177. P. K. Ku, *Diss. Abstr. Int. B*, *31*, 6717 (1971).

178. H.-P. Roth and M. Kirchgessner, in *Trace Element Metabolism in Man and Animals*, Vol. 4 (J. McC. Howell, J. M. Gawthorne, and C. L. White, eds.), Australian Academy of Science, Canberra, 1981, p. 334ff; distributed by Springer-Verlag, Heidelberg.

179. F. Bischoff, *Am. J. Physiol.*, *117*, 182 (1936).

180. F. Bischoff, *Am. J. Physiol.*, *121*, 765 (1938).

181. L. C. Maxwell, *Am. J. Physiol.*, *110*, 458 (1934).

182. M. J. Millar, P. V. Elcoate, M. I. Fischer, and C. A. Mawson, *Can. J. Biochem. Physiol.*, *38*, 1457 (1960).

183. N. Sato and R. I. Henkin, *Am. J. Physiol.*, *225*, 508 (1973).

184. M. H. Briggs, M. Briggs, and J. Austin, *Nature (Lond.)*, *232*, 480 (1971).

185. J. A. Halsted, B. M. Hackley, and J. C. Smith, *Lancet*, *2*, 278 (1968).

186. J. A. Halsted and J. C. Smith, *Lancet*, *1*, 322 (1970).

187. J. G. Schenker, W. Z. Palishuk, and E. Jungreis, *Fertil. Steril.*, *22*, 229 (1971).

188. A. S. Prasad, D. Oberleas, K. Y. Lei, K. S. Moghissi, and J. C. Stryker, *Am. J. Clin. Nutr.*, *28*, 377 (1975).

189. L. D. McBean, J. C. Smith, Jr., and J. A. Halsted, *Proc. Soc. Exp. Biol. Med.*, *137*, 543 (1971).

190. J. Apgar, *J. Nutr.*, *100*, 470 (1970).

191. W. J. Pories, E. G. Mansour, F. R. Plecha, A. Flynn, and W. H. Strain, in *Trace Elements in Human Health and Disease* (A. S. Prasad, ed.), Academic Press, New York, 1976, p. 115ff.

192. A. Flynn, W. J. Pories, W. H. Strain, and F. L. Weiland, *Naturwissenschaften, 60,* 162 (1973).

193. M. Kirchgessner and U. Schneider, *Arch. Tierernaehr. 28,* 211 (1978).

194. U. Schneider and M. Kirchgessner, *Nutr. Metab., 23,* 241 (1979).

195. S. Gombe, J. Apgar, and W. Hansel, *Biol. Reprod., 9,* 415 (1973).

196. M. Kirchgessner, D. Schams, and H.-P. Roth, *Z. Tierphysiol. Tierernaehr. Futtermittelkd., 37,* 151 (1976).

197. R. Hartoma, *Acta Physiol. Scand., 101,* 336 (1977).

198. K. Y. Lei, A. Abbasi, and A. S. Prasad, *Am. J. Physiol., 230,* 1730 (1976).

199. A. Abbasi, A. S. Prasad, and J. Ortega, *Ann. Int. Med., 85,* 601 (1976).

200. A. S. Prasad, A. Abbasi, and J. Ortega, in *Zinc Metabolism: Current Aspects in Health and Disease* (G. J. Brewer and A. S. Prasad, eds.), Alan R. Liss, New York, 1977, p. 211ff.

201. K. H. Neldner, L. Hagler, W. R. Wise, F. B. Stifel, E. G. Lufkin, and R. H. Herman, *Arch. Dermatol., 110,* 711 (1974).

202. R. Cash and C. K. Berger, *J. Pediatr., 74,* 717 (1969).

203. H. B. White and J. M. Montalvo, *J. Pediatr., 83,* 999 (1973).

204. E. J. Moynahan, *Lancet, 1,* 399 (1974).

205. G. W. Evans, C. I. Grace, and H. J. Votava, *Am. J. Physiol., 228,* 501 (1975).

206. M. K. Song and N. F. Adham, *Fed. Proc., 35,* 1667 (1976).

207. G. W. Evans and P. E. Johnson, *Lancet, 2,* 52 (1977).

208. M. K. Song and N. F. Adham, *Fed. Proc., 36,* 4583 (1977).

209. M. K. Song and N. F. Adham, *Am. J. Physiol., 234,* E 99 (1978).

210. B. L. O'Dell, G. Reynolds, and P. G. Reeves, *J. Nutr., 107,* 1222 (1977).

211. M. Kirchgessner and H.-P. Roth, in *Trace Element Metabolism in Man and Animals,* Vol. 4 (J. McC. Howell, J. M. Gawthorne, and C. L. White, eds.), Australian Academy of Science, Canberra, 1981, p. 327ff; distributed by Springer-Verlag, Heidelberg.

Chapter 10

ZINC DEFICIENCY SYNDROME DURING PARENTERAL NUTRITION IN HUMANS

Karin Ladefoged and Stig Jarnum
Medical Department P
Division of Gastroenterology
Rigshospitalet
Copenhagen, Denmark

1. INTRODUCTION

Zinc deficiency was first produced artificially in a mammalian
species in 1934 [1]. Growth retardation and loss of hair were
observed in rats fed a low-zinc diet. Twenty years later Tucker
and Salmon [2] found that zinc cures and prevents parakeratosis in
pigs. Human zinc deficiency in a chronic form was reported in the
early 1960s by Prasad et al. [3,4]. They observed a syndrome of
iron deficiency anemia, hepatosplenomegali, hypogonadism, and
dwarfism in patients whose intake of animal protein was negligible.
Acrodermatitis enteropathica has been known since 1936 [5], but it
was not related to zinc deficiency until 1973 [6]. A syndrome of
acute zinc deficiency during total parenteral nutrition with symptoms identical to those found in acrodermatitis enteropathica [7]
was reported in 1976. The findings have been confirmed in several
reports [8-18]. Zinc deficiency with acrodermatitis has also been
observed in a patient with chronic alcohol abuse, cirrhosis, and
malnutrition [19], and recently it was reported in patients with
Crohn's disease [20], possibly caused by zinc malabsorption [21].

2. METABOLIC ASPECTS OF ZINC
IN HUMAN NUTRITION

2.1. Zinc Absorption and Excretion

A normal human adult consumes about 200 μmol of zinc per day, but
large variations, from 70 to 450 μmol per day, have been reported
[22,23]. Absorption of dietary zinc has been estimated at 30-40%
[24,25]. Absorption depends on the source of zinc intake. Zinc
from animal sources is better absorbed than that from plant products
[26,27]. Phytate from the plants binds zinc in the intestinal lumen
and renders it unavailable for absorption. Fibers exert a similar
effect.

According to animal studies zinc absorption occurs throughout
the small intestine by a carrier-mediated process with little or no

zinc absorbed from the colon [28,29]. Animal studies suggest that
the duodenum contributes most to the overall zinc absorption [28],
although the ileum may have the greatest absorptive capacity [30].
Gastrointestinal secretion represents the main excretory route for
zinc [31,32]. Spencer et al. [25] found that fecal excretion of an
intravenous [65]Zn dose averaged 12.8% in 12 days, and urinary excretion averaged 1.6%. Assuming a normal renal zinc excretion (5-10
µmol per day), it yields an endogenous fecal zinc loss of 40-80 µmol
per day. Animal studies suggest that zinc homeostasis is maintained
by endogenous zinc excretion in the gut rather than by regulation of
dietary zinc absorption [33].

Urinary zinc excretion normally amounts to 5-10 µmol per day
[34]. It is within wide ranges unaffected by dietary zinc intake
[35-37], but reduced in chronic zinc-deficiency states [4]. Zinc
concentration in sweat has been estimated at 18 µmol per liter [38].
In studies of preadolescent children, estimates of whole-body sweat
loss of zinc have ranged from 2 µmol per day [39] to 20 µmol per day
[37]. Zinc secretion by sweat is also reduced in zinc deficiency
states [38].

In the normal human adult zinc balance can be obtained by daily
intake of 200-250 µmol of zinc when this intake is derived from a
mixed diet [40].

2.2. Body Stores of Zinc

The total body content of zinc approximates 30,000 µmol, with the
majority present in bone and muscle [41,42]. The large quantities
of zinc found in bone do not seem readily available for mobilization
[41]. Zinc concentration in plasma ranges from 10 to 20 µmol per
liter [34], 30-40% is tightly bound by an α_2-macroglobulin, 60%
loosely bound to albumin, and a minor fraction is chelated by amino
acids [43]. Zinc concentration in erythrocytes is 10-fold that of
plasma, and leucocytes contain up to 25 times the zinc concentration
of erythrocytes [43].

2.3. Biological Function

Zinc plays an important metabolic role as a component of numerous
enzyme systems, among which are alkaline phosphatase, carbonic
anhydrase, aminopeptidases, and several dehydrogenases [41,42].
The physiological functions and biochemical actions of zinc are not
clearly elucidated, but among other things it seems essential for
fatty acid metabolism, collagen synthesis, the cell division cycle,
and stabilization of cellular and organelle membranes [41].

3. ZINC DEFICIENCY DURING PARENTERAL NUTRITION

In the late 1960s Dudrick et al. [44] showed that long-term survival,
growth, and positive nitrogen balance was possible by total parenteral
nutrition. Since then more and more patients with severe gastrointestinal dysfunction have been nourished partly or completely by
the veins. With this development came a well-controlled research
field to define the body requirements for all nutritional substances
necessary to induce anabolism. It is in patients on intravenous
feeding that we have become aware of acute syndromes related to
deficiency of several trace elements: copper (anemia, leucopenia,
skeletal lesions) [45-47], chromium (glucose intolerance, neuropathy
encephalopathy) [48,49], selenium (cardiomyopathy, skeletal myopathy)
[50,51], and zinc (see below) [7-18].

3.1. Pathogenesis of Zinc Deficiency

Several factors may contribute to zinc deficiency.

3.1.1. Increased Losses

During catabolism renal zinc excretion is increased [52-54] and the
increase is related to the degree of catabolism as evaluated from
urinary excretion of nitrogen and creatine [53]. In severe catabolic

states the zinc loss may be extremely large: 60-120 μmol per day
[55]. The zinc loss presumably derives from breakdown of the zinc-
rich skeletal muscles [56].

Abnormally high zinc excretion from the gastrointestinal tract
occurs in patients with diarrheal stools, gastric suction, intestinocutaneous fistulas, or high-output enterostomies [57]. The zinc
loss is related to the volume and source of gastrointestinal discharge. In patients on total parenteral nutrition mean zinc concentration has been measured to 230 μmol per liter in fecal output from
an intact small bowel, and to 55 μmol per liter in excreta from small
intestinal fistulas and after extensive small-bowel resection [57].

It is possible but undocumented that cutaneous zinc loss is
increased in patents with excessive sweat [38] or extensive burns.

3.1.2. Increased Requirements

Restoration of body cell mass in a malnourished patient implies
increased tissue demand for zinc, and zinc deficiency syndrome often
precipitates upon induction of weight gain [7,9,10,13,14,55].

3.1.3. Inadequate Supply

Most nutrient solutions contain small amounts of zinc as a contaminant, especially protein hydrolysates and amino acid solutions [58].
Zinc concentration in mixed nutrient solutions has been measured to
5-15 μmol per liter, corresponding to a daily supply of 15-45 μmol,
which seldom suffices to maintain zinc homeostasis as evidenced by
a gradual decrease of plasma zinc [59,60]. Supplementary zinc therefore is warranted.

3.2. Symptoms of Zinc Deficiency

Symptoms of zinc deficiency have appeared from 3 weeks to 7 months,
most often 4-8 weeks, after the institution of parenteral nutrition
[7,8,10-18]. Diarrhea, dermatitis, and psychological alterations
are the classical features.

Diarrhea may be the first symptom [7,15], but is not invariably present [14].

The mental changes described are: irritability, apathy, depression, and confusion [7,8,13,14,18].

Dermatitis is the most commonly reported clinical manifestation, possibly because it is the most specific symptom and therefore leads to the diagnosis. The skin lesions are quite similar to those found in acrodermatitis enteropathica, a hereditary zinc deficiency disorder [61] possibly caused by zinc malabsorption [62]. The dermatitis may be preceded by pruritus [15]. It characteristically begins in the nasolabial folds and progresses rapidly to widespread lesions, predominantly involving the face, anogenital region, heels, elbows, toes, and fingers [7-18]. The skin lesions may resemble seborrhoic eczema with erythematous, scaling efflorescences [9,14,17,18]. Crusted psoriasis-like plaques with an erythematous base may also occur [7, 10,16-18]. Vesiculopustular lesions or larger bullae surrounded by an erythematous ring are frequently present [7,8,12,17,18].

Stomatitis [7,13,14,16], conjunctivitis [14], blepharitis [18], and paronychial inflammation [8,18] have been reported.

Pronounced hair loss usually occurs [7,9,11,12,14-18], often with a delay of a few days to a couple of weeks after the appearance of the dermatitis [7,14,15]. It may proceed for some time despite zinc therapy [7,8] and may progress to total or almost total alopecia.

Skin biopsies have shown hyperkeratosis or parakeratosis of the epidermis and infiltration with polymorphnuclear leucocytes [17,18, 55].

3.3. Diagnosis of Zinc Deficiency

Zinc deficiency syndrome may be accompanied by a decline of serum alkaline phosphatase, a zinc-containing enzyme [9,10,12,15,18], but normal enzyme levels have also been reported [15,16].

Most cases of zinc deficiency syndrome have been associated with very low serum zinc concentrations: 1.2-8.0 µmol per liter (normal range: 10-20 µmol per liter) [7-13,15-18]. However, symptoms typical for zinc deficiency and susceptible to zinc supply may occur despite normal serum zinc [14] or maybe even high zinc levels [63]. The diagnostic significance of serum zinc is further questioned by the fact that a low zinc level may occur in serum in several conditions without symptoms of zinc deficiency [43].

Urinary zinc excretion may decrease considerably during zinc depletion but may still exceed normal ranges in a frank zinc deficiency syndrome [7,18]. It is thus of no diagnostic value unless consecutive measurements are made.

Hair zinc content is more reflective of total body stores, but since hair is slow to grow, low levels are more indicative of long-term deficiencies than of acute shortcomings.

It is likely but not firmly established that zinc concentration in rapidly renewing cells such as leukocytes may provide a qualified diagnostic approach to zinc depletion.

So far the diagnosis rests mainly on a characteristic symptomatology in association with proper response to zinc therapy.

3.4. Therapy of Zinc Deficiency

Symptoms of zinc deficiency improve dramatically within 2-3 days upon proper zinc therapy and complete healing of the skin lesions is usually obtained within 10 days [7-11,13,14,18]. Hair loss is slower to revert and may even proceed for some time after the institution of zinc supply, but subsequently all symptoms subside completely. The zinc doses used in adults have been 765-2100 µmol per day for oral supply [7,8,18] and 75-300 µmol per day intravenously [9,11,14,16,18]. Infants have received 90-350 µmol per day orally [12,13].

3.5. Prevention of Zinc Deficiency

A zinc deficiency syndrome may develop within 3 weeks upon institution of parenteral nutrition. Furthermore, insufficient zinc supply
impairs insulin response and the utilization of glucose and amino
acids [57]. Parenteral nutrition therefore should always include
sufficient amounts of zinc. Balance studies have shown that a
positive zinc balance can be obtained in patients with high gastrointestinal zinc losses by parenteral supply of 100-200 μmol of zinc
per day [57]. Patients during extreme catabolic stress possibly
require larger amounts since their urinary zinc excretion may exceed
these levels [7,55]. However, caution should be taken not to exceed
toxic levels. Death from renal, hepatic, and cardiopulmonary failure
has followed parenteral supply of 46 mmol zinc over 60 h due to a
prescribing error [64] and symptoms of acute zinc intoxication has
occurred upon rapid zinc administration: 150 μmol of zinc infused
over 1 h [9].

In general, supplementary zinc should always be given to
patients on parenteral nutrition. A total of 70-100 μmol of zinc
will be sufficient unless extraordinary large losses take place
through the gastrointestinal tract (fistulas, gastric suction, etc.)
or through the kidneys (severe catabolic stress).

The following three case reports illustrate the development
of zinc deficiency syndrome during parenteral nutrition.

4. CASE REPORTS

4.1. Case 1

A 49-year-old woman had suffered from Crohn's disease for 8 years.
Because of an exacerbation, she underwent extensive bowel resection
in 1975. A jejunocolonic anastomosis was performed between 100 cm

jejunum and the left half of the colon. Postoperatively, she had
septicemia, wound rupture, and an intestinocutaneous fistula. Total
parenteral nutrition was initiated with a daily supply of 100 g
of amino acids, 3200 kcal (13,400 kJ), essential fatty acids
(Intralipid), vitamins, and minerals (Na, K, Ca, Mg, PO_4). Zinc
was not supplied except for what might be included as a contaminant
of the nutrient solutions. A surgical attempt to close the fistula
was unsuccessful. Once again wound dehiscence occurred, split-skin
grafts from the thighs failed to heal, and a duodenal and jejunal
leakage with four intestinocutaneous fistulas appeared. Two months
after the start of total parenteral nutrition the patient was increasingly poor and apathetic with high fistular output (1.5-2 liters
per day) and with no signs of healing of the abdominal defect or of
the split-skin donor areas on the thighs (Fig. 1). Within 3 days
she developed a widespread dermatitis with crusted skin lesions
along the nasolabial folds and on the elbows (Fig. 1), and pustules
in the finger creases (Fig. 2) and on the heels bullae, which broke
down to large eroded areas (Fig. 3). Some loss of hair occurred.
Within a month serum alkaline phosphatase decreased from 850 to 265
U/liter (normal range: 50-275). Serum zinc was measured to 4.1 μmol/
liter (normal range: 10.6-17.7). Urinary zinc excretion was 31 μmol
per day. Parenteral supply of 210 μmol of zinc per day was started.
It was followed by rapid improvement of both the general condition
and the dermatitis. Within a week the skin lesions healed completely
(Figs. 4 and 5), body weight increased 4 kg on unchanged parenteral
energy and nitrogen supply, fistular secretion decreased to about
1 liter per day, serum zinc became normal, and serum alkaline phosphatase rose to 800 U/liter. However, despite zinc supply, hair loss
proceeded to almost total alopecia (Fig. 6). A slight increase of
urinary zinc excretion to about 40 μmol per day followed zinc supply.
Within 2 weeks the abdominal defect and the split-skin donor areas
healed (Fig. 7).

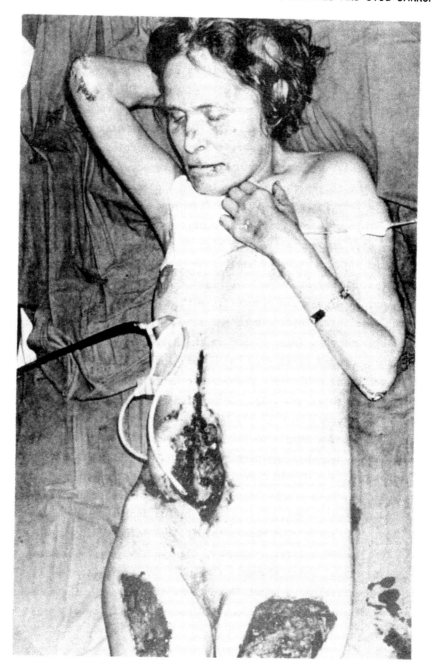

FIG. 1. Case 1: crusted dermatitis periorbitally, periorally, on
the nose and on the elbows. No healing of abdominal wound dehiscence
or of split-skin donor areas on the thighs.

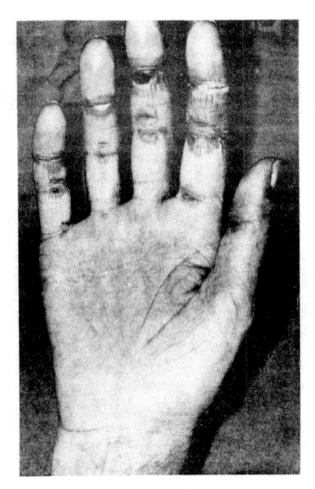

FIG. 2.   Case 1: pustules in the finger creases.   (From Ref. 18.)

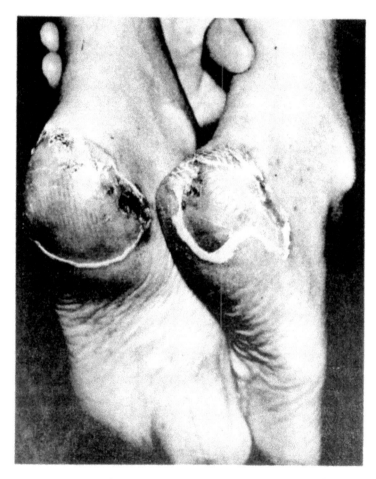

FIG. 3.   Case 1: erosions on the heels.   (From Ref. 18.)

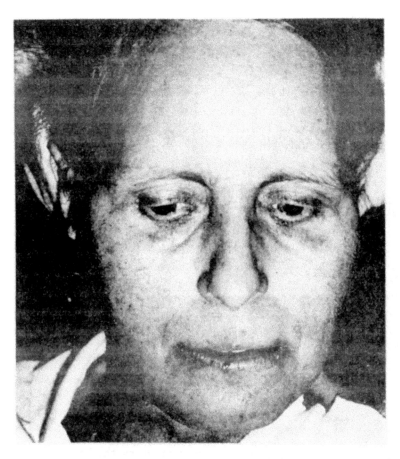

FIG. 4. Case 1: healing of the facial dermatitis upon parenteral
zinc supply.

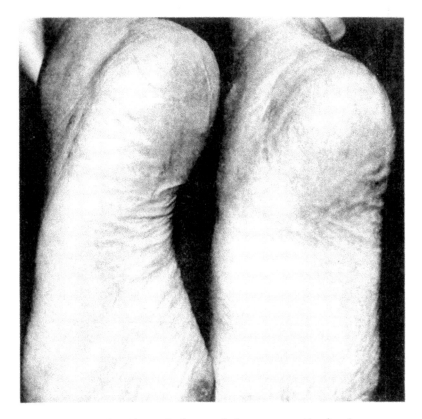

FIG. 5.   Case 1: healing of the eroded areas on the heels.

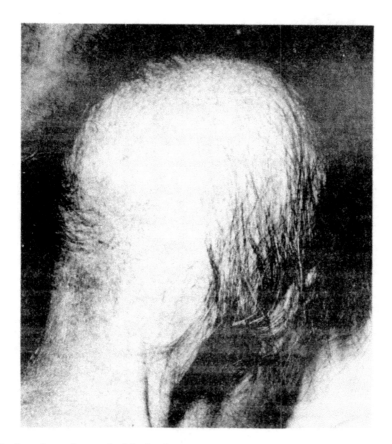

FIG. 6.   Case 1: marked hair loss.

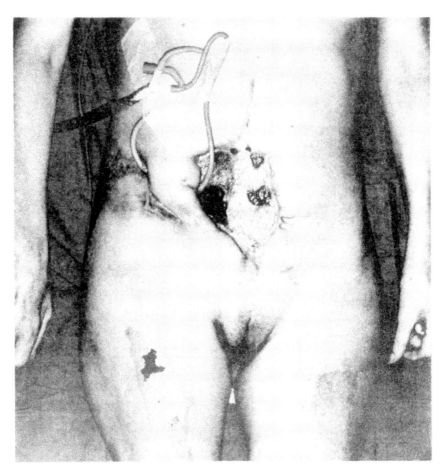

FIG. 7. Case 1: healing of the abdominal wound dehiscence and of
the split-skin donor areas on the thighs.

4.2. Case 2

A 43-year-old man had undergone a gastric resection for ulcer 13
years previously. Three years later a Roux anastomosis was performed
and, after another 2 years, an abdominal vagotomy. He had suffered
from symptoms of Crohn's disease (abdominal pain, malabsorption,
diarrhea) for 10 years and had radiographic lesions of the terminal
ileum and the whole colon. He had been on continuous glucocortico-
steroid therapy (prednisone) for 4 years. He was persistently malnourished, with biochemical signs of inflammatory activity (low
plasma albumin, elevated plasma orosomucoid) and disabled by diarrhea and tiredness. Parenteral nutrition had been administered for
short periods. In late 1975 his condition deteriorated, with weight
loss, tiredness, aggravated diarrhea, and vomiting. In the course
of a month a generalized scaly dermatitis appeared (Figs. 8-10).
Parenteral nutritive support with daily supply of 1500 kcal (6,300
kJ), 100 g of amino acids, fat (Intralipid), minerals, and vitamins
but without supplementary zinc was initiated. During the following
month his condition remained poor. No weight gain was obtained
(body weight: 49 kg, height: 173 cm). Serum alkaline phosphatase
decreased from 501 to 66 U/liter. Serum zinc was not measured. An
exploratory laparotomy was performed, but the intestinal tract was
found to be normal. Parenteral nutrition was discontinued, but had
to be resumed a month later because of a further weight loss of 4 kg,
increasing fatigue, diarrhea, and vomiting. The skin lesions had
deteriorated. Four months after commencement of the dermatitis
serum zinc was measured to 5.4 μmol/liter (normal range: 11.4-18.9).
Parenteral supply of 210 μmol of zinc per day was initiated and was
followed by dramatic improvement of the dermatitis and of the general
condition, a slow weight gain, and a gradual decrease of the gastrointestinal symptoms. Serum zinc increased to 10.4 μmol/liter.
Plasma albumin was 184 μmol/liter (normal range: 532-813).

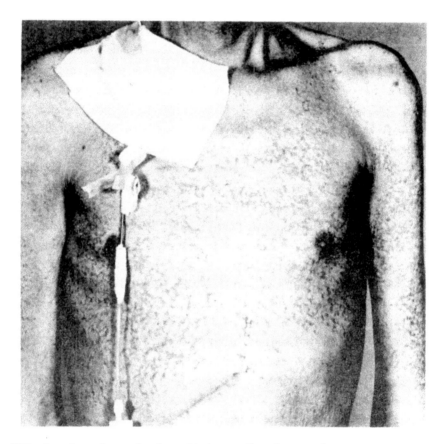

FIG. 8.   Case 2: scaly dermatitis on the chest and arms.

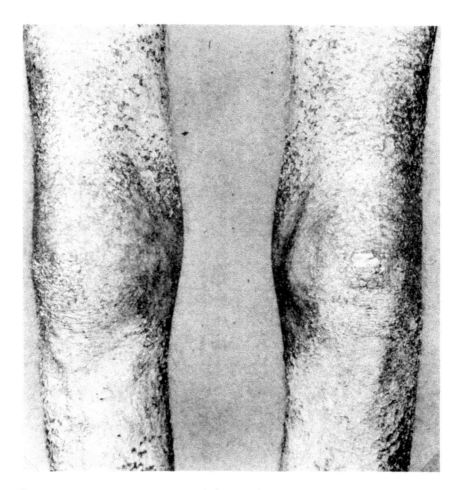

FIG. 9. Case 2: scaly dermatitis on the legs.

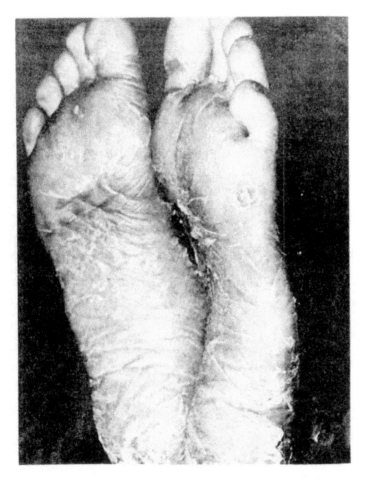

FIG. 10.   Case 2: pronounced desquamation of the soles.

4.3. Case 3

A 56-year-old man had suffered from Crohn's disease for 8 years.
He had undergone a resection of 130 cm terminal ileum 2 years previously. In 1975 a subtotal colectomy was performed and an ileos-
tomy was constructed. Postoperatively he developed a subphrenic
abscess and a jejunocutaneous fistula. Total parenteral nutrition
was initiated with a daily supply of 3000 kcal (12,500 kJ), 100 g
of amino acids, fat (Intralipid), minerals, and vitamins. No
supplementary zinc was administered. During the following months
body weight showed a gradual increase from 55 to 61 kg (height:
178 cm). Two operations with drainage of subphrenic abscesses were
performed. The fistula remained open with variable output. After
4 months of parenteral nutrition and a minimal oral intake his
general condition deteriorated and he contracted a seborrhoic
dermatitis of the face with crusted perioral lesions, and a scaly
erythematous affection of the anogenital region, bullae in the
finger creases, and paronychial inflammation. In a week serum
alkaline phosphatase decreased from 1056 to 394 U/liter (normal
range: 50-275). Serum zinc was measured to 4.9 µmol/liter (normal
range: 11.4-18.9). Following a parenteral supply of 300 µmol of
zinc per day the dermatitis subsided within 3-4 days, serum alkaline
phosphatase rose to 991 U/liter, and serum zinc increased to 9.6
µmol/liter (plasma albumin: 322 µmol/liter, normal range: 532-813).

REFERENCES

1. W. R. Todd, C. A. Elvehjelm, and E. B. Hart, *Am. J. Physiol.*,
   107, 146 (1934).

2. H. F. Tucker and W. D. Salmon, *Proc. Soc. Exp. Biol. Med.*, *88*,
   613 (1955).

3. A. S. Prasad, J. A. Halsted, and M. Nadimi, *Am. J. Med.*, *31*,
   532 (1961).

4. A. S. Prasad, A. Miale, Jr., Z. Farid, H. H. Sandstead, and
   A. R. Schulert, *J. Lab. Clin. Med.*, *61*, 537 (1963).

5. T. Brandt, *Acta Dermato-Venerol.*, *17*, 513 (1936).

6. P. M. Barnes and E. J. Moynahan, *Proc. R. Soc. Med.*, *66*, 327 (1973).

7. R. G. Kay, C. Tasman-Jones, J. Pybus, R. Whiting, and H. Black, *Ann. Surg.*, *183*, 331 (1976).

8. B. Bernstein and J. J. Leyden, *Arch. Dermatol.*, *114*, 1070 (1978).

9. L. P. Bos, W. A. van Vloten, A. F. D. Smit, and M. Nubé, *Neth. J. Med.*, *20*, 263 (1977).

10. I. de Leeuw, R. Peeters, and A. Croket, *Acta Clin. Belg.*, *33*, 227 (1978).

11. S. Jarnum and K. Ladefoged, *Scand. J. Gastroenterol.*, *16*, 903 (1981).

12. N. Principi, A. Giunta, and A. Gervasoni, *Acta Paediatr. Scand.*, *68*, 129 (1979).

13. M. N. Srouji, W. F. Balistreri, M. H. Caleb, M. A. South, and S. Starr, *J. Pediatr. Surg.*, *13*, 570 (1978).

14. J. W. Steger and G. T. Izuno, *Int. J. Dermatol.*, *18*, 472 (1979).

15. S. Suita, K. Ikeda, A. Nagasaki, and Y. Hayashida, *J. Pediatr. Surg.*, *13*, 5 (1978).

16. S. B. Tucker, A. L. Schroeter, P. W. Brown, and J. T. McCall, *JAMA*, *235*, 2399 (1976).

17. W. A. van Vloten and L. P. Bos, *Dermatologica*, *156*, 175 (1978).

18. K. Weismann, N. Hjorth, and A. Fischer, *Clin. Exp. Dermatol.*, *1*, 237 (1976).

19. R. I. Ecker and A. L. Schroeter, *Arch. Dermatol.*, *114*, 937 (1978).

20. C. McClain, C. Soutor, and L. Zieve, *Gastroenterology*, *78*, 272 (1980).

21. G. C. Sturnioli, M. M. Molokhia, R. Shields, and L. A. Turnberg, *Gut*, *21*, 387 (1980).

22. K. A. Haeflein and A. I. Rasmussen, *J. Am. Diet. Assoc.*, *70*, 610 (1977).

23. D. Osis, L. Kramer, E. Wiatrowski, and H. Spencer, *Am. J. Clin. Nutr.*, *25*, 582 (1972).

24. J. C. King, W. L. Raynolds, and S. Margen, *Am. J. Clin. Nutr.*, *31*, 1198 (1978).

25. H. Spencer, V. Vankinscott, I. Lewin, and J. Samachson, *J. Nutr.*, *86*, 169 (1965).

26. B. Sandström, B. Arvidsson, Å. Cederblad, and E. Björn-Rasmussen, *Am. J. Clin. Nutr.*, *33*, 739 (1980).

27. B. Sandström and Å. Cederblad, *Am. J. Clin. Nutr.*, *33*, 1778 (1980).

28. N. T. Davies, *Br. J. Nutr.*, *43*, 189 (1980).

29. A. H. Methfessel and H. Spencer, *J. Appl. Physiol.*, *34*, 58 (1973).

30. D. L. Antonson, A. J. Barak, and J. A. Vanderhoof, *J. Nutr.*, *109*, 142 (1979).

31. H. Spencer, B. Rosoff, and A. Feldstein, *Radiat. Res.*, *24*, 432 (1965).

32. R. L. Aamodt, W. F. Rumble, G. S. Johnston, D. Foster, and R. I. Henkin, *Am. J. Clin. Nutr.*, *32*, 559 (1979).

33. G. W. Evans, E. C. Johnson, and P. E. Johnson, *J. Nutr.*, *109*, 1258 (1979).

34. S. Kiilerich, M. Sanvig Christensen, J. Naestoft, and C. Christiansen, *Clin. Chim. Acta*, *105*, 231 (1980).

35. J. L. Greger, R. P. Abernathy, and O. A. Bennett, *Am. J. Clin. Nutr.*, *31*, 112 (1978).

36. C. R. Meiners, L. J. Taper, M. K. Korslund, and S. J. Ritchey, *Am. J. Clin. Nutr.*, *30*, 879 (1977).

37. S. J. Ritchey, M. K. Korslund, L. M. Gilbert, D. C. Fay, and M. F. Robinson, *Am. J. Clin. Nutr.*, *32*, 799 (1979).

38. A. S. Prasad, A. R. Schulert, H. H. Sandstead, A. Miale, Jr., and Z. Farid, *J. Lab. Clin. Med.*, *62*, 84 (1963).

39. M. E. Harrison, C. Walls, M. K. Korslund, and S. J. Ritchey, *Am. J. Clin. Nutr.*, *29*, 842 (1976).

40. H. Spencer, D. Osis, L. Kramer, and C. Norris, in *Trace Elements in Human Health and Disease*, Vol. 1 (A. S. Prasad and D. Oberleas, eds.), Academic Press, New York, 1976, p. 345ff.

41. P. J. Aggett and J. T. Harries, *Arch. Dis. Child.*, *54*, 909 (1979).

42. P. A. Walravens, *West. J. Med.*, *130*, 133 (1979).

43. N. W. Solomons, *Am. J. Clin. Nutr.*, *32*, 856 (1979).

44. S. J. Dudrick, D. W. Wilmore, H. M. Vars, and J. E. Rhoads, *Surgery*, *64*, 134 (1968).

45. R. M. Heller, S. G. Kirchner, J. A. O'Neill, A. J. Hough, Jr., L. Howard, S. S. Kramer, and H. L. Green, *J. Pediatr.*, *92*, 947 (1978).

46. J. T. Karpel and V. H. Peden, *J. Pediatr.*, *80*, 32 (1972).

47. R. W. Vilter, R. C. Bozian, E. V. Hess, D. C. Zellner, and H. G. Petering, *N. Engl. J. Med.*, *291*, 188 (1974).

48. H. Freund, S. Atamian, and J. E. Fischer, *JAMA, 241,* 496 (1979).

49. K. N. Jeejeebhoy, R. C. Chu, E. B. Marliss, G. R. Greenberg, and A. Bruce-Robertson, *Am. J. Clin. Nutr., 30,* 531 (1977).

50. R. A. Johnson, S. S. Baker, J. T. Fallon, E. P. Maynard, J. N. Ruskin, Z. Wen, K. Ge, and H. J. Cohen, *N. Engl. J. Med., 304,* 1210 (1981).

51. A. M. van Rij, C. D. Thomson, J. M. McKenzie, and M. F. Robinson, *Am. J. Clin. Nutr., 32,* 2076 (1979).

52. A. Askari, C. L. Long, and W. S. Blakemore, *JPEN, 3,* 151 (1979).

53. D. P. Cuthbertson, G. S. Fell, C. M. Smith, and W. J. Tilstone, *Br. J. Surg., 59,* 68 (1972).

54. R. D. Lindeman, R. G. Bottomley, R. L. Cornelison, Jr., and L. A. Jacobs, *J. Lab. Clin. Med., 79,* 452 (1972).

55. C. Tasman-Jones, R. G. Kay, and S. P. Lee, *Surg. Annu., 10,* 23 (1978).

56. G. S. Fell, D. P. Cuthbertson, C. Morrison, A. Fleck, K. Queen, R. G. Bessent, and S. L. Husain, *Lancet, 1,* 280 (1973).

57. S. L. Wolman, G. H. Anderson, E. B. Marliss, and K. N. Jeejeebhoy, *Gastroenterology, 76,* 458 (1979).

58. E. C. Hauer and M. V. Kaminski, *Am. J. Clin. Nutr., 31,* 264 (1978).

59. C. R. Fleming, R. E. Hodges, and L. S. Hurley, *Am. J. Clin. Nutr., 29,* 70 (1976).

60. N. W. Solomons, T. J. Layden, I. H. Rosenberg, K. Vo-Khactu, and H. H. Sandstead, *Gastroenterology, 70,* 1022 (1976).

61. E. J. Moynahan, *Lancet, 2,* 399 (1974).

62. K. Weismann, S. Hoe, L. Knudsen, and S. Sølvsten Sørensen, *Br. J. Dermatol., 101,* 573 (1979).

63. C. T. Strobel, W. J. Byrne, W. Abramovits, V. J. Newcomer, R. Bleich, and M. E. Ament, *Int. J. Dermatol., 17,* 575 (1978).

64. A. Brocks, H. Reid, and G. Glazer, *Br. Med. J., 1,* 1390 (1977).

AUTHOR INDEX

Numbers in parentheses are reference numbers and indicate that an author's work is referred to although his name may not be cited in the text. Underlined numbers give the page on which the complete reference is listed.

A

Aamodt, R. L., 417(32), 437
Aasa, R., 113(68), 148
Abadi, P., 333(74), 357
Abbasi, A., 277(15), 297(15), 311; 367(12), 370(12), 373(12), 403(198,199), 404(200), 406, 414
Abbound, M. M., 162(24), 163(24), 189
Abdallah, M. A., 40(137), 53
Abdulla, M., 276(13), 311
Abernathy, R. P., 330(55,56), 356; 417(35), 437
Abraham, E. P., 105(75,77,78, 210), 109(36), 115(75), 116(75, 77,78), 117(75), 147, 149, 155
Abramovits, W., 421(63), 438
Abrams, R., 373(51), 407
Abramson, R., 158(3), 159(3), 161 (3), 162(3), 183(3), 188; 371 (41), 407
Adam, J. D., 133(164), 153
Adams, K. F., 278(52), 312
Adham, N. F., 327(46), 356; 404 (206,208,209), 414
Adler, M., 343(118), 359
Adman, E. T., 260(87), 273
Aggett, P. J., 327(45), 356; 417 (41), 418(41), 437
Ahmad, F., 107(19), 146

Åkeson, Å., 35(125,128,129,133), 39(128,135), 40(135), 45(129), 53; 76(115), 95
Alben, J. O., 304(150), 308(150), 317
Alberti, G., 121(107), 150
Albrecht, G., 394(141), 411
Allain, P., 276(12), 311
Allen, L. C., 65(75), 94
Altmann, H., 186(90), 191
Ambler, R. P., 123(120), 151
Ament, M. E., 421(63), 438
Amiconi, G., 300(145), 304(149), 317
Amiel, S., 195(3,5), 209
Anders, H. P., 394(140), 411
Anderson, B., 85(171), 98
Anderson, D., 45(158), 54
Anderson, D. R., 288(104,105), 315
Anderson, G. H., 419(57), 422(57), 438
Anderson, P. M., 276(8), 311
Anderson, R. A., 83(158,159), 97; 116(83), 117(83), 120(83), 124 (127), 125(83,127), 138(127), 140(83), 141(83), 149, 151; 185 (82), 191
Andersson, A., 394(142), 411
Andersson, B., 8(12), 10(12), 48
Andersson, I., 42(152), 54; 112 (58), 114(58), 115(58), 118(58), 120(99), 121(58), 126(99), 129

439

R

Rabbani, P., 278(50), 313; 367
(12), 370(12), 373(12), 406
Rabenstein, D. L., 277(22), 300
(22), 304(22), 305(22), 311
Rabinowitz, R., 286(83), 314
Radomski, M. W., 279(53), 313
Raff, M., 367(11), 405
Rajogopalan, K. V., 88(177), 98;
104(202), 111(55), 148, 155
Rajan, K. S., 277(33), 312
Ralston, R. B., 289(114), 315
Ramaiah, A., 281(69), 285(69),
313
Ramaswamy, B. S., 60(50), 93
Ranhotra, G. S., 331(59), 356
Rao, K. S. J., 398(156), 412
Rasmussen, A. I., 331(65), 357;
416(22), 436
Raulin, J., 1(1), 47
Rayburn, C. S., 18(54), 19(54),
50
Raynolds, W. L., 416(24), 436
Raynor, J. B., 104(122), 123
(122), 151
Re, P. D., 348(132), 360
Read, R. J., 62(64), 93
Reba, R. C., 336(95), 358
Reber, K., 396(153), 412
Recant, L., 385(103), 398(103),
410
Reck, S. J., 333(80), 353(155),
357, 361
Reedijk, J., 89(184), 99
Rees, D. C., 67(84), 68(84), 94;
108(24,26), 120(24), 140(24),
141(24), 146
Reeves, P. G., 401(176), 404
(210), 405(210), 413, 414
Reichlmayr-Lais, A. M., 335(84),
358
Reid, H., 422(64), 438
Reid, T. W., 109(30), 146
Reina, J. C., 339(101), 358
Reinhold, J. G., 332(72), 333
(74), 357; 375(60), 389(60),
408
Renzi, P., 106(9), 145
Repke, K. R. H., 239(59), 240
(59), 271

Reynolds, G., 404(210), 405(210),
414
Rhee, M. J., 111(49), 147
Rhoads, J. E., 418(44), 437
Rich, A., 31(113), 52
Rich, W. E., 88(177), 98; 104
(202), 155
Richards, M. P., 214(6), 269
Richardson, C., 186(92), 191
Richardson, D. C., 87(174,175),
98; 103(5), 106(5), 111(55),
116(5), 118(5), 127(5), 145,
148
Richardson, J. S., 87(174,175),
98; 103(5), 106(5), 116(5), 118
(5), 127(5), 145
Rifkind, J. M., 186(92), 191; 277
(18,21), 278(21,46,47), 279(64),
281(21), 282(21), 285(80), 287
(47), 288(21,47), 291(21,80),
294(18,64,129), 295(18,125-129),
296(18), 297(18,125,129,137,138),
298(18,64,129), 299(129), 300(18,
146), 301(18), 302(18,64), 303
(18,129,146), 304(18), 305(146),
306(18), 308(18,146), 311, 312,
313, 314, 316, 317
Rigo, A., 120(101), 150
Rinaldi, R. A., 186(85), 191
Ring, J., 179(57), 190
Ringold, H. J., 35(125), 53
Riordan, J. F., 40(138,139), 41
(138,139), 53; 375(59), 408
Ritchey, S. J., 330(54), 340(107),
351(149,150), 356, 359, 361; 417
(36,37,39), 437
Ritzl, F., 327(44), 356; 367(13),
404(13), 406
Robberecht, P., 343(118), 359
Robertson, G. B., 85(171), 98
Robin, E. D., 2(5), 47
Robinson, M. F., 351(150), 361;
417(37), 418(51), 437, 438
Rodriguez, O. G., 195(8), 210
Roginski, E. E., 392(127,128), 411
Romanelli, P., 115(88), 117(88),
121(88), 149
Romano, S., 65(73,74), 66(73,74),
94
Romans, A. Y., 13(38), 49
Root, C. A., 113(69,70), 115(69,
70), 119(69,70), 139(69,70), 148

SUBJECT INDEX

A

Absorption bands and spectra (and spectrophotometry) (*see also* Extinction coefficients and UV absorption spectra), 14, 15, 17, 26, 43, 44, 58, 60-62, 64, 65, 67, 76, 78, 79, 81-83, 112-118, 123, 125, 127, 135, 138, 139, 141, 143, 165, 167, 170, 172, 173, 177, 202, 205, 206, 220-223, 241, 242, 244, 253-259, 261, 263, 264, 267
Acetaldehyde, 59, 121
Acetamide, N-methyl-, 75
Acetate (or acetic acid), 65, 130, 280
  as ligand, 16, 17, 117, 144
  bromo-, 41
  buffer, 42, 121
  iodo-, 41, 321
  methoxy-, 130
  p-nitrophenyl-, 59
  oxalo-, 110, 395
Acetolamide, 18, 124
Acetylcholine esterase, 194
Acidity constants, 7, 15, 45, 57-60, 65, 66, 69, 71, 74, 84, 86, 106, 130, 135, 142-145, 235, 236, 262, 265, 303
  apparent, 13, 19, 58, 63
Acrodermatitis enteropathica, 2, 325, 327, 333, 334, 367, 368, 372, 404, 416, 420
ACTH, *see* Corticotropin
Actin, 291
Active site (*see also* Coordination spheres), 19, 23, 25, 28, 32, 33, 56, 58, 67, 68,

[Active site], 75-77, 84, 87-89, 121, 130, 131, 133, 134, 140, 145, 209
Adenosine 5'-diphosphate, *see* ADP
Adenosine monophosphate, *see* AMP
Adenosine 5'-triphosphate, *see* ATP
ADP, 111, 176
  -ribose, 40
Adrenal
  glands, 399
  steroid hormones, 399
Adrenocorticotropic hormone, *see* Corticotropin
Affinity constants, *see* Stability constants
α-Alanine (and residues), 21, 30
  phenyl-, *see* Phenylalanine
Albumins, 400, 417
  human serum, 282-284
  plasma, 326, 431, 435
Alcohol dehydrogenase, 3, 5, 6, 8, 35-47, 103, 104, 112, 115, 118, 121, 125, 129, 185, 209, 365, 370, 374
  amino acid sequence, 36, 37
  horse, 5, 6, 47, 76, 82, 129, 185, 209, 242, 244, 248
  human, 5, 47
  inhibition, 126, 133
  liver, 76, 81, 112, 114, 120, 126, 129, 133, 134, 138, 139, 185, 209, 242, 244, 245, 248
  models for, 76-82
  stereodiagram, 38
  vertebrates, 6
  yeast, 6, 47, 76, 81
Alcoholic groups, *see* Hydroxyl groups and Phenolates
Alcohols (*see also* individual names), 76, 83, 103

473

Milton Keynes UK
Ingram Content Group UK Ltd.
UKHW020007071024
449327UK00031B/2692